T0251996

MEDICAL
INTELLIGENCE
UNIT

Organ Preservation for Transplantation
Third Edition

Luis H. Toledo-Pereyra
Departments of Research and Surgery
Michigan State University
Kalamazoo Center for Medical Studies
Kalamazoo, Michigan, USA

CRC Press
Taylor & Francis Group
Boca Raton London New York

CRC Press is an imprint of the
Taylor & Francis Group, an **informa** business

ORGAN PRESERVATION FOR TRANSPLANTATION
THIRD EDITION

Molecular Intelligence Unit

First published 2010 by Landes Bioscience

Published 2018 by CRC Press
Taylor & Francis Group
6000 Broken Sound Parkway NW, Suite 300
Boca Raton, FL 33487-2742

First issued in paperback 2018

© 2010 by Taylor & Francis Group, LLC
CRC Press is an imprint of Taylor & Francis Group, an Informa business

No claim to original U.S. Government works

ISBN-13: 978-1-138-11545-3 (pbk)
ISBN-13: 978-1-58706-337-4 (hbk)

Visit the Taylor & Francis Web site at
http://www.taylorandfrancis.com

and the CRC Press Web site at
http://www.crcpress.com

Library of Congress Cataloging-in-Publication Data

Organ preservation for transplantation / [edited by] Luis H. Toledo-Pereyra. -- 3rd ed.
    p. ; cm. -- (Medical intelligence unit)
Rev. ed of: Organ procurement and preservation for transplantation / edited by Luis H. Toledo-Pereyra. 2nd ed. c1997.
Includes bibliographical references and index.
ISBN 978-1-58706-337-4
1. Preservation of organs, tissues, etc. 2. Transplantation of organs, tissues, etc. I. Toledo-Pereyra, Luis H. II. Organ procurement and preservation for transplantation. III. Series: Medical intelligence unit (Unnumbered : 2003)
[DNLM: 1. Organ Transplantation. 2. Organ Preservation. 3. Tissue and Organ Procurement. WO 660 O65665 2010]
RD129.O73 2010
617.9'54--dc22
                                2009048359

Dedication

To Felix T. Rapaport
Great friend, generous supporter, committed advisor.
Thanks for all Felix.

About the Editor...

LUIS H. TOLEDO-PEREYRA—Transplant Surgeon, Researcher and Educator. He is currently Professor of Surgery and Director of Research at Michigan State University, Kalamazoo Center from Medical Studies where he teaches medical students, residents and fellows in the essentials of medical and surgical research. He is particularly interested in promoting scholarly activities among the faculty, residents and medical students. His work has ranged from the basis of organ preservation, the mechanisms of the molecular biology of ischemia and reperfusion and the role of medical humanities on the improvement of health care.

CONTENTS

EDITOR

Luis H. Toledo-Pereyra
Departments of Research and Surgery
Michigan State University
Kalamazoo Center for Medical Studies
Kalamazoo, Michigan, USA
Email: toledo@kcms.msu.edu
Chapters 1, 2, 4, 7, 10 and 13

CONTRIBUTORS

Note: Email addresses are provided for the corresponding authors of each chapter.

Roberto Anaya-Prado
Department of Research
Western Medical Center, IMSS
Guadalajara, Mexico
Email: robana@prodigy.net.mx
Chapters 1 and 2

William A. Baumgarnter
Division of Cardiac Surgery
Johns Hopkins Hospital Blalock
Baltimore, Maryland, USA
Email: wbaumgar@csurg.jhmi.jhu.edu
Chapter 12

Sufan Chien
Health Sciences Campus
University of Louisville
Louisville, Kentucky, USA
Email: sufanc@netscape.net
Chapter 6

Angelika Gruessner
IPTR, Diabetes Institute for Immunology
    and Transplantation
Department of Surgery
University of Minnesota
Minneapolis, Minnesota, USA
Chapter 8

James V. Guarrera
Division of Abdominal Organ
    Transplantation,
Columbia University
College of Physicians and Surgeons
New York, New York, USA
Email: jjg46@columbia.edu
Chapter 9

Mitchell L. Henry
Division of Transplantation
    and Comprehensive Transplant Center
The Ohio State University Medical Center
Columbus, Ohio, USA
Email: mitchell.henry@osumc.edu
Chapter 3

Naoya Kobayashi
Department of Surgery
Okayama University Graduate School
    of Medicine and Dentistry
Okayama, Japan
Chapter 8

Fernando Lopez-Neblina
Department of Biological Sciences
University of Baja California
Mexicali, BC, Mexico
Email: flneblina@yahoo.com
Chapter 4

Shinichi Matsumoto
Kawahara-cho Shogoin Sakyo-ku
Kyoto University Hospital
    Transplantation Unit
Kyoto, Japan
Chapter 8

Debra McKeehen
Veterans Affairs Medical Center
Minneapolis, Minnesota, USA
Email: debramckeehen@va.gov
Chapter 10

John S. Najarian
Minneapolis, Minnesota, USA
Foreword

Hirofumi Noguchi
Kyoto University Hospital
    Transplant Surgery
Kyoto, Japan
Chapter 8

Francis D. Pagani
University of Michigan Medical School
    Cardiac Surgery Division
Ann Arbor, Michigan, USA
Email: fpagani@umich.edu
Chapter 11

Juan M. Palma-Vargas
Texas Transplant Institute
Methodist Hospital
San Antonio, Texas, USA
Email: juan.palma-vargas@mhshealth.com
Chapters 1 and 7

Ronald P. Pelletier
Division of Transplantation
    and Comprehensive Transplant Center
The Ohio State University Medical Center
Columbus, Ohio, USA
Email: ronald.pelletier@osumc.edu
Chapter 3

Amer Rajab
Division of Transplantation
    and Comprehensive Transplant Center
The Ohio State University Medical Center
Columbus, Ohio, USA
Email: amer.rajab@osumc.edu
Chapter 3

Francisco Rodríguez-Quilantán
Department of Surgery
Medical Center
San Luis Potosí, S.L.P., Mexico
Chapter 2

Ashish S. Shah
Division of Cardiac Surgery
Johns Hopkins Hospital Blalock
Baltimore, Maryland, USA
Chapter 12

Roberta E. Sonnino
University of Minnesota
    Medical School
Minneapolis, Minnesota, USA
Chapter 10

David E.R. Sutherland
IPTR, Diabetes Institute for Immunology
    and Transplantation
Department of Surgery
University of Minnesota
Minneapolis, Minnesota USA
Email: dsuther@umn.edu
Chapter 8

Shohachi Suzuki
Second Department of Surgery
Hamamatsu University School
    of Medicine
Hamamatsu, Japan
Email address: shohachi@hama-med.ac.jp
Chapter 5

Yoshikazu Suzuki
University of Michigan Medical School
    Cardiac Surgery Division
Ann Arbor, Michigan, USA
Email: ysuzuki@umich.edu
Chapter 11

Alexander H. Toledo
Division of Abdominal Transplantation
University of North Carolina
Chapel Hill, North Carolina, USA
Email: ahtoledo@med.unc.edu
Chapters 1, 2, 4 and 7

Jason A. Williams
Division of Cardiac Surgery
Johns Hopkins Hospital Blalock
Baltimore, Maryland, USA
Chapter 12

FOREWORD

The Foreword for the first edition (1982) of this book, which was entitled *Basic Concepts of Organ Procurement, Perfusion, and Preservation for Transplantation*, was written by the late Folkert O. Belzer, one of the real giants in the field of organ preservation. In his Foreword, he described his serendipitous discovery of cryoprecipitated plasma. One evening in 1966, he had inadvertently stored dog plasma in the freezer—instead of the refrigerator—for the next day's experiment. When he noted the flocculation in the frozen perfusate the next day, he reasoned that filtering it might still net him usable fresh plasma for perfusion. He proceeded to use this filtered plasma and realized that, without the lipoproteins, no pressure rise occurred when perfusing dog kidneys: thus was born cryoprecipitated plasma. When used in a perfusion machine, cryoprecipitated plasma would allow for a 3-day kidney preservation time.

Two years after Belzer's 1967 publication detailing his discovery, Geoffrey Collins described in 1969 a solution that could be used for cold storage by suppressing the loss of intracellular potassium in exchange for sodium. Although both Belzer's and Collins' methods of preserving kidneys made a major impact in clinical kidney transplantation, the simpler method of cold storage is currently the primary technique used by most transplant teams.

The Foreword for the second edition (1997) of this book was written by Felix T. Rapaport, who had a long interest in perfusion and preservation. In his Foreword, he focused on the essential role of organ procurement and preservation in the success of solid organ transplantation. Rapaport stressed that the University of Wisconsin solution was the most significant event that occurred since the first edition and could allow extended long-term preservation of clinically transplantable kidneys to 72 hours and livers up to 30 hours. In addition, the use of this solution for preservation has stimulated interest in the use of "marginal" donors such as asystolic donors for transplantation. Finally, clinicians will find in this volume an impressive review of the relevant literature.

Sadly, both Belzer and Rapaport have now passed away, but the legacy of sincere caring and scientific curiosity that they passed on to their patients and colleagues will endure forever.

In this third edition (2009), more than a quarter-century after this book was first published, Dr. Luis H. Toledo-Pereyra has once again engaged experts in the field to review, update, and rewrite each of the chapters. Crafted in collaboration with Toledo-Pereyra, the text is eminently informative and easily understandable. What I consider most intriguing is that each solid organ that can be transplanted has an entire chapter devoted to the particular methods of its preservation. One chapter apiece, in other words, features insights on the preservation of the kidney, the liver, the pancreas, the small bowel, the heart, and the heart-lung.

I was also captivated by Toledo-Pereyra's own chapter on the future of procurement and preservation, especially his emphasis on the continuing significance of these two crucial and complex steps. It is with a great deal of pride that I have been asked by my former student, Toledo-Pereyra, to offer this Foreword for the third edition of his comprehensive accounting of *Organ Preservation for Transplantation, Third Edition*. Luis has always been one of my favorite students who not only was an outstanding clinical surgeon but devoted himself to research and as a resident published over 60 articles in medical literature. His inexhaustible writing efforts extend into many other books on transplantation and are primarily devoted to historical biographies. His efforts have made him one of the more notable transplant surgeons both nationally and internationally.

This valuable, well-organized book should be in the library of all solid organ transplant programs.

John S. Najarian
Minneapolis, Minnesota

PREFACE

The first edition of this book was published 27 years ago, in 1982[1] when organ procurement and preservation began to advance in the study of the best ways to preserve organs for transplantation. The second edition followed 15 years later, in 1997,[2] with the goal of finding common denominators in the best preservation techniques for transplantation. In our current third edition, 11 years after the second edition, we are still pursuing similar goals of defining the best preservation methods, but now we have more evidence, including results, and new advances have reached publication and are being incorporated into ischemia and reperfusion techniques and organ preservation studies.[3] Needless to say, many preservation solutions have been introduced, important preservation solution components have been better defined, and improved perfusion methods are being considered, especially for the increasing number of organ donors with non-beating hearts who are being sought for transplantation.

I could not agree more with the advanced concepts that the preservation pioneer, Fred Belzer (1930-1994), so wisely incorporated into the Foreword of the first edition:[1]

> *New discoveries are continually being made concerning the mechanism of cell death, the mechanism of action of pharmacological agents, and the role of newly discovered, naturally occurring regulators of hemodynamics and metabolism (calmodulin, prostaglandins, etc.). These discoveries may some day become integrated into the scope of methods to optimize long-term preservation of organs.*

These accurately crafted scientific predictions continue to be true today as we review the advances made since the first edition of this book more than two and a half decades ago.

Another preservation pioneer, Felix Rapaport (1929-2001), clearly analyzed the field of organ preservation in the Foreword to the second edition of this book[2] when he wrote:

> *The art and science of organ procurement and preservation have made significant progress in understanding the mechanisms of cell damage induced by ischemia, and in the development of new physiological and pharmacological approaches for the protection of cells and tissues from ischemia.*

He then considered the development of the University of Wisconsin solution as the most significant achievement between the first and second editions of this book (1982-1997). Furthermore, he reviewed pharmacological, biochemical and molecular mechanisms associated with organ ischemic injury as valuable tools for organ preservation.

Both committed students of preservation, Belzer and Rapaport were able to summarize in an elegant manner the important principles defining preservation, and to forecast the possibilities of new developments in this exciting field of transplantation science.

In this third edition (2009),[3] I have invited new transplant and preservation specialists, who have modified many of the chapters to the point of rewriting them from scratch. As much as it is desirable to maintain a uniform approach in revised editions, I saw it differently as quite clearly better to present different approaches to transplantation. We encouraged all contributors to manifest their own personal points of view, since this approach would present different views and alternatives and definitely enhance our knowledge of this challenging field of organ preservation.

I welcome previous and new contributors to this edition. I will leave their chapters as demonstrated proof of their knowledge and expertise. When necessary, I offered suggestions but, fundamentally, the chapters remain the same as submitted. The chapter on cryopreservation has been completely changed. This time, it has been given a more clinical approach by studying each one of the organs pertaining the results of cryopreservation attempts. The other chapters are self-explanatory regarding content and authorship. We anticipate a good response to all the chapters presented in this book.

The title of the present book has been shortened to better represent its content. Thus, *Organ Preservation for Transplantation, Third Edition* demonstrates a more accurate depiction of its current status. Nevertheless, the goal of this book remains exactly the same as in the first edition, that is "to analyze the most important aspects of organ procurement, perfusion and preservation for transplantation".[1]

My heartfelt thanks go to the authors and coauthors of the various chapters, who evidenced great patience in completing the text. Without their support it would have been difficult to carry out our mission of finalizing this book. Debra A. (Gordon) McKeehen, former fellow worker while at Detroit's Henry Ford and Mount Carmel Mercy Hospitals, was a faithful and dedicated supporter in the elaboration of this work. In addition, she helped me diligently and effectively with the first edition of this book. I am grateful for her caring work and excellent assistance. Special consideration and appreciation goes to Ron Landes, editor and founder of Landes Bioscience, and their superb staff as well. They were extremely helpful in providing the last details for the publication of this text. Thank you all.

Finally, I would like to thank my esteemed and appreciated mentor, Dr. John S. Najarian, who exhaustively reviewed this text and provided insightful and very worthwhile information in the Foreword of this third edition. Thanks so much Dr. Najarian for your generous support and invaluable time on the making of this edition.

Luis H. Toledo-Pereyra

References

1. Toledo-Pereyra LH. Basic Concepts in Organ Procurement, Perfusion and Preservation for Transplantation. New York: Academic Press, 1982.
2. Toledo-Pereyra LH. Organ Procurement and Preservation for Transplantation. Second Edition. Austin/New York: Landes Bioscience/Chapman & Hall, 1997.
3. Toledo-Pereyra LH. Organ Preservation for Transplantation. Third Edition. Austin: Landes Bioscience, 2009

CHAPTER 1

Science of Organ Preservation

Roberto Anaya-Prado, Alexander H. Toledo, Juan M. Palma-Vargas
and Luis H. Toledo-Pereyra*

Introduction

The goal of organ preservation is to obtain perfect preservation for as long as needed. Unfortunately, this has not been accomplished for the majority of organs; although the use of different techniques and solutions for preservation has modified and improved the results observed both clinically and experimentally. Hypothermia has long been considered to be an essential component of organ preservation. Therefore, it has been used both alone and in combination with perfusion techniques in order to protect organs from ischemia/reperfusion injury. The use of various preservation solutions has modified and improved some of the results observed in experimental and clinical organ transplantation. In this chapter, we analyze recent advances in organ preservation both clinically and experimentally; it also explores the basic mechanisms involved during hypothermia and reperfusion organ preservation, with special attention to efforts being done in refining solutions and preservation techniques.

The Effects of Hypothermia

Clinical hypothermia, ideally maintained between 6-10°C, significantly reduces the tissue metabolic rate and is the primary reason why preserved organs remain viable. Although the metabolic demand continues at about 10% of normal during hypothermia, the activity of the different enzymatic catalytic reactions is not uniformly affected by reduced temperatures. Most enzymes of normothermic animals show a 1.5 to 2.0 fold decrease in activity for every 10°C decrease in temperature. Thus, the metabolic rate is suppressed by about 12 to 13 fold when the temperature is reduced form 37°C (body temperature) to 0°C, probably the lowest point of preservation temperature.[1-3] Furthermore, membrane-bound enzymes systems. This difference is apparently due to the temperature-induced decrease in membrane fluidity via the solidification of the membrane-bound lipids.[1]

Hypothermia Works through Different Mechanisms

The organ's metabolic activity is not entirely abolished even at 0°C and this fact imposes a limit on the duration of organ viability under hypothermic storage.[3,4] Hypothermia decreases the rate at which intracellular enzymes degrade essential cellular components necessary for organ viability. Yet this protective effect is incomplete since hypothermia does not completely stop metabolism and ultimately the organ can cease to function under these conditions as well. However, in slowing the reactions and cell death, hypothermia greatly extends the preservation time.

In transferring the organ from normothermic to hypothermic conditions, the optimum rate of cooling has been a topic of much discussion. The rapid cooling of the kidney can lead to a loss of glycolytic enzymes, an event which could be prevented by a more gradual reduction of temperature.[5,6]

*Corresponding Author: Luis H. Toledo-Pereyra—Departments of Research and Surgery, Michigan State University, Kalamazoo Center for Medical Studies, Kalamazoo, Michigan 49008, USA. Email: toledo@kcms.msu.edu

Organ Preservation for Transplantation, Third Edition, edited by Luis H. Toledo-Pereyra.
©2010 Landes Bioscience.

The mechanism of organ injury may have been cold-induced vasoconstriction.[7] At temperatures in the range of 0° to 6°C, surface cooling has been shown to results in good preservation of renal function for periods of six to eight hours.[2,8-9] But deterioration becomes evident as preservation with this method is progressively extended. Ultrastructural changes can be found in the proximal and to a lesser extent in the distal tubules after 12 hours.[2,10] Surface cooling, however, has been consistently utilized in clinical transplantation for storage times considerably longer than 12 hours, probably due to better protective methods utilized with current preservation.

Hypothermia is not simply a process of slowing normothermic metabolic pathways. Indeed, the benefits of hypothermia in sharply reducing metabolic demands are counterbalanced by a parallel blockade of a number of key enzymatic reactions. For example, glycolysis, which normally supplies the major energy requirements of the renal cortex, is blocked partially at several levels as a consequence of the Hypothermic inactivation of key enzymes, such as glycogen phosphorylase and glyceraldehydes-3-phosphate dehydrogenase. Furthermore, fatty acid metabolism is also impaired by inactivation of pyruvate carboxylase by hypothermia.[11]

Glycolytic Enzymes and Energy Stores

It has been demonstrated that hypothermia produces conformational changes in the tertiary structure of a number of enzymes (pyruvate kinase, glutamate dehydrogenase, argininosuccinase), there by inactivating them. Although this process usually is reversible by rewarming, a number of key glycolytic enzymes accepting sulfhydryl groups such as glyceraldehyde dehydrogenase, lactate dehydrogenase and lactase may form disulfide bonds, with resulting irreversible deactivation.[12]

Glycolysis also can be blocked by the action of intermediate products of fatty acid metabolism which may accumulate as a consequence of hypothermia. For example, it has been shown that citrate inhibits the enzyme phosphofructokinase, a primary glycolytic regulating enzyme and accumulation of acetyl coenzyme A has been shown to inhibit the pyruvate dehydrogenase that prevents the final degradation of pyruvate in the Krebs cycle, diminishing then the production of lactate (Fig. 1).

While hypothermia can block energy production at various stages, a number of cold-resistant enzymatic reactions will persist, with a shunting of glycolytic intermediates and energy stores to other reactions that are not required to maintain cellular integrity. Persistence of the cold-resistant glycerophosphate pathway, for example, stimulates the formation of triglycerides from glucose in cold preserved organs. In this regard, for instance, the principal source of energy for the renal cortex during hypothermia proceeds from free fatty acid metabolism. Octanoid acids (especially caprylic acid) are degraded to acetyl coenzyme A and enter the Krebs cycle. Unlike octanoic acids, however, long-chain free fatty acids such as palmitic and myristic acids cannot be degraded into energy-producing cycles, but, instead, they are incorporated into tissue triglycerides through an energy-consuming process.[13] However, during hypothermia, phosphorylation is suppressed because of the inability of adenosine diphosphate to penetrate the inner mitochondrial membrane following hypothermic inactivation of adenosine diphosphate translocase (a rate limiting adenosine diphosphate transfer enzyme). The adenosine diphosphate remains in the cytosol and becomes degraded into adenosine monophosphate and ultimately into hypoxanthine, which readily diffuses out of the cell. Recovery of the resulting depleted pool of adenine nucleotide by the novo synthesis may take several hours after restoration of normal temperatures and oxygen levels.

Another important enzymatic system deactivated by hypothermia is the sodium potassium adenosine triphosphate (ATP), which is important in maintaining all membrane integrity.[14,15] The cold has been found to deactivate sodium-potassium adenosine triphosphatase by disassembling the enzyme into its 3, 5, and (S) subunits. Upon rewarming, reassembly occurs but irreversible damage develops after prolonged hypothermia. Deactivation of this system results in a massive redistribution of intracellular and extracellular electrolytes, with cell swelling and rapid loss of intracellular potassium; the later is essential to maintain many vital enzymatic reactions within the cell.[16] The combined loss of adenine nucleotides and intracellular potassium; significantly hinders the replenishment of needed stores after restoration of blood flow, at the time when energy demands

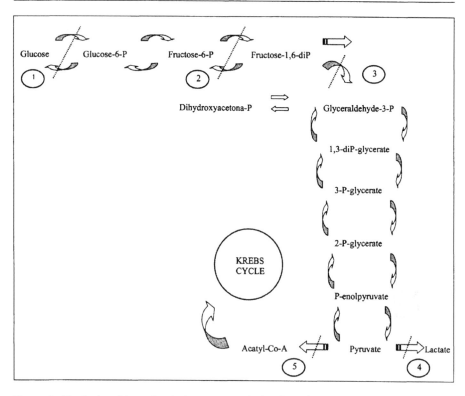

Figure 1. Blockade of key glycolytic enzymes during hypothermic organ preservation. 1, hexokinase; 2, phosphofructokinase; 3, glyceraldehyde phosphate dehydrogenase; 4, pyruvate carboxylase; 5, pyruvate dehydrogenase.

are particularly great. The amelioration of damage in regard to the key enzymatic reactions during hypothermic ischemia remains one of the most critical prerequisites to improvement in the quality and duration of organ preservation.

Calcium-Calmodulin Complex

The calcium-calmodulin complex has a central role in the regulation of many key enzymes responsible for mitochondrial respiration, adenosine triphosphate transport and regulation of ionic transport and membrane potential (Fig. 2).[17] These effects raise the possibility that regulation of cytosolic calcium may restore or preserve the enzymatic reactions needed for the integrity of the hypothermic ischemic cell. Calcium activates several enzymatic systems including glycogenolysis, lipase and phospholipase. It inhibits other enzymatic systems such as pyruvate kinase and phospholipid synthesis and it is involved ultimately in hormonal regulation, including the release of insulin, steroids, vasopressin and catecholamines and the binding of prostaglandins to membranes.[18] Calcium may exert its functions by activation and deactivation of other intracellular messengers and also possesses regulatory effects of adenyl cyclase and phosphodiesterase.[19]

In the cytosol, calcium exists bound to calmodulin, this calsium binding protein was discover by Cheung in 1967.[20] It is a prime mediator of important cellular functions such as membrane phosphorylation, microtubule disassembly, activation of phosphorylase A and B kinase and release of neurotransmitters.[19] The precision with which this protein regulates these enzymes may due to conformational changes of its alpha helical structure, caused by successive binding of calcium to its four calcium binding sites.[21]

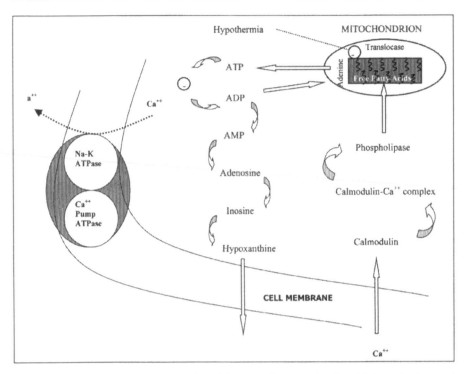

Figure 2. Possible role of calcium-calmodulin complex on mitochondrial respiration and phosphorylation during hypothermic preservation.

The cellular reactions mediated by calcium may require that calcium be kept in the cytosol at low levels of activity and that it be allowed to fluctuate widely and rapidly around this low level switch intracellular targets on and off alternately. During ischemia and specifically during hypothermic ischemia the enzymatic systems primarily responsible for the efflux of calcium across plasma membranes (calcium specific adenosine triphosphatase and the sodium calcium exchange system) are deactivated.[22] The rapid depletion of energy stores during hypothermic ischemia results in the deactivation of calcium-specific adenosine triphosphatase and the deactivation of sodium potassium adenosine triphosphatase during hypothermia which causes massive influx of sodium into the cytosol, resulting in failure of the sodium calcium exchange system. The resultant massive influx of sodium into the cytosol affects numerous cellular enzymes adversely, which produces relentless deterioration of cellular function and ultimately cell death (Fig. 3).

Alterations in the calcium-calmodulin complex can cause mitochondrial and membrane dysfunction by damaging the phospholipids moiety of these structures.

Phospholipids not only form the backbone structure of the bilayer membrane but they provide sites of attachment and optimum environment for energy transducing units. Activation of phospholipase A mediated by the calcium-calmodulin complex results in fusion of the inner and outer membranes of the mitochondria in a pentamellar structure, with a decrease in membranes function.[23] Prolongation of the ischemic insult then produces swelling of the inner membrane compartments, sometimes massively, leading to rupture of the outer membrane and further dilution of the inner membrane compartament.[18]

The principal limitation of the safe use of cold preservation techniques is due to mitochondrial dysfunction and particularly to the failure of the adenosine translocase system, a rate-limiting step in the transfer of adenosine diphosphate from cytosol to mitochondria. As a consequence

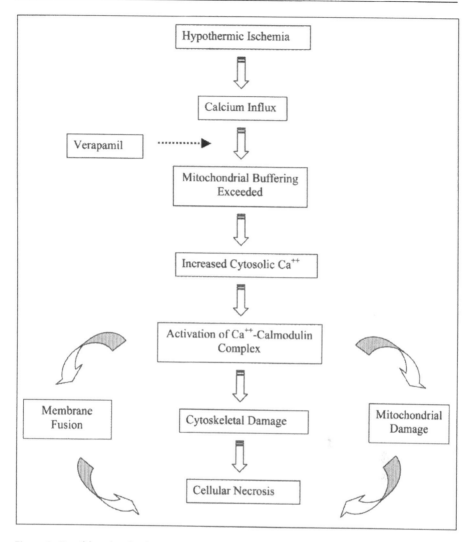

Figure 3. Possible role of calcium-calmodulin complex in cellular deterioration during hypothermic ischemia.

and despite an adequate supply of oxygen, the substrate energy stores cannot be maintained by phosphorylation and cellular deterioration ensues. The inability to phosphorylate adenosine diphosphate results in dephosphorylation of adenosine diphosphate to adenosine monophosphate to hypoxanthine; the later is freely diffusible and thus provides a mechanism for loss of adenosine nucleotides from the cell.[24,25]

Damage to cellular membranes during hypothermic ischemia also may be mediated by the action of calmodulin on the cytoskeleton.[26] Calmodulin besides mediating the calcium-dependent assembly and disassembly of microtubules, activates a specific myosin light-chain kinase, which in turn, regulates the activity of the microfilamental myosin.[27]

Cell separation may be another important mechanism underlying organ failure after ischemic damage. Endothelial cell separation due to cytoskeleton damage may result in exposure of collagen, which then can lead to platelet aggregation and intravascular coagulation. Together, these data suggest that cellular necrosis following hypothermic ischemia may be consequence of a rapid influx of calcium into cytosol. This influx then exceeds the buffering capacity of the mitochondria, with activation of the calcium-calmodulin complex. Activation of this complex then causes phospholipase activation with membrane breakdown, cytoskeleton damage and adverse effects on mitochondrial respiration. If allowed to progress, this event will result in irreversible cell damage.

Hypothermia-Induced Cell Swelling

Normally, the cells are bathed in an extracellular solution which is high in sodium (Na^+) and low in potassium (K^+). This ratio is maintained by the Na^+/K^+ ATPase pump, which uses much of the energy derived from ATP during oxidative phosphorylation. The Na^+ pump effectively makes Na^+ an impermeant anion outside the cell that counteracts the colloidal osmotic pressure force derived from the intracellular proteins and other impermeable anions (Fig. 4). The calculated osmotic force derived from intracellular proteins and impermeable anions is about 110-140 mOsm/kg.[28] The interstitial fluid bathing the cell is relatively low in protein and therefore water tends to enter the cell to equalize the pressure difference between the two compartments. The presence of nondiffusible anions on one side of a semipermeable membrane results in a redistribution of ions that favors increasing intracellular osmolarity with a concomitant redistribution of tissue water. Finally, the inside of the cell is electronegative relative to the outside and this orientation of the resting membrane potential prevents the accumulation of permeable anions in the interstitial fluid.

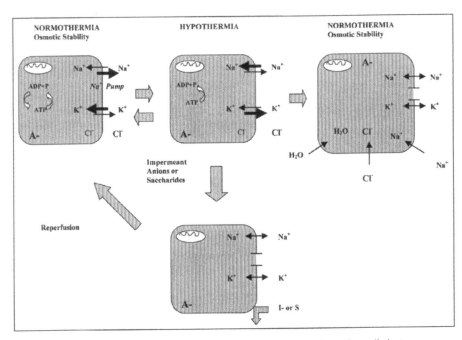

Figure 4. Hypothermic-induced cell swelling. The osmotic stability of a cell during normothermia is lost at hypothermia due to plasma membrane ion pump activity depression and loss of energy (ATP) for these energy-linked pumps. Continuous exposure to hypothermia leads to Chloride (Cl-) influx due to lack of membrane potential. This Cl- influx must go along with a cation; which in turn increases osmotic strength inside the cell. The combination of the already present impermeant anions (A-) within the cell, along with the increased osmotic strength, leads to water uptake and cell swelling. This whole process can be lethal to cell viability. Modified from Southard JH and Belzer FO. Principles of Organ Preservation: Part I, Surg Rounds 1993; May:353

Unfortunately, hypothermia per se has some adverse effects at the present time and could be a major unknown obstacle in organ preservation.[29] Anaerobic hypothermia preservation suppresses the activity of the Na^+/K^+ pump and decrease the membrane potential of the plasma membrane. Consequently, Na^+ and Cl^- enter the cell against a low concentration gradient and the cell swells because it takes up water due to an increase in the intracellular osmotic strength. The tendency for cell swelling is due to the lack of the intracellular position of sodium ions; this situation is normally maintained by Na^+/K^+ pump located in the outer membrane of all cells. These pumps continuously extrude sodium from the interior of the cell as rapidly as sodium enters the cells by fusion from a high extracellular to a low intracellular concentration. This pump mechanism is an energy requiring transport mechanism. The active extrusion of sodium ions and uptake of potassium under normal condition produces a membrane potential in which the cell interior is relatively negatively charged ions such as chloride from entering the cell. In addition, the extracellular position of sodium ions can counterbalance the oncotic effect of intracellular colloids. Thus, the sodium pump balances solute distribution and stabilizes the cell volume. It is apparent that any condition which adversely affects the sodium pump mechanism will result in cell swelling.

Cell swelling occurs in two phases (Fig. 5). If the sodium pump is inhibited, the initial result is an exchange of sodium for potassium and a redistribution of ions. Thus, more sodium enters the cell than potassium leaves because the internal proteins also pull water into the cells resulting in some degree of swelling (Phase 2).[29] This tendency to swell can be counteracted by adding 110-140

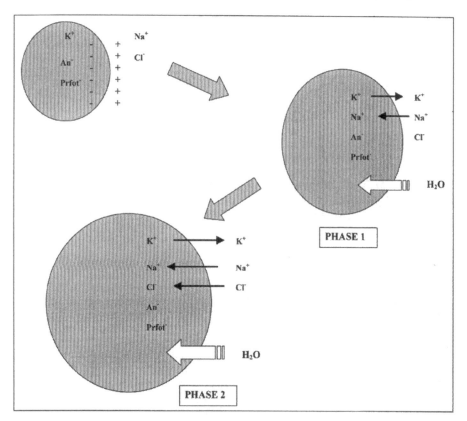

Figure 5. Theoretical stages of cellular swelling. Major swelling occurs after membrane potential disappears, allowing chloride to enter the cell. Modified from Flye MW. Principles of Organ Transplantation. Philadelphia: WB Saunders, 1989.

mmol/l (110-140 mOsm/kg osmotic force) of substances that are impermeant to the cell. This concentration approximately equals the impermeant concentration in most of the cold storage solutions. In successful organ preservation, maintenance of the membrane potential would then appear to be another major prerequisite. However, this appears to be extremely difficult without an active pumping mechanism to extrude positively charged sodium ions in exchange for potassium ions.

If cell water can be maintained in the intravascular system by very high but nontoxic concentration of substances such as mannitol, glucose or sucrose, why then are these substances not acceptable for long term storage of organs? The problem here is that these substances, although poorly diffusible, will diffuse in time into the cell. This produces two problems: First, a progressive increase in intracellular osmolarity during perfusion might produce detrimental effects. Second and perhaps more important, severe harmful effects will occur after reimplantation of these organs following the period of storage. On implantation of the organ, these cells are now subject suddenly to a normal extracellular osmolarity which is relatively hypotonic to the intracellular osmolarity. The sodium pump may not be immediately functional, especially if ATP levels are low and thus accumulated sodium cannot be rapidly extruded. Furthermore, the accumulated intracellular mannitol cannot be metabolized and also is extruded slowly. This high intracellular osmolarity will results in increased swelling. Swelling of the endothelial cells may prevent restitution of normal bloodflow to the organ, which by itself leads to warm ischemia. Cell swelling of the parenchyma cells will also involve the mitochondria resulting in structural and functional derangement. Thus, although the organs may have been viable at the end of the storage period, destruction of multiple components will occur on re-implantation. Although mainteinance of normal intracellular volume appears to be prerequisite for successful organ preservation, many other physiological and biochemical parameters may need to be controlled.

Hypothermia and pH

Cold ischemia can also stimulate glycolysis and glycogenolysis, as well as increase the production of lactic acid and the concentration of hydrogen ions. It is clear that acidosis develops rapidly during normothermic ischemia and that the magnitude of this pH shift is much reduced by cooling. Tissue acidosis can damage cells, induce lysosomal instability, active lysosomal enzymes and alter mitochondrial properties. However the damage associated with decline in cellular pH during ischemia is self-limiting, owing to the fact that key regulatory glycolytic enzymes such as phosphofructokinase are inhibited below a pH of 6.7.[30-32] However, the prevention of intracellular acidosis is an important element necessary for good preservation. The effect buffering of cold storage solutions or the use of a flush-out solutions with an alkaline pH are some methods suggested to improve the storage of solid organs.[33,34]

The Effects of Preservation Solutions

A basic problem of cold preservation is that when the sodium pump is turned off by ischemia and by cold, cells progressively lose potassium and magnesium and gain water and sodium with prolonged storage. Similar effectiveness of flushing solutions of varied composition suggests that one reason for success may be that poorly diffusible ions or solutes delay cellular swelling an edema. Since the first descriptions of intracellular composition for hypothermic preservation, flushing solutions have generally been formulated with high potassium content such as Collins, Euro-Collins and UW solutions (Table 1). Because the high potassium content tends to cause vasospasm and may lead to cardiac irregularities on reperfusion of a transplanted organ, a reduction in the potassium concentration of the perfusate is desirable; however, the question of the optimal cation content is still in debate. Multiple studies have been carried out to elucidate this significant question.[35,36]

An effective cold flush-out solution may also prevent the expansion of the interstitial space, which would occur during the in situ flushing of the donor organ. If the capillary system is affected by an ineffective solution, this will produce poor distribution of the solutions do not contain substances that exert oncotic support (albumin or other colloids) except for the UW and previously

Table 1. Essential components of various preservation solutions*

	Coll	HTK	UW	Celsior	Col	UP
KH$_2$PO$_4$ (mmol/L)	15.1					
K$_2$HPO$_4$ (mmol/L)	42.5		25.0		25.0	
KCL (mmol/L)	15.0	16.0		15.0		
NaHCO$_3$ (mmol/L)	10.0					
Glucose (mmol/L)	140.0				67.4	11.0
MgSO$_4$ (mmol/L)	30.0		5.0		5.0	
Histidine (mmol/L)		180.0		30.0		
Histidine-HCl (mmol/L)		18.0				100.0
Tryptophane (mmol/L)		2.0				
α-ketoglurate (mmol/L)		1.0				
NaCl (mmol/L)		15.0				
MgCl$_2$ (mmol/L)		4.0		13.0		
Mannitol (mmol/L)		30.0		60.0		20.0
Lactobionate (K) (mmol/L)			100.0	80.0		
Glutathione (mmol/L)			3.0	3.0		
Adenosine (mmol/L)			5.0		5.0	5.0
Allopurinol (mmol/L)			1.0			
Raffinose (mmol/L)			30.0			
HES (g/dL)			5.0			
Glutamic acid (mmol/L)				20.0		
CaCl2 (mmol/L)				0.25		
Lidocaine (mg/L)						100.0
Insulin (U/L)						10.0
K gluconate (mmol/L)					95.0	
Heparin (U/mL)					10.0	
Dextran (g/dL)					5.0	
Nitroglycerin (mg/L)					0.1	
Verapamil (:g)					10.0	
N-acetylcysteine (mmol/L)					0.5	
Db-cAMP (mmol/L)					2.0	

*Coll: Collins (originally developed in 1964 for the use in kidneys, later applied to livers); HTK: Histidine-Tryptophane-Ketoglutarate (it was introduced in 1980 and is safe for kidney, liver and pancreas transplants); UW: University of Wisconsin (first developed in 1986. Used for the storage of pancreas, later introduced to safely preserve livers and kidneys); Col: Columbia (used for storage of both heart and lungs); UP: University of Pittsburgh (used for the preservation of hearts).

the silica gel formulated solutions.[34-39] The components of the flush-out solutions, therefore, rapidly diffuse into the interstitial space and could produce tissue edema.[1] Thus, the ideal in situ flush-out solution should contain substances that create colloidal osmotic pressure and allow the free exchange of essential constituents of the flush-out solution without expanding the interstitial space.

Hypothermia and Microcirculation

The ischemic damage following organ transplantation is predominantly located at the endothelial cell level and is a major cause for disturbance of the microcirculation.[40] Thus, in order to improve the quality of hypothermic preservation, the study of preserved organs should be focused in the microcirculation.[41]

Experimental studies in hepatic microcirculation have shown that after reperfusion the number of blood cells adhered to endothelial cells increased as the preservation time increased, sinusoidal blood flow also decreased as the preservation time increased and the phagocytic activity of Kupffer cells increased with the preservation time.[42,43] These features may cause cell death and contribute to primary graft nonfunction after transplantation. Because of these effects, several investigators have addressed the impact that different storage solutions might have in the disturbances of the microcirculation.[44-46] For instance the hepatic graft immediately before revascularization with a specially designed rinse solution such as the Carolina rinse solution, has been reported to improve early function after prolonged cold ischemia.[47,48] From the microvascular changes observed in the experimental liver preserved for up to 24 hours with UW solution, it has been concluded that the UW solution can prevent endothelial damage and microcirculatory injury.[45,49]

Since temperature increases membrane fluidity and decreases vascular resistance, Takei et al, investigated the effect of a brief rinse of liver grafts with a warm buffer.[50] This idea markedly improved the hepatic microcirculation, leading to a dramatic improvement in graft survival.

Injury Caused by Preservation

There are multiple changes induced by prolonged hypothermia in the metabolism of parenchymal cells that compromise the survival of the organ. To obtain successful long-term preservation will require correcting, simultaneously, all the various limiting factors extensively mentioned throughout this chapter. Preservation injury, which represents the anatomical and functional alterations seen after cold storage or hypothermic perfusion and reperfusion, needs to be overcome to improve the results after transplantation.

When one deals with hypothermic perfusion, the possibility of damage to endothelial cells is more apparent. This damage is more pronounced when organs are perfused at hypothermic high pressures. In the kidney, hypothermic high pressures are considered to be higher than 60 mmHg. Consequently, less endothelial cell damage as well as good quality short-term preservation has been obtained with lower perfusion pressures of 20-30 mmHg.

Endothelial cell damage may result from either mechanically or metabolically induced changes. The mechanical damage induced by continuous hypothermic perfusions is the major cause. One way to avoid perfusion-induced damage may be to combine the benefits of cold storage to those of perfusion by intermittent perfusion. In addition to mechanical damage to endothelial cells, the accumulation of toxic metabolites may react with endothelial cells as well as other elements. Two classes of toxic substances may accumulate that include lipid peroxides and/or denatured proteins. The presence of oxygen may promote the formation of peroxides. Although at hypothermia this reaction may be slow, there is evidence that it occurs too. Other preservation injury changes are described in the upcoming sections.

Free Radicals and Preservation Injury

During cold storage, the organ could undergo certain injury that might be exacerbated by the effect of oxygen free radicals during the reperfusion period. This could occur either by the induction of the activity of various enzymes such as xanthine oxidase, mitochondrial oxidases and others that produce cytotoxic free radical products, or the depletion of endogenous factors necessary for the scavenging of oxygen radicals such as glutathione, vitamin E, NADH, superoxide dismutase, gluthatione peroxidase and catalase.[51] The injury to the vascular system that results in attraction of macrophages to the preserved and reperfused organ with activation and production of oxygen free radicals is another important factor. The basic source of oxygen- derived free radicals is xanthine oxidase.[52] During ischemia, xanthine dehydrogenase (an enzyme that uses nicotinamide adenine dinucleotide (NAD) as an electron acceptor and does not produce superoxide anions) is converted to xanthine oxidase, which uses oxygen as electron acceptor and generates superoxide anions. The mechanism of conversion from dehydrogenase to oxidase may depend on the calcium activation of a protein kinase. Substrates for xanthine oxidase are generated by the ischemia-induced breakdown of adenine nuecleotide to hypoxanthine. After reperfusion, oxygen is supplied to xanthine oxidase, which convers hypoxanthine to uric acid and generates superoxide anions (Fig. 6).[53]

The preservation of organs for transplantation induces cell injury similar in some ways to the one observed after warm ischemia. Organ failure or delayed function after preservation may, therefore, be related to tissue damage caused by oxygen derived free radicals. The damaging metabolites (superoxide anion, hydroxyl radical or H_2O_2) may originate during either reperfusion or preservation or both.[54] Most likely, some organs are more sensitive to damage from oxygen derived free radicals because of the abundance of xanthine oxidase.

The primary goal of organ preservation is to maintain the integrity of all cellular systems so that there is minimal opportunity for oxygen free radical injury to develop upon reperfusion. This appears to be possible under conditions of cold storage of organs by utilization of an appropriate

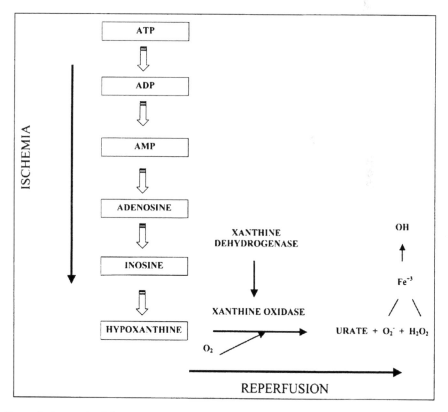

Figure 6. Free radical formation (OH, O_2^-, H2O2) after ATP breakdown, following ischemia and subsequent reperfusion.

organ preservation solution.[55] However, it is less likely under conditions of warm ischemia. Organs exposed to warm ischemia may be more prone to injury during reperfusion and this injury may be partially due to the generation of oxygen free radicals.[56] Certain types of oxygen free radicals may be of little significance in human liver and kidney because endogenous xanthine oxidase has a relatively low activity compared with the high endogenous activity of superoxide dismutase, which scavenges superoxide anions.[57] More studies are being re-appraised on this topic. However, the injury induced by oxygen free radicals may be extremely important in lungs and intestine, which have high activity of the enzymes able to generate these compounds.

Gluthatione appears to play a critical role as an antioxidant in the liver and heart and it is relatively in high concentrations in these organs.[58] During cold storage, there is a nearly 50% loss of liver cell gluthatione and previous studies have shown that the depletion of gluthathione from the liver sensitizes the cells to oxidative injury.[59] The normal endogenous rate of oxygen free radical generation could prove to be cytotoxic of the cellular antioxidant defense systems which would be compromised. Thus, the loss of endogenous antioxidants can lead to injury to the preserved organ by cytotoxic oxygen metabolites. The rapid regeneration of the naturally occurring antioxidants may be a requisite for good return of function following organ transplantation.[60] These naturally occurring antioxidants may be more important in the protection of tissue than exogenously added scavengers, especially if they do not rapidly penetrate the cell membrane.

Endothelial Cell Adhesion Molecules and Preservation Injury

Ischemia is associated with upregulation of the three principal families of adhesion molecules (integrins, immunoglobulins and selectins),[61] as well as with increase expression of major histocompatibility antigens.[62,63] These processes are thought to increase the organ damage and the immunogenicity of the graft loss from rejection under this setting.[64] With the design of specific monoclonal antibodies against the aforementioned upregulated adhesion molecules or mimetic of its counter ligands, such as Sialyl Lewis^x, several investigators have been able to overcome some of the damage occurring after warm ischemia and reperfusion; with the possible application of this principle to future therapeutic interventions for cold preservation.[65-70]

The Future of Organ Preservation

Transpant centers and research teams will continue to search for the perfect preservation solution. However, it may be that we are nearing our peak of solution development and all that remains to be done is modifications of solutions already in use. New solutions often prove to be superior to existing solutions.[71] However, it is important to realize the point when solutions may be near the best that they can be and to look at other ways for improving organ viability e.g., preparation of the donor before procurement. In the future, we may see a greater use of perfusion machines, although this may be unjustified when preservation solutions are satisfactory.[72] With increased multiple-organ procurement from single donors, the new challenge is to find solutions that are universally optimum for each organ. This may be performed through hybridization of existing solutions.

Conclusion

Significant changes have occurred in the field of organ preservation during the last 30 years. The hypothermic perfusion, initially development as the procedure of choice, is today limited in its use and cold storage for now remains as the conventional method of preservation. Crystalloid solutions and later colloid preservation solutions have been successful in improving organ function. Currently, there are four principles to bear in mind in order to improve organ preservation: hypothermia, maintenance of the physical as well as the biochemical environments and modulation of the reperfusion response. Each one of these factors may affect, individually or as a whole, the success of organ preservation. Finally, prevention of preservation injury with carefully planned therapeutic manipulations might be of increased significance in attaining near normal organ function after transplantation.

References

1. Belzer FO, Southard JH. Principles of solid organ preservation by cold storage. Transplantation 1988; 45:673.
2. Calne RY, Pegg DE, Pryse-Davies J et al. Renal preservation by ice cooling. An experimental study relating to the kidney transplantation from cadavers. Br Med J 1963; 2:651.
3. Levy MN. Oxygen consumption and blood flow in the hypothermic, perfused kidney. Am J Physiol 1959; 197:111.
4. Burg M, Orloff JM. Active cation transport by kidney tubules at 0°C. Am J Physiol 1964; 207:983.
5. Francavilla A, Brown TH, Fiore R et al. Preservation of organs for transplantation: evidence of detrimental effect of rapid cooling. Eur Srg Res 1973; 5:384.
6. Jacobsen IA, Kemp E, Buhl MR. An adverse effect of rapid cooling in kidney preservation. Transplantation 1979; 27:135.
7. Fonteles MC, Karow AM Jr. Vascular alpha adrenotropic responses of the isolated rabbit kidney at 15 degrees C. Arch Int Pharmacodyn Ther 1977;227(2):195-205
8. Kerr W, Kyle V, Keresteci A et al. Renal hypothermia. J Urol 1960; 81:236.
9. Collins G, Bravo-Shugarman M, Novom S et al. Kidney preservation for transportation 12 hours storage in rabbits. Transplant Proc 1969; 1:801.
10. Fisher E, Copeland C, Fisher B. Correlation of ultrastructure and function following hypothermic preservation of canine kidneys. Lab Invest 1967; 17:99.
11. Lundstam S, Claes G, Johnson O et al. Metabolism in the hypothermically perfused kidney. Production and utilization of acetate in the dog kidney. Eur Surg Res 1976; 8:300.
12. Chilson O, Costello L, Kaplan N. Effects of Freezing on enzymes. Fed Proc 1965; S55:24.
13. Pettersson S, Claes G, Scherstein T. Fatty acid and glucose utilization during continuous hypothermia perfusion of dog kidney. Eur Surg Res 1974; 6:79.
14. Martin DR, Scott D, Downes G et al. Primary cause of unsuccessful liver and heart preservation: cold sensitivity of the ATPase system. Am Surg 1972; 175:111.
15. Leaf A. Maintenance of concentration gradients and regulation of cell volume. Ann NY Acad Sci 1959; 72:396.
16. Lubin M. Intracellular potassium and control of protein synthesis. Fed Proc 1964; 23:994.
17. Carafoly E. Membrane transport and regulation of the cell calcium levels. In: Cowley R, Trump BF, eds. Pathophysiology of Schock, Anoxia and Ischemia. Baltimore: William and Wilkins, 1982; 7:5.
18. Trump B, Berezesky I, Cowley R. The cellular and subcellular characteristics of acute and chronic injury with emphasis on the role of calcium. In: Cowley R, Trump BF, eds. Pathophysiology of shock, anoxia and ischemia. Baltimore: William and Wilkins, 1982; 1:6.
19. Cheung W. Calmodulin plays a pivotal role in cellular regulation. Science 1980; 207:19.
20. Cheung W. Cyclic 3'5'-nucleotide phosphodiesterase: pronounced stimulation by snake venom. Biochem Biophys Res Comm 1967; 29:478.
21. Richman P, Klee C. Specific perturbation by Ca++ of tyrosyl residue 138 of calmodulin. J Biol Chem 1979; 254:5372.
22. Gmaj P, Murer H, Kinee R. Calcium ion transport across plasma membranes isolated from rat kidney cortex. Biochem 1979; 178:549.
23. Wakabayashi T, Green D. Membrane Fusion in Mitochondria ultrastructural basis for fusion. J Electron Micros 1977; 26:305.
24. Mittnacht S, Sherman S, Farber J. Reversal of ischemic mitochondrial dysfunction. J Biol Chem 1979; 254:9871.
25. Feinberg H. Energetics and Mitochondria. In: Pegg D, Jacobsen I, Halasz N, eds. Organ preservation: Basic and Applied Aspects Lancaster: Eng. MIT Press LTD, 1982; 1:3.
26. Welsch M, Debman J, Brinkley B et al. Tubulin and calmodulin: effects of microtubule and microfilament inhibitors on localization in the mitotic apparatus. J Cell Biol 1979; 81:624.
27. Marcum J, Debman J, Brinkley B et al. Control of microtubule assembly-disassembly by calcium dependent regulator protein. Proc Natl Acad Sci 1978; 75:3771.
28. Jac-Knight A, Leaf A. Regulation of cellular volume. Physiol Rev 1977; 57:510.
29. Belzer F, Hoffman R, Southard J. Kidney preservation. Surg Clin North Amer 1978; 2:621.
30. Hardi I, Clunie G, Collins G. Evaluation of simple methods for assessing renal ischemic injury. Surg Gynecol Obstet 1973; 136:143.
31. Rehncrona S, Siesjo B, Smith D. Reversible ischemia of the brain: biochemical factors influencing restitution. Acta Physiol Scand 1979; 492:135.
32. Bock P, Frieden C. pH induced cold liability of rabbit skeletal muscle phosphofructokinasa. Biochemistry 1974; 13:4191.
33. Lie T, Ukikusa M. Significance of alkaline preservation solutions in transplantation. Transplant Proc 1984; 16:34.

34. Abouna GM, Sutherland DE, Florack G, Heil J, Najarian JS. Preservation of human pancreatic allografts in cold storage for six to 24 hours. Transplant Proc. 1987;19(1 Pt 3):2307-9.

35. Moen J, Claesson K, Pienaar H et al. Preservation of dog liver, kidney and pancreas, using the Belzer-UW solution with high sodium and low potassium content. Transplantation 1989; 47:490.

36. Collins GM, Warren et al. Canine and cadaver kidney preservation with sodium lactobionate sucrose solution. Transplant Proc 1993; 25:1588.

37. Toledo-Pereyra LH, Condie RM, Malmberg R et al. Long-term Kidney preservation with a new plasma perfusate. Forum Proc Clin Dial Transplant 1973; 3:88.

38. Toledo-Pereyra LH. A new generation of colloid solutions for preservation. Dial Transplant 1985; 14:143.

39. Walhberg JA, Southhard JH, Belzer FO. Development of a cold storage solution for pancreas preservation. Cryobiology 1986; 23:477.

40. Manner M, Shult W, Senninger N et al. Evaluation of preservation damage after porcine liver transplantation by assessment of hepatic microcirculation. Transplantation 1990; 50:940.

41. Holloway C, Harver P, Mullen J et al. Evidence of cold preservation induced microcirculatory injury in liver allografts is not mediated by oxygen free radicals or cell swelling in the rat. Transplantation 1989; 48:579.

42. Teramoto K, Browers J, Kruskal J et al. Hepatic microcirculatory changes after reperfusion in fatty and normal liver transplantation in the rat. Transplantation 1993; 56:1076.

43. Marzi I, Walcher F, Menger M et al. Microcirculatory disturbances and leukocyte adherence in transplant livers after cold storage in Euro Collins, UW and HTK solutions. Transpl Int 1991; 4:45.

44. Furukawa H, WU Y, Zhu Y et al. Disturbance of microcirculation asociated with prolongued preservation of dog livers under UW solution. Transplant Proc 1993; 25:1591.

45. Haba T, Hayashi S, Haschisuka T et al. Microvascular changes of the liver preserved in UW solution. Pathological and inmunoitohistochemical examination. Criobiology 1992; 29:310.

46. Palma P, Post S, Rentsh M et al. Effects of Carolina rinse on hepatic microcirclarion an leukocyte endothelium interaction in rat liver transplantation. Transplant Proc 1993; 25:2536.

47. Post S, Rentsh M, Gonzalez A et al. Efects of Carolina rinse and adenosine on microvascular perfusion and intrahepatic leukocyte endothelium interation after liver transplantation in the rat. Transplantation 1993; 55:972.

48. Post S, Palma P, Menger M et al. Differential impact of Carolina rinse and UW solutions on microcirculation, leukocyte adhesion, kupffer cell activity and biliary excretion after liver transplantation. Hepatology 1993; 18:1490.

49. Schlumpf R, Morel P, Loveras J et al. Examination of the role of the colloids hydrosyethylstarch, dextran, human albumin and plasma proteins in a modified UW solution. Transplant Proc 1991; 23:2362.

50. Takei Y, Gao W, Hijioka T et al. Increase insurvival of liver grafts after rinsing with warm Ringer's solution due to improvement of hepatic microcirculation. Transplantation 1991; 52:225.

51. Ballmer P, Reinhart W, Gey K. Antioxidant vitamins and disease risk of a suboptimal supply. There-Umsh 1994; 51:467.

52. Southard J, Den-Butter B, Marsh D et al. The role of oxygen free radicals in organ preservation. Klin Wochenschr 1991; 69:1073.

53. Southard J, Marsh D, Mc Anulty J et al. Oxygen derived free radical damage inorgan preservation: Activity of superoxide dismutase and xanthine oxidase. Surgery 1987; 5:566.

54. Fuller B, Grower J, Green C. Free radical damage and organ preservation: fact or fiction. A review of the interrelationship between oxidative stress and physiological ion disbalance. Cryobiology 1988; 25:377.

55. Toledo-Pereyra LH. Effect of warm ischemic time on kidney transplant. Contrib Nephrol 1989; 71:129.

56. Mc-Cord JM. Oxygen-derived free radicals in post ischemic tissue injury. N England J Med 1985; 312:158.

57. Keith F. Oxygen free radicals in cardiac transplantation. J Card Surg 1993; 8:245.

58. Vreugdehil P, Belzer F, Southard J. Effect of cold storage on tissue and cellular gluthathione. Cryobiology 1991; 28:143.

59. Scaudutuo R, Gattone V, Grotyohann L et al. Effect of altered gluithathione content on renal ischemic injury. Am J Physiol 988; 255:F911.

60. Colomer J, Fernandez-Cruz L, Saenz A et al. Oxygen free radicals and organ transplants. Rev Quir Esp 988; 15:305.

61. Bergese SD, Pelletier RP. Treatment of mice with anti-CD3 mAB induces endothelial vascular cell adherion molecule 1 expression. Transplantation 1988; 57:711.

62. Shackleton CR, Ettionger SI L, Mc Loughlin MG et al. Effect of recovery form ischemic injury on class I and II MHC antigen expression. Transplantation 1990; 49:641.

63. Shoskes DA, Halloran PF. Ischemic injury induces altered MHC gene expression in Kidney by an interferon gamma dependent pathway. Transplant Proc 1991; 23:599.

64. Halloran PF, Aprile MA, Farewell V et al. Early function as the principal correlate of graft survival. A multivariate analysis of 200 cadaveric renal transplants treated with a protocol incorporating antilymphocyte globulin and cyclosporine. Transplantation 1988; 46:223.
65. Winn RK, Liggit D, Vedder NB et al. Anti-P-selectin monoclonal antibody attenuates reperfusion injury to the rabbitear. J Clin Invest 1993; 92:2042.
66. Lefer AM, Xin-Liang MA. PMN adherence to cat ischemic reperfused mesenteric vascular endothelium under flow role of P-selectin. J Appl Physiol 1994; 76:33.
67. Okada Y, Coppeland BR, Mori E et al. P-selectin and intracellular adhesión molécula I Expresión alter focal brain inschemia and reperfusion. Stroke 1994; 76:33.
68. Garcia-Criado FJ, Toledo Pereyra LH, Lopez-Neblina F et al. Role of P-selectin in total hepatic ischemia and reperfusion. J Am Coll Surg 1996; 181:327.
69. Seekamp A, Till GO, Mullingan MS et al. Role of selectin in local and remote tissue injury following ischemia and reperfusion. Am J Pathol 1994; 144:592.
70. Misawa K, Toledo-Pereyra LH, Phillips ML et al. Role of Sialyl Lewis^x in total hepatic ischemia and reperfusion. J Am Coll Surg 1996; 182:251.
71. Badet L, Petruzzo P, Lefrancois N et al. Kidney preservation with IGL-1 solution: a preliminary report. Transplant Proc 2005; 37:308-311.
72. Southard JH, Blezer KO. Organ Preservation. Annu Rev Med 1995; 46:235-247.

CHAPTER 2

Organ Donation and Procurement

Francisco Rodríguez-Quilantán, Alexander H. Toledo,
Roberto Anaya-Prado and Luis H. Toledo-Pereyra*

Introduction

The use of appropriate techniques of organ procurement is the hallmark of successful preservation and transplantation. As indicated in previous editions of this book, organ procurement has varied depending on the organs to be recovered for transplantation or research according to each particular circumstance. In the last few years, the transplantation activity has increased to reach 28,357 organ transplants in the United States in 2007 alone according to the statistics of the US Department of Health and Human Services.[1] The number of patients dying in the waiting list in the same year climbed to 6,554. Because of these numbers and the imperative of maximizing the organs recovered for transplantation, the technical and logistic characteristics of organ procurement have become a rather significant aspect of this specialty. In this chapter, we will review the standard procedures utilized for organ procurement, as well as the most frequent indications for organ donation.

Criteria for Organ Donation: Brain Death

Since 1968 when the Ad Hoc Committee of Harvard Medical School defined brain death, its diagnosis became a requirement prior to the consideration of organ donation for transplantation.[2] The basic diagnostic findings of brain death should be satisfactorily fulfilled to allow for expeditious organ recovery. When organs are recovered from brain dead donors with intact circulation and rapidly cooled following the arrest of circulation, the preservation-reperfusion injury is diminished significantly, therefore cellular and whole organ viability can be achieved.

According to the American Medical Association, brain death is established when "permanent and irreversible cessation of the function of the brain" is reached. Table 1 lists the brain death criteria.[3,4] Exclusion parameters for organ donation are delineated separately in the organs studied. Allografts from non-heart-beating cadaveric donors[5] are emphasized in the next section of this chapter.

Non-Heart-Beating Organ Donors

Non-heart-beating donors (NHBDs) have been used mainly as kidney donors since the 1960s by few isolated transplant centers.[6] Recently, because of the severe shortage of organ donors the number of centers using organs from NHBDs has considerably increased.[2,6,7] In fact, some liver transplant centers have considered these donors as potential source for liver transplantation, if the time of complete ischemia is limited to 30 minutes maximum, the liver looks normal at the time of exploration and the donor does not have any other associated contraindications for liver use. Additionally, evaluating the effluent when initiating cold preservation is critical. Successful

*Corresponding Author: Luis H. Toledo-Pereyra—Departments of Research and Surgery, Michigan State University, Kalamazoo Center for Medical Studies, Kalamazoo, Michigan 49008, USA. Email: toledo@kcms.msu.edu

Organ Preservation for Transplantation, Third Edition, edited by Luis H. Toledo-Pereyra.
©2010 Landes Bioscience.

Table 1. Brain death criteria

- Irreversible deep coma in absence of hypothermia or nervous system depressant drugs
- No spontaneous movement and no decerebrate of decorticate posture
- Apnea with no spontaneous respiratory movements for at least three minutes without ventilatory assistance with a $PaCO_2$ greater than 55 mmHg without muscle relaxants
- No response to deeply painful stimuli
- No cranial reflexes:
  Not responsive, fixed and dilated pupils
  No oculocephalic reflex, external eye movements or response to head turning
  No oculovestibular reflex
  No corneal reflex
  No response to upper and lower airway stimulation (pharyngeal and endotracheal suctioning)

pancreas transplantation from NHBDs has been established at several centers. Further advances in the management of ischemia by antioxidants, nitric oxide, anti-selectins, calcium channel blockers and many other compounds might facilitate the possibility of increasing the donor pool through the use of NHBDs (see other chapters on Organ Preservation and Ischemia and Reperfusion).

According to Merion[2], NHBDs, also known as DCD (donation after cardiac death) donors, are a good opportunity for increasing the donor pool. In fact, he referred to the positive outlook given by the Institute of Medicine which emphasized that these donors are "an important medically effective and ethically acceptable approach".

The circumstances for the consideration of NHBDs are presented in the following way: once the donor does not meet the criteria for brain death, he/she can be considered as a potential NHBD, decision to withdraw life support is discussed and obtained, then, of course, with proper informed consent, the donor is moved to the operating room where is extubated and 60 to 90 minutes are alloted for the donor to expire, depending on the hospital specific regulations. If expiration occurs, the procurement of kidneys or liver and pancreas under the best conditions is started after additional 5 minutes after pronouncement of death is reached. It is worth noting that time of hypotension with systolic pressure of less than 60 mmHg for more than 20-30 minutes is more meaningful than the time of extubation to cross-clamp. Immediately thereafter, the NHBD abdomen and chest (if needed) are opened through a standard kidney and liver procurement procedure approach. At the earliest time possible when entering the abdomen, either one of the distal aorta or common iliac or external iliac is cannulated, perfusion starts and the distal inferior vena cava or common iliac vein is incised for venous drainage. Continuous suction for the draining of the abdominal venous blood is maintained and ice slush is placed around the organs to be dissected.[8,9]

Organ Donation and Procurement

Cardiac Donor

The decision to procure the heart depends not on the heart function at the time of procurement, but on the potential of the heart to recover full function after transplantation. The main criteria depends on the existence of an ABO blood group compatibility with a difference between donor and recipient size no greater than 20%. However, in infants with high pulmonary artery pressure, the difference in size could be bigger. Crossmatching is done retrospectively.[10] On the other hand, smaller hearts for otherwise dying patients can be heterotopically transplanted. The heart is assessed also by direct visualization during the recovery of the donor's heart (see Chapter 11).

Most hearts are orthotopically transplanted and all are obtained from cadaveric donors. Nevertheless, as mentioned above, in some selected cases, the heart can be placed heterotopically. Some centers report also the use of hearts obtained after one hour of cardiac arrest.[11]

Table 2. Heart donor acceptance criteria*

ABO Compatibility
  Age:
    Men less than 45, women less than 50 (flexible upper limit and in both cases could be older according to the results of coronary angiography)
  Normal cardiac function by ECG, echocardiogram
  Pulmonary function:
    Evaluated by gas exchange and chest x-ray studies
  No evidence of sepsis or malignancy other then intracranial tumor
  Negative serology for hepatitis B and C, HIV and CMV in pediatric recipients
  Up to 25% of difference in size (smaller donor)
  Estimated cold ischemic time 4 hours ideally
  Cause of death:
    Blunt and penetrating trauma: Rule out myocardial injury
    Carbon monoxide
  Medical history:
    Heart disease
    Hypertension
    Drug abuse
    Smoking
    Alcoholism
  Hemodynamic evolution
    Cardiac arrest, prolonged hypotension
    Vasoactive drug administration
    High central venous pressure

*Review Chapter 11 for more detail.

Table 2 describes the main issues considered for potential cardiac donors. In the United States, 2,210 hearts were recovered for transplantation in 2007. Cold ischemic time must not ideally exceed 4 hours. Age criteria are established in order to minimize silent coronary atherosclerosis. Tissue matching remains controversial as a means of organ distribution. Blunt trauma victims are candidates in whom it is necessary to evaluate myocardial contusion or direct penetrating injury, along with the fact of history of drug or alcohol abuse. Patients with primarily neurologic brain death must be evaluated for sequelae of severe hypertension and cardiac lesion. Prominent ST-T changes might be associated with cerebro-vascular disease, hypothermia or electrolyte abnormalities. Positive cytomegalovirus (CMV) donors are used for CMV positive recipients with prophylactic treatment.[10]

Heart Procurement

The donors' heart excision is performed through a median sternotomy in combination with the liver-pancreas-kidney procurement team whose main approach is through a midline laparotomy (Review chapter 11 for more detail) (Figs. 1A and 6A). The donor is heparinized (30,000 units i.v.) and the infusion of cardioplegia solution directly into the aorta or through the innominate artery is started while cold solution surrounds the heart and it is followed by ligation and division of the superior vena cava and pulmonary veins in order to decompress the heart once the aorta has been clamped proximally to the innominate artery (Fig. 1A). If the lungs are procured also, a cannula is placed into the pulmonary artery and lung preservation is initiated. Otherwise, the aorta is sectioned along with the pulmonary artery (Fig. 1B), completing the division of the left atrium (Fig. 1C) and the heart is taken out and placed in plastic bags containing ice-cold normal saline or University of Wisconsin preservation solution (UW solution) or HTK (Hystidine-tryptophan-ketoglutarate)

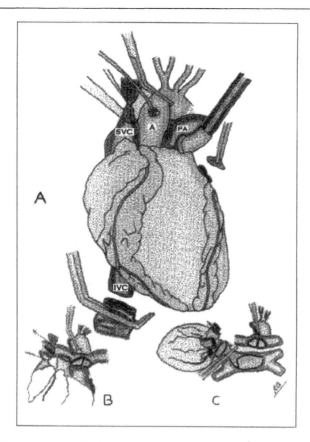

Figure 1. Cardiac recovery (and pulmonary preservation). A) Preliminary steps in cardiac recovery. The cannula of cardioplegia is inserted, the aorta clamped, the pulmonary veins vented, the inferior vena cava sectioned and the superior vena cava ligated. The cannula in the pulmonary artery is placed when both lungs are harvested concomitantly. SVC: Superior vena cava. IVC: Inferior vena cava. A: Aorta. PA: Pulmonary artery. B) Section of the superior vena cava, aorta and pulmonary artery is complete. C) Completion of the section of the left atrium.

or the selected preservation solution, and the graft is ready for implantation. If the block of heart-lungs was removed, the heart can be dissected out on the back table and the lungs left with the main bronchus, pulmonary veins and main branch of the pulmonary artery; otherwise the entire block (heart-lung) is transplanted, or the bilateral sequential lung transplantation (see: Lung Procurement) or separate transplantation (Figs. 1 and 2).

Lung Donor

By UNOS guidelines, a lung donor should be <55 years old, have no significant smoking history or active pulmonary disease, a clear chest x-ray and a PaO_2 >30 mg Hg, when measured on 100% oxygen and 5 cm positive end expiratory pressure for 5 minutes. The donor airways should be free of purulent secretions or aspirated material on bronchoscopy. Contraindications to double-lung recovery include unilateral infiltrate, pneumothorax, or pulmonary contusion, but these would allow for a single-lung recovery. Unfortunately, only 5-25% of the cases of multiple-organ donation have lungs that are suitable for transplantation.[12-17] In patients diagnosed as brain dead, lung injury occurs quickly due to a number of factors such as mechanical ventilatory support, i.v. fluids used for resuscitation, neurogenic interstitial edema and iatrogenic pulmonary infections. There is also lung trauma in many patients. Only 30% of heart donors are suitable for lung donation after lung trauma related injuries.

Size matching avoids atelectasias (bigger size donor) or repetitive pleural effusions or pneumothorax (smaller size donor) and also permits and adequate size of bronchial anastomosis. However, the telescoping technique might be used when small differences are evident in size matching.[15] Size matching in cases of heart-lung transplants is crucial because mediastinal shift is not always reliable to compensate. Recipients with emphysema, cystic fibrosis or other hyperinflated lungs need a donor with a vital capacity one liter greater than the recipient. The oxygenation on arterial oxygen tension must be ≥ 300 mmHg on inspired oxygen fraction of 1.0 and 5 cm H_2O positive end-expiratory pressure.

Lung Procurement

It has become standard practice to recover both lungs for transplant whenever feasible without affecting the use of the heart for transplant, through the median sternotomy. The determination of suitability of the heart and lungs (bronchoscopy, open assessment) is made only early on, after which the abdominal team continues their procurement. The pericardium is sectioned and pericardial stay sutures are placed. On the venous side, the intrapericardial superior vena cava is circled with sutures and the inferior vena cava with an umbilical tape. On the arterial side, the ascending aorta and main pulmonary artery are separated and circled with an umbilical tape each. The posterior pericardium is incised and the distal trachea exposed. Before the installation of the cannulas, the donor is heparinized and in coordination with the other teams, all the perfusion cannulas are installed. The cardioplegia cannula is also placed (Fig. 1A). It is particularly important to leave enough left atrial cuffs in the donor and the recipient, taking care of the atrial grove in the donor (Fig. 1C). When cardiac arrest occurs, the cross-clamp of the aorta, ligature of superior vena cava, section of inferior vena cava and amputation of the tip of the left atrial appendage takes place (Fig. 1A). Bolus administration of 500 mg of prostaglandin E_1 is drawn up in 10 ml of saline solution into the main pulmonary artery alongside the cannula (not shown in Fig. 1A) and the infusion of the preservation solution begins (3-5 liters of modified Euro-Collins, UW, HTK solution or Celsior or other special solution) at 4°C and the chest is flooded with ice-cold saline solution. A few minutes later the heart is extracted dividing the superior vena cava between the ligatures, the aorta is transected proximal to the cross-clamp, the distal main pulmonary artery is also transected (Fig. 1B). The roof of the left atrium is then divided toward the inter-atrial grove and the inferior side must be away from the mitral valve and the coronary sinus (Fig. 1C) (also see Heart Procurement).

Following this, the trachea is encircled digitally one or two rings above the carina and with the lungs moderately inflated (10-20 cm H_2O pressure) the trachea is double-stapled and divided. The same is done with the esophagus proximally and distally (Fig. 2). The esophagus and the aorta are extracted with the specimen to preclude injury to the vessels or the airway. The pericardium near the diaphragm is incised taking care to stay superior to the open inferior vena cava; the pulmonary ligaments are divided and the block is extracted, placed on ice and ready for implantation. The division of lungs might be done at this time, with the section of the main pulmonary artery, the trachea (bronchus) and the pulmonary veins with a left atrial cuff each (not shown in Fig. 2) (also Heart-Lung Procurement).[13,14]

Lung-Heart Donor

In general, similar indications utilized for individual heart and lungs are used for lung-heart donors (Table 2).

Lung-Heart Procurement

Through the above mentioned median sternotomy, the cardiothoracic cavity is approached. Dissection of the vessels proceeds as described for lung procurement. Systemic heparin is administered, the cardioplegia solution is infused and after the cardiac arrest, prostaglandin E_1 injection into the pulmonary artery is used regularly. The pulmonary circulation is not disturbed in block heart-lung harvesting, as well as the left atrial appendage, otherwise, only preserved along with

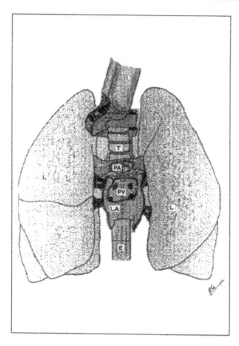

Figure 2. Extraction of the lungs en bloc. The figure shows the lines of division of the esophagus (proximally and distally), as well as tracheal division. When the lungs are transplanted separately, the pulmonary veins are divided in the middle and the pulmonary artery at the bifurcation. LA: Left atrium. PA: Pulmonary artery. PV: Pulmonary veins. E: Esophagus. T: Trachea. L: Lungs.

cardiac preservation (Fig. 1A). After the conventional dissection of the heart and lungs, the main difference is the handling of the trachea, which is double-stapled two or three rings above the carina and excised (Fig. 2); the esophagus is also stapled and transected. The rest of the technique is presented in lung procurement, leaving the whole block ready to be implanted.

Liver Donor

Patients with irreversible neurologic damage are the most common donors of cadaveric livers and 60% of them are associated with multi-organ donation.[2,18] In 2007, 265 living liver transplants were performed in US centers[1] (HSRA, 2007) in addition to those performed in Japanese and European institutions. In Table 3, the criteria to accept liver transplant donors are listed. The age ranges between neonates and donors 70 years old or more. Size matching is less strict than for lung; if the graft is bigger, it can sometimes be reduced on the back table—an option systematically used in pediatric liver transplantation. ABO blood group compatibility is the rule, but nevertheless, ABO incompatible liver grafts occasionally are used despite the survival rate of 20% at 1 year. ABO incompatible liver transplants in neonates and infants have much better results. Probably owing to their naive immune systems. Tissue typing is not used regularly in the decision to allocate them.

The circumstances surrounding the donor's death, past history of associated diseases, cardiopulmonary status and liver and renal function are the most reliable factors for acceptable liver donation. Acute and severe hemorrhage in trauma victims must be carefully assessed and in cases of definite cardiovascular instability, they are not considered as liver donors. More centers are now using NHBDs for liver transplantation. While some centers have shown equivalent results with well selected NHBDs, an increased risk of biliary complications is acknowledged.

Active alcoholism and drug abuse, hepatitis, chronic liver disease, surface antigen for hepatitis B or C positive, HIV positive, sepsis and cancer other than brain or skin tumors preclude liver donation.

Table 3. Liver donor acceptance criteria

Age: Neonates to 70 years or above
Hemodynamically stable
No sepsis
No Neoplastic disease other than skin or brain tumors
Acceptable cardiopulmonary and renal function
No history of chronic liver disease
Negative serology for human immunodeficiency virus (HIV), hepatitis B and syphilis, as well as risk factors such as prostitution
Direct (or conjugated) bilirubin less than 3.0 mg/dl
Aspartate and alanine transaminase levels less than two to three times normal
Alkaline phosphatase less than twice normal
Prothrombin time and partial thromboplastin time no more than twice normal
Steatosis involving less than 30-50% of hepatocytes

However, some transplant centers might accept donors with positive serology for hepatitis B, C or both as long as the donor's liver biopsy report is normal and the liver is transplanted into recipients with positive serology for hepatitis B, C or both.[19,20]

Abnormalities in prothrombin time and partial thromboplastin time might result from brain damage, rather than from liver dysfunction. If the liver biopsy shows 30-50% macrovesicular hepatocyte fat degeneration, the liver will not be used for transplantation. Microvesicular fat is not as relevant as an independent factor. Similar parameters are considered for living related donation of segmental grafts.

Liver Procurement

The body is opened from the suprasternal notch to the pubic symphysis, and a large Balfour retractor placed for exposure. Alternatively a cruciate incision can be used. After taking down the falciform and left triangular ligaments, the abdomen is inspected for infection and malignancy. The right colon is mobilized fully to expose the great vessels (Fig. 3-6A,B). Heavy silk or umbilical tape should be placed around the distal most aorta early, in case instability of the donor necessitates early clamping and flushing. This is also true of the IMV (inferior mesenteric vein) if a portal flush will be performed in situ. The liver hilum should be identified. Specifically, the hepatic artery must be identified and traced distally towards but not beyond the GDA (gastroduodenal artery) junction. Careful inspection for accessory or replaced hepatic arteries should be done prior to crossclamp. Near the duodenum, the common bile duct should be divided and the gallbladder flushed with cold saline (Fig. 6C). An appropriate clamp and window for supraceliac crossclamp, in the chest or abdomen, should be sought. 30,000 units of IV heparin should be administered in coordination with the chest team. A 24 French canula is then placed in the aorta, with occlusion of the distal aorta. Aortic crossclamp, initiating the flush and opening the IVC should again be coordinated with the chest team.

Ice should be immediately available to cool the organs. The portal can be done in situ via the IMV or on the backtable. Two to four L of flush via the aorta and 1-2 L via the portal vein should be sufficient. Failure for the effluent to clear, or clots in the effluent, is concerning for a poor flush, and canula and clamp location should be verified. The celiac axis is taken off the aorta for the liver procurement. The splenic artery is cut a few centimeters beyond its takeoff if the pancreas is to be used. Similarly, the portal vein is divided at the coronary vein if the pancreas is being procured. The vena cava is divided just above the entry of the renal veins. Great vigilance is needed to avoid cutting replaced hepatic arteries. The left replaced hepatic artery can come off the left gastric, celiac trunk or aorta. The right replaced hepatic artery frequently arises fromt he superior mesenteric artery and can course laterally or posterior to the head of the pancreas, and is often posterior and lateral in the hilum (See refs. 2 and 21 for further anatomical detail). Sometimes gently retracting on a loosely

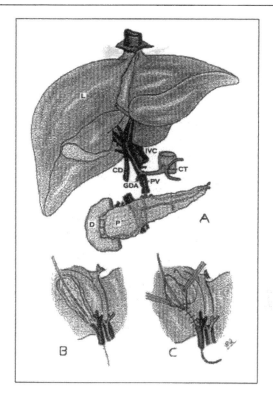

Figure 3. A) Final steps in liver and pancreas recovery. CD: Common duct. CT: Celiac trunk. PV: Portal vein. GDA: Gastroduodenal artery. P: Pancreas. D: Duodenum. L: Liver. IVC: Infrahepatic vena cava. B) Submucosal cholecystectomy and insertion of the stent with a guidewire. C) Suturing of the gallbladder flaps.

placed vessel loop around the SMA can help identify an accessory/replaced right hepatic artery in the liver hilum.

In hemodynamically compromised cases and in pediatric donors (where the cannula is not placed in the inferior mesenteric vein), instead of complete retroperitoneal dissection, a rapid preservation technique is used. Once the abdominal cavity is opened, a catheter is placed in the inferior mesenteric vein at the ligament of Treitz, the iliac arteries clamped, the aorta's cannula installed and the infusion of chilled electrolyte solution started. The rest of the dissection is done after this procedure. On the back table, a submucosal cholecystectomy is performed, a stent is inserted in the donor liver cyst with a guide wire, is fixed with a suture around the cystic duct and the gallbladder flaps are closed around the catheter (Fig. 3B,C).[20]

The reduced-size liver donor is commonly used in pediatric recipients due to the shortage of small or pediatric donors This is done by taking segments II and III, or the whole left or right lobes of the donor's liver, allowing the implementation of grafts into recipients who weigh seven to ten times less than the donor and preserving the hilum vessels. The "split" liver transplantation (one-half to one recipient and the other one to a second recipient) has not gained popularity because its complications and graft failure have been high, unless the right lobe hilum vessels were preserved.[22] In living-related donation, the approach is directly to segments II and III of the donor's liver, without dissecting the donor's hepatic hilum (Fig. 4A). Recently, laparoscopic assisted right lobe donor hepatectomy has been described.[23] A similar organ procurement technique is done in cases of auxiliary liver transplants, where the liver can be implanted in the recipient either orthotopically (removing segments II and III of the recipient's liver) or heterotopically (subhepatic).

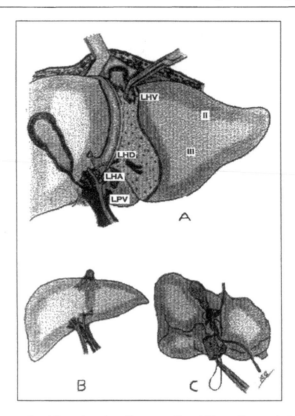

Figure 4. A) Living-related liver donation. Segments II and III are dissected. A clamp is placed in the left hepatic vein. LVH: Left hepatic vein. LHA: Left hepatic artery. LPV: Left postal vein. LHD: Left hepatic duct. II and III: Liver segments. B) "Piggy back" technique for preservation of the inferior vena cava (see text). C) End-to-end cavo-caval anastomosis for preservation of the inferior vena cava (see text).

Some transplant groups are performing the recipient vena cava preservation technique, either the so-called "piggy back" technique, using the suprahepatic veins in the donor and the recipient to re-establish the venous circulation (Fig. 4B)[24] or the variant technique of suturing the donor's infrahepatic and suprahepatic vena cava (upside 1 cm above the hepatic veins, downside 5-10 cm above the hepatic veins, downside 5-10 cm above the renal veins) and performing the anastomosis in the recipient (side to side cavocaval anastomosis) without occlusion of the recipient's vena cava and without by-pass (Fig. 4C).[25]

Pancreas Donor

The criteria in selecting pancreas donors are in many ways similar to those for liver, renal or thoracic donors (Tables 2, 3; see also: Lung, Kidney and Small Bowel donors).[26] Pancreases are recovered usually along with other thoracic and abdominal organs (see Pancreas Procurement). Most pancreases are obtained from cadaveric donors.[3,27] However, in a multivariate analysis, the age of the donor (older than 45), the gender (female) and the blood group (non-A blood group) have shown some negative influence on graft survival in the recipient.[28] It seems that pancreases procured from donors who died of cardio-cerebro-vascular disease, grafts obtained and preserved more than 30 hours, or donors after more than 15 hours of brain death might also affect graft survival related to vascular graft thrombosis. Graft implantation has shown better results when placed on the right side due to the length of the iliac vein.[29]

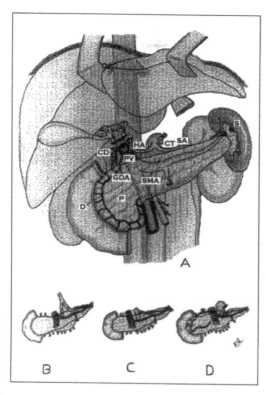

Figure 5. A) Final dissection of pancreas recovery. CT: Celiac trunk. HA: Hepatic artery. GDA: Gastroduodenal artery. SMA: Superior mesenteric artery. SA: Splenic artery. S: Spleen. PV: Portal vein. CD: Common duct. P: Pancreas. D: Duodenum. B) Extension arterial graft (see text). C) End-to-end spleno-superior mesenteric artery anastomosis. D) Aortic patch (see text).

Living-related donations of segmental pancreas grafts have not gained widespread acceptance because of the potential risk of donor surgery and the possibility of a positive glucose tolerance test in the donor two years or more after the hemipancreatectomy.[30] For a more current analysis of this topic, Gruessner and Sutherland[31] reviewed this area in a more detailed fashion in their recently updated book on Transplantation of the Pancreas (2004).

A donor history of type I diabetes is an absolute contraindication of donation; relapsing acute pancreatitis and chronic pancreatitis are relative contraindications.[32] Hyperglycemia and hyperamylasemia have not been absolute contraindications. Pancreas examination at the time of recovery is possibly one of the best ways to confirm the suitability of pancreas procurement. In those cases of living-related pancreas donation, the donor must be at least 10 years older than the age of onset of diabetes in the recipient; for sibling donors, no family member other than the recipient can be diabetic and their post-intravenous glucose stimulatory first-phase insulin level must be above the 30th percentile of the normal range.[30]

Pancreas Procurement

Through the above-mentioned large midline abdominal incision, multiple organ retrieval is not accepted as standard since better exposure is required for multiple organ recovery (Fig. 5A). Inspection and palpation can be achieved by opening the lesser sac. The pancreas should be soft and pinkish as opposed to firm or fatty. The exposure of the pancreatic vessels starts with the dissection of the hepatoduodenal ligament. In most circumstances the splenic and superior mesenteric artery remain with the pancreas graft and the hepatic and celiac trunk are left with the liver graft. The gastroduodenal artery is divided only in cases of liver-pancreas procurement, otherwise it is preserved with the pancreas (Fig. 5A,D). The common bile duct is divided as it enters the pancreas and the donor is systemically heparinized (see Liver Procurement).

Arterial abnormalities in some cases preclude the possibility of harvesting the liver and pancreas simultaneously (see Liver Procurement). The decision is made based on the expertise of the harvesting team and priority is always given to the liver in such situations.

After the dissection of the heart and lung vessels by the thoracic team, the abdominal surgeons divide the duodenum, which had been instilled in advance with 300 ml of amphotericin/antibiotic solution. The distal abdominal aorta is clamped and the perfusion cannula is placed into it or into the inferior mesenteric artery (Fig. 7, top). At the same time the supraceliac aorta is clamped, along with the cardiac-lung harvesting team clamping the proximal aorta; the portal vein is divided and the suprahepatic vena cava, too. In case of liver harvesting, at this moment the hepatic perfusion cannula would be inserted into the portal vein (Fig. 3A). Usually UW or HTK preservation solution is utilized in liver-pancreas procurement (1 liter into the portal vein and 2 or more liters into the aorta). The liver and the pancreas might be separated in situ or ex vivo (the spleen conventionally is dissected out in the latter manner), both organs are surrounded by the same preservation solution (UW or HTK) and ready for implantation (Fig. 5A,D).

It is accepted worldwide, as described above, to procure the pancreas with a segment of duodenum, the exocrine pancreatic secretion is directed to the 2d.portion of the duodenum and the final drainage is nowadays from a pancreaticoduodenal graft to the proximal jejunum without a Roux-en-Y.[31,32] For solitary pancreas transplants, some centers still prefer bladder drainage.

The pancreatic vessels might be reconstructed using a graft extension (Fig. 5B), side-to-side splenic-superior mesenteric artery anastomosis (Fig. 5C), or directly to the recipient's vessels through the aortic patch including the celiac trunk and the superior mesenteric artery (Fig. 5D). The resection of the redundant duodenum is also done on the back table, it is stapled at both ends and the spleen is taken out.

In pancreas living-related donation, the celiac trunk is exposed initially, the spleen is left in situ only with the short gastric vessels, the body and tail of the pancreas are transected with the scalpel, the pancreatic duct is identified and marked with prolene suture and the proximal duct is suture-ligated with 5-0 prolene. The patient is heparinized (4-units/kg); vascular clamps are placed on the splenic artery and across the splenic vein. The graft is removed, immediately flushed with UW or HTK and kept on ice until implantation.[26,27]

Kidney Donor

Generally, patients are not considered for organ donation if they have a history of disease or trauma involving the organ considered for donation. Criteria similar to those proposed for the recipient are used for the donor. Neoplastic disease, sepsis, severe fungal infection, HIV positive, or serious systemic viral infection and a history of end-stage renal disease preclude kidney donation. The donor's age even though it used to be ideally between 5-55 years, today it has been increased to 65 years and older depending on the rest of the donor's history.[2] Donor's below 5 years of age tend to exhibit lower graft survival than older donors.[2] In order to expand the number of older donors (≥60) or to use kidneys with a prolonged cold ischemia time (> 30 hours), estimated creatinine clearance between 40-80 ml/min, or biopsy specimen with <40% glomerulosclerosis, a double adult renal allograft has been proposed recently. Here it is possible to increase the nephron dosing, similar to the principle used to transplant both kidneys in adult recipients coming from pediatric donors.[33] This approach of using dual adult renal allografts is rarely employed. Criteria suggested for donors within the conventional age limits of donation such as diabetics, hypertensive or hypotensive, infected, non-heart beating, subnormal organ function, with a high risk of viral infection or malignancy are expanded for recipients with short-life expectancy without transplant. For recipients requiring inotropic support, graft survival is around 70% at 1 year and they have higher incidence of acute tubular necrosis.[34] The goal in such cases is the optimization of kidneys with normal function transplanted in recipients with longer life expectancy.[35,36] Nevertheless, these criteria are clearly defined by some transplant groups.[37] If doubt arises, it is advisable to perform biopsy of a donor's kidney at the time of procurement. Cytomeglovirus (CMV) positive donors are used for CMV-positive recipients, but also for CMV-negative recipients, with a prophylactic treatment posttransplantation with gancyclovir for two weeks, followed by acyclovir for three months.[3]

Living donors are crossmatched at the onset and just before surgery in order to reconfirm that transplantation is immunologically feasible. Preoperative examination involves a thorough history and physical examination, including examination of serum creatinine, urinary sediment and a creatinine and protein in a 24 hour urine collection, completed by noninvasive digital subtraction angiography. The short-term results are good (over 90% 1 year graft survival), but the long-term results are especially better than for those with a postmortem kidney.[38]

Kidney Procurement

There are three ways to procure kidneys: isolated cadaveric kidney procurement, multi-organ procurement and kidney living-related donation.

In the first case, the abdomen is approached through a midline incision from the xiphoid appendix to above the pubis, the routine abdominal initial exploration is done and if there is no contraindication for donation, the right colon is mobilized and lifted upward to the left along with the small bowel (Fig. 6A,B). The duodenum and the pancreas are also retracted upward. The connective tissue is dissected around the proximal aorta to expose the celiac trunk and the superior mesenteric artery, which are ligated and divided. The proximal aorta is isolated above the celiac trunk, with special attention to the renal vessels (Fig. 6C). The distal aorta and inferior vena cava are also isolated and

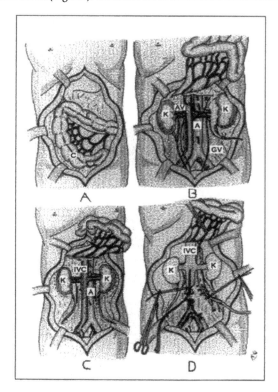

Figure 6. Cadaveric kidney recovery. A) Initial exposure—Line of incision. C: Colon. B) Retroperitoneal exposure: K: Kidneys. IVC: Inferior vena cava. A: Aorta. GV: Gonadal vein. AV: Adrenal vein. C) Kidney dissection:— Line of incision. D) Preservation flush (see text).

circled with umbilical tapes. At this time systemic heparin (300 units/kg) is administered. Proximal aorta occlusion is performed, a perfusion cannula is placed into the distal aorta and the preservation solution is infused, usually 2-3 liters of UW or HTK solutions maintained at 0-4°C. The distal vena cava is incised for free drainage or cannulated for direct drainage into a container, immediately after the preservation solution starts.

The final dissection of the kidneys is carried out through the Gerota's fascia without disturbing the renal hilum and the ureters are divided as distally as possible, maintaining the fat around them to preserve their vascularization. The distal vena cava and aorta are divided and their lumbar vessels ligated, the proximal aorta and vena cava are sectioned, much above the emergence of both renal vessels, the block of both kidneys is extracted and placed on a basin of cold saline with the preservation solution Complementary dissection is performed on the back table that allows one to take both kidneys with the complete length of both renal veins and arteries with an aortic cuff, the right renal vein comes with the inferior vena cava attached to it and the left renal vein with a cuff of the vena cava (the last step not shown in Fig. 6D). If the kidneys will be placed in the perfusion machine, the aortic patch should be carefully cut and fashioned not to impair the placement of the renal artery into the device and the connection to the machine. This step is better carried out with full cooperation and direct participation of the perfusion technologist.

In multiple organ donation, the kidneys are recovered after the extraction of heart, lungs, liver and pancreas (see: Liver and Pancreas Procurement) and, in some cases, even from non-heart-beating cadaveric donor organs that otherwise are not considered for donation.[2,21,34,35]

In living donor nephrectomy, the main issues are adequate exposure, minimal hilar dissection to preserve the periureteral fat to ensure sufficient blood supply to the ureter and acceptable diuresis during the operation. Most transplant surgeons prefer to use the left kidney because of its longer renal vein. The patient is placed in a lateral position with the table flexed to raise the flank; the incision is carried out from the tip of the twelfth rib to the external border of the rectus muscle and the posterior auxiliary line dorsally The external oblique muscle is divided and the twelfth rib might be removed; the internal oblique muscle, the transverse abdominis and the underlying transversalis fascia are divided to facilitate entry into the retroperitoneal space. Gerota's fascia is dissected and the perinephritic fat is removed. The ureter is isolated with a portion of its fat and divided close to the common iliac artery bifurcation. The renal vein is dissected to its junction with the vena cava and in the left side gonadal and adrenal veins are ligated and divided. Finally, the kidney is lifted upward and the renal artery is dissected all the way to its origin from the aorta, diuretics are given i.v. and the donor is temporarily heparinized (action is contra-arrested with protamine as soon as the kidney is extracted). Two clamps are placed; one in the renal artery just in the margin of the aorta and another one in the renal vein just in the margin of the inferior vena cava; they are sectioned and the kidney is removed. The stumps of both vessels are sutured with 5-0 prolene and the clamps are taken out. The wound is closed in the regular fashion.

Over the last fifteen years, laparoscopic or laparoscopic assisted donor nephtectomy has largely replaced the open approach. This is due to nearly equivalent complication rates and decreased morbidity and length of hospital stay for the laparoscopic approach. With the patient in a semi-lateral position a hand-port, approximately 7 cm, is placed in the upper midline. A camera and dissecting port are placed infraumbilically and in the left lower quadrant. After reflecting the colon medially, the gonadal vein and ureter are identified, and the vein traced back to the renal vessels. Vascular staplers are used to divide the vessels.

Small Bowel Donor

Most small bowels have been recovered from cadaveric donors. In Starzl's early experience, the majority of small bowel transplants have been performed simultaneously with a liver transplant and some are done as part of a multi-visceral transplant.[21] The least frequent small bowel transplants were performed as an isolated organ.

The main criteria for the donation of small bowels are the conventional ones; no malignancy, sepsis, positive serology for HIV or other serious systemic viral diseases, or already compromised small bowel. The graft should be ABO-blood type compatible and lymphocytic crossmatch test negative. Size matching is a questionable issue; intestines of pediatric donors are better allocated to pediatric recipients. The same is true to the intestines of adult donors, which are better allocated to adult recipients. In cases in which small bowels are transplanted with the liver or multi-viscerally, liver function and the function of the other organs must be assessed according to protocols (see Liver, Pancreas and Kidney Donors).

Living-related donation is not commonly performed, but it is preferable to transplant a segment of ileum instead of jejunum because of its absorptive capacity, acceptable control of rejection and its less vascular technical complexity (Fig. 7, bottom). One and a half meters is enough to re-establish and maintain the nutritional needs of an adult with minimal morbidity.[39]

Small Bowel Procurement

Through a midline celiotomy the small bowel is harvested; all the non-intestinal organs to be procured are mobilized and removed before the intestinal lumen is entered. In many cases, the liver and kidney will be recovered along with the intestine. The lesser sac is entered through the gastrocolic ligament. The omentum is lifted upward with the stomach in which the short gastric vessels were previously divided. The connective tissue, ventral to the celiac trunk is divided and the vessels are exposed; spleen and pancreas are lifted to the right (Fig. 7, top) and the ligament of

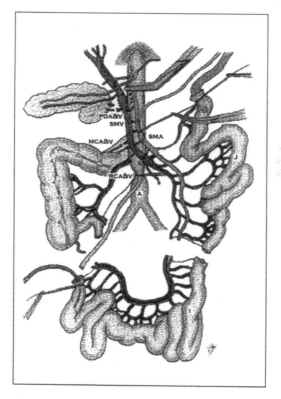

Figure 7. Small bowel procurement. Top: Cadaveric small bowel procurement along with pancreas. SMA: Superior mesenteric artery. MCA and V: Middle colic artery and vein. RCA and V: Right colic artery and vein. PDA and V: Pancreatocuodenal artery and vein. A: Aorta. J: Jejunum. I: Ileum. Bottom: Living-related small bowel procurement (see text).

Treitz is divided; the superior mesenteric artery and vein are isolated, ligating two to four duodenal branches, as well as the right and middle colic vessels (artery and vein) (Fig. 7, top). The pancreatic vessels are divided unless the pancreas will be harvested. If the liver is going to be harvested separately, the portal vein is not mobilized. If the liver and/or the pancreas will be procured, the superior mesenteric artery and vein are dissected a few centimeters underneath the pancreatic border (Fig. 7, top). The donor's iliac grafts are used as necessary.

The donor is heparinized, the aorta is cross-clamped proximal to the distal insertion of the perfusion cannula and the infusion of UW or HTK solution (Fig. 7, top) is started. Liver, pancreas and kidneys are extracted at this time directly into the superior mesenteric artery. The flush of small bowel vasculature is done with the same UW or HTK solution, until the venous effluent is clear. The intestinal graft is removed and immersed in iced saline to keep it cold. Later the lumen of the intestine is flushed with preservation solution plus antibiotics (Fig. 7), so the graft is ready to be implanted.[21] As in the case of heart and lungs, the sooner it is implanted the better the result will be.

In living related donation, the length of small bowel that allows for the maintenance of normal absorptive and nutritional function in the recipient is around 1.5 meters in length. Shorter segments lead to an inadequate nutritional state and need a preserved ileo-cecal valve. On the other hand, this length of small bowel does not affect the donor. As mentioned above, the ileum is preferred to the jejunum, due to its acceptable physiologic properties, equal degree of rejection and less technical complexity—the ileocolic artery can be divided without jeopardizing the donor's mesenteric circulation. The approach is through conventional midline-celiotomy and the difference lies in the preservation method, which is done ex vivo, as well as the lumen intestinal flush (Fig. 7, bottom). The donor is temporarily and moderately heparinized and reversed with protamine as soon as the intestine is out.

Multivisceral Donation for Multivisceral Transplantation

An interesting approach emerging in treatment for patients with malignancy or multiple abdominal organ failure when it is not enough to replace one or two organs is the transplantation of the entire block of abdominal viceras (so called Cluster Transplantation).[40] Ethical dilemmas are always present in these cases, because of this high morbidity and mortality, intensive medical care, high cost and short life-expectancy in patients with malignant tumors.[40,41] Considering all tumor cases with the triad of negative lymph nodes, only one organ compromised and no vascular invasion by tumor, makes the results more acceptable. The longest survival has been achieved with endocrine tumors and the shortest with colon, pancreas or gallbladder adenocarcinoma. Sarcoma patients have a high rate of survival, but also a high recurrence. Patients with cholangiocarcinoma are not good candidates unless the primary tumor is confined to the liver and bile ducts. In cases of hepatocellular carcinoma, this option is just applied to the fibrolamellar variant.

Other indications for multi-visceral transplantation, besides gastrointestinal tumors, are Crohn's disease, abdominal trauma, celiac artery occlusion, superior mesenteric artery thrombosis, surgical adhesions, Budd-Chiari syndrome in adults and gastroschisis, necrotizing enterocolitis, vovulus, intestinal atresia, microvillus disease, pseudo-obstruction and multiple polyposis in children, with much better patient survival.[42] The alternative of immunologic tolerance when performing bone marrow transplantation along with the cluster transplant needs to be further studied in order to elucidate the best means to modulate the rejection process.

Multi-visceral organ donors must meet the criteria for donation for liver, kidney, pancreas and small bowel individually in order to be candidates for the cluster procedure. Organs, also, need to be obtained from and ABO blood type identical cadaveric donor, ranging from less than 1 year to 40 years of age. It is not necessary to have good HLA matching. Some transplantations have been performed even with a positive lymphocyte crossmatch—an alternative that is not considered desirable.

Multivisceral Procurement for Multvisceral Transplantation

Multivisceral procurement is done en bloc by dissecting mainly the retroperitoneal structures of the liver (intrahepatic vena cava, suprahepatic veins and suprehepatic vena cava) (Figs. 3A, 4B), pancreas (celiac trunk, superior mesenteric artery) (Fig. 5A,D), spleen, duodenum, small bowel (superior mesenteric artery and vein) (Fig. 7, top) and kidneys (aortic and vena cava segments including renal vessels) (Fig. 6B-D). Another important aspect is the transection of the gastrointestinal tract and the decontamination of the small bowel (see Pancreas and Small Bowel Procurement). Other aspects of dissection and procurement for multi-visceral transplantation have already been presented in other sections of this book regarding individual organs and discussed in other important publications.

Authors' Note

Excellent reviews that analyze the topic of organ donation and procurement have appeared in the literature in the last few years. We recommend the readers to complement this chapter with the publications by Merion[2] and Humar, Matas and Payne[21] that covered these topics extremely well. They also presented especially well-designed artistic drawings of the various steps seen through all the techniques included in this chapter.

References

1. HRSA (Health Resources and Services Administration) Division of Transplantation, 2007. National Breakthrough Collaborative. UNOS/OPTN Data, 2007. www.donatelifeny.org/organ/2007.
2. Merion RM. Organ Preservation. In: Mulholland MW, Lillemoe KD, Doherty GM et al. Greenfield's Surgery. Scientific Principles and Practice. 4th Ed. Lippincott Williams and Wilkins, Philadelphia, 2006: 563-580.
3. Van Buren CT, Barakat O. Organ donation and retrieval. Surg Clin N Am 1994; 74:1055-81.
4. Yanaga K, Podesta LG, Broznick B et al. Multiple organ recovery for transplantation. In: Starzl TE, Shapiro R, Simmons RI, eds. Atlas of Organ Transplantation. New York: Gower Medical Publishing, 1992; 3.2-3.49.
5. Xu H-S, Scott Jones R. Study of a rat liver transplantation from non-heartbeating cadaver donors. J Am Coll Surg 1995; 181:322-26.
6. Bos MA. Ethical and legal issues in non-heart-beating organ donation. Transplant Proc 2005; 37:574-576.
7. Brook NR, Waller JR, Nicholson JR. Non-heart-beating kidney donation: Current practice and future developments. Kid Int 2003; 63:1516-1529.
8. Klintmalm GB, Levy MF. Organ Procurement and Preservation. Austin: Landes Biosciences, 1999.
9. Peterson TP, Johnson J, Fleming A et al. Surgical recovery of organs. In a clinician's guide to donation and transplantation. Cleveland: ASAIO Book, 2008:857-866.
10. Simmons RL, Ilstad ST, Smith CR et al. Heart transplantation. In: Schwartz S, Shires T, Spencer FC, eds. Principles of Surgery. 6th ed. New York: McGraw Hill Inc., 1994:419-28.
11. Takagaki M, Misamochi K, Morimoto T et al. Successful transplantation of cadaver hearts harvested one hour after hypoxic cardiac arrest. J Heart Lung Transplant 1995; 15:527-31.
12. Zenati M, Dowling Rd, Armitage IM et al. Organ procurement for pulmonary transplantation. Ann Thorac Surg 1989; 48:882-86.
13. Sundaresan S, Trachiotis GD, Aoe M et al. Donor lung procurement: assessment and operative technique. Ann Thorac Surg 1993; 56:1409-13.
14. Simmons RL, Ildstad ST, Smith CR et al. Lung and heart lung transplantation. In: Schwartz S, Shires T, Spencer FC, eds. Principles of Surgery; 6th ed. New York: McGraw Hill Inc, 1994:426-28.
15. Davis DR, Pasque MK. Pulmonary transplantation. Ann Surg 1995; 221:14-28.
16. Kshetty VR, Kroshus TJ, Burdine I et al. Does donor organ ischemia over four hours affect long term survival after lung transplantation? J Heart Lung Transplant 1996; 15:169-174.
17. D'Armini AM, Roberts CS, Lemasters H et al. Lung retrieval from cadaver donors with nonbeating hearts; optimal preservation solution. J Heart Lung Transplant 1996; 15:496-505.
18. Starzl TE, Miller C, Broznik B et al. An improved technique for multiple organ harvesting. Surg Gynecol Obstet 1987; 165:343-48.
19. Wright TL. Liver transplantation for chronic hepatitis C viral infection. Gastroenterol Clin N Am 1993; 22:231-42.
20. Turcotte JG, Campbell DA Jr, Bromberg JS et al. Hepatic transplantation. In: Zuidema GD et al. Hepatic transplantation and In: Zuidema GD, ed. Shackelford's Surgery of the Alientary Tract, 4th ed. Vol III. Philadelphia: WB Saunders Company, 1996:600-15.
21. Humar A, Matas AJ, Payne WD. Atlas of Organ Transplantation. Springer Verlag, New York, 2006.
22. Kalayoglu M, D'Alessandro AM, Sollinger HW et al. Experience with reduced-size liver transplantation. Surg Gynecol Obstet 1990; 171:138-47.

23. Koffran AJ, Kung R, Baker T et al. Laparoscopic assisted right lobe donor hepatectomy. Am J Transplant 2006; 10:2522-2525.
24. Tzakis A, Todo S, Starzl TE. Orthotopic liver transplantation with preservation of the inferior vena cava. Ann Surg 1989; 10:649-52.
25. Belghiti J, Panis Y, Sauvanet A et al. A new technique of side to side caval anastomosis during orthotopic hepatic transplantation without inferior vena caval occlusion. Surg Gynecol Obstet 1992; 175:271-72.
26. Benedetti E, Sileri P. Donor Selection and Management. In: Gruessner RWG, Sutherland DER. Transplantation of the Pancreas. New York: Springer Verlag, 2004:111-118.
27. Lloyd DM. Pancreas transplantation. In: Kremer B, Christoph E, Broelsch CE et al. eds. Atlas of Liver, Pancreas and Kidney Transplantation. New York: Thieme Medical Publishers Inc, 1994:107-27.
28. Douzdjian V, Gugliuzza KG, Fish JC. Multivariate analysis of donor risk factors for pancreas allograft failure after simultaneous pancreas-kidney transplantation. Surgery 1995; 118:73-81.
29. Troppmann C, Gruessner AC, Benedetti E et al. Vascular graft thrombosis after pancreatic transplantation: univariate and multivariate operative and nonoperative risk factor analysis. J Am Coll Surg 1996; 182:285-316.
30. Gruessner RWG, Sutherland DER. Simultaneous kidney and segmental pancreas transplants from living related donors—the first two successful cases. Transplantation 1996; 61:1265-68.
31. Gruessner RWG, Sutherland DER. Recipient Procedures. In: Gruessner RWG, Sutherland DER. Transplantation of the Pancreas. New York: Springer Verlag, 2004:119-178.
32. Toledo-Pereyra LH, Rodríguez-Quilantán FJ. Evolución histórica de las técnicas quirúrgicas del transplante de pancreas. Gac Med Mex 1994; 3:487-94.
33. Johnson LB, Kuo PC, Dafoe DC et al. Double adult renal allografts: a technique for expansion of the cadaveric kidney donor pool. Surgery 1996; 120:580-84.
34. Marshall R, Ahsan N, Dhillon S et al. Adverse effect of donor vasopressor support on immediate and one-year kidney allograft function. Surgery 1996; 120:663-66.
35. Orloff MS, Reed AI, Erturk E et al. Non-heartbeating cadaveric organ donation. Ann Surg 1994; 220:578-85.
36. Ratner LE, Kraus E, Magnuson T et al. Transplantation of kidneys from expanded criteria donors. Surgery 1996; 119:372-77.
37. Kiberd BA. Should hepatitis C-infected kidneys be transplanted in the United States? Transplantation 1994; 57:1068-72.
38. Lowell JA, Brennan DC, Shenoy S et al. Living-unrelated renal transplantation provides comparable results to living-related renal transplantation: a 12-year single-center experience. Surgery 1996; 119:538-43.
39. Hoffman AL, Lee KKW, Schraut WH. Small bowel and multivisceral organ transplantation. In: Starzl TE, Shapiro R, Simmons RI, eds. Atlas of Organ Transplantation. New York: Gower Medical Publishing, 1992:9.2-9.16.
40. Allessiani M, Tzakis A, Toto S et al. Assessment of five-year experience with abdominal organ cluster transplantation. J Am Coll Surg 1995; 180:1-9.
41. Todo S, Tzakis A, Abu-Elmagd K et al. Abdominal multivisceral transplantation. Transplantation 1995; 59:234-49.
42. Abu-Elmagd D, Todo S, Tzakis A et al. Three years clinical experience with intestinal transplantation. J Am Coll Surg 1994; 179:385-400.

CHAPTER 3

Perfusion and Storage Techniques

Amer Rajab, Ronald P. Pelletier and Mitchell L. Henry

Introduction

The clinical practice of living-related renal transplantation became firmly established by the late 1960s to early 1970s. However, it quickly became apparent that the need for transplantable kidneys far exceeded the availability. Subsequent legislation made it possible for people meeting the criteria of "brain death" to donate their organs to chronic renal failure patients who had easier access to these organs. As a result of these changes, a need arose for the development of organ preservation methods that would facilitate the successful transplantation of these cadaveric organs. In 1935, Lindbergh and Carrel were the first to describe organ preservation, utilizing a normothermic, pressurized and oxygenated solution system.[1] Nearly 30 years later, Calne et al reported adequate kidney preservation for 8-12 hours using simple cold storage in 1963.[2] Soon thereafter, Folkert Belzer et al reported a successful human kidney transplant following preservation with a hypothermic, cryoprecipitated-plasma perfusion method refined in the laboratory.[3] To date, there continues to be debate among transplant physicians regarding simple cold storage versus machine perfusion as the preferred method of kidney preservation.

Goals of Preservation

The goal of organ preservation is to minimize metabolic processes at the cellular level, reducing the rate of cellular energy utilization. Ongoing cellular energy use in the absence of oxygen and appropriate substrate nutrients results in ischemic injury and irreversible cell damage and death. Accumulation of metabolic end-products induced by ongoing aerobic and anaerobic metabolism results in reperfusion injury when blood flow is re-established following transplantation. The resulting end-products of anaerobic metabolism leads to the formation of intracellular toxic compounds such as superoxide and hydroxyl radicals and hydrogen peroxide, that exacerbate ischemia induced cellular and microvascular (endothelial) injury. The preservation process of solid organs incorporates the preservation solutions, the aim being to minimize the ischemia/reperfusion injury that occurs during the transplant procedure.

Ischemic Phase

The ischemic phase of transplantation includes periods when the organ is relatively warm (warm ischemia) and periods when the organ is cold (cold ischemia). Some warm ischemia is inevitable in organ transplantation. The warm ischemia begins even before organ procurement in the hemodynamically unstable donor. Further warm ischemic damage occurs during implantation due to technical factors and can be exacerbated after reperfusion by a delay in establishing normal microcirculation (vasospasm). Cold ischemia occurs during organ cold storage prior to implantation. During warm and cold ischemia a lack of oxygen precipitates a series of cellular physiologic disturbances that lead to impaired function and can result in cell death. A drop in

*Corresponding Author: Mitchell L. Henry— The Ohio State University Medical Center, Division of Transplantation and Comprehensive Transplant Center, 1654 Upham Drive, Columbus, OH 43210 USA. Email: mitchell.henry@osumc.edu

Organ Preservation for Transplantation, Third Edition, edited by Luis H. Toledo-Pereyra.
©2010 Landes Bioscience.

the tissue oxygen content will interfere with oxidative phosphorylation within the mitochondria and the anaerobic glycolytic pathway remains the only available metabolic pathway for energy production.[4,5] Consequently the cellular level of adenosine triphosphate ATP will drop.[5] This depletion of ATP plays a central role in the development of ischemic injury.

Hypothermia is an essential component of organ preservation. Reducing the temperature of an organ dramatically decreases the rate of cellular metabolism. Some biologic cellular functions are reduced more effectively by cooling than are others. Ion trans-membrane passive diffusion is not appreciably effected by temperature reduction while energy-dependent, ion trans-membrane transport mechanisms are inhibited below $10°C$.[6] One result of this differential effect is that permeable substances equilibrate across the plasma membranes, leading to cellular swelling and injury. Preservation solutions have therefore been designed to inhibit cell swelling by the addition of impermeants. Low-level metabolism persists even though hypothermia significantly decreases oxygen consumption at the tissue level (approximately 5% of normal at $5°C$).[7] Therefore, the accumulation of damaging metabolic end products will eventually occur, even at low temperatures. Many preservation solutions contain additives that will inhibit metabolism either during hypothermia or upon reperfusion. Preservation solutions are also formulated to control the pH of the extracellular fluid in the organ. Ongoing reduced metabolism during hypothermia allows for the accumulation of waste products, i.e., lactic acid, that lower the pH. Buffers in the preservation solution ameliorates the adverse events of acidosis at the local cellular environment.

Organ Preservation and Injury

Organ damage during procurement, preservation and reperfusion predisposes to graft dysfunction and in extreme circumstances to transplanted organ failure. To achieve successful organ preservation, the pathophysiology of ischemic and reperfusion injuries has been carefully studied. Loss of organ viability during preservation and reperfusion is the end result of a sequence of events that culminate in cell death. The combination of cell swelling, calcium influx, loss of ATP precursors and production of the destructive oxygen free radicals (OFR) contribute to organ damage. The role of OFR is felt to be especially important in ischemia-reperfusion injury during lung and intestine transplantation.

Organ Injury: Pathophysiology of Ischemic Phase

Cell Swelling

Lack of ATP results in cell swelling.[5,8] In a normal situation, an active extrusion of sodium ions by the Na/K-ATPase pump creates a Donnan equilibrium balancing the cell membrane impermeant intracellular anions, mainly proteins. During ischemia and lack of ATP, this pump is inhibited. Sodium enters the cell, which results in the cellular accumulation of electrolytes and water causing swelling of the cell and its organelles.[4,5]

Calcium Inflow

A critical reduction in cellular ATP level results in the elevation of intracellular calcium. Normally, the cytosolic Ca^{2+} concentration is 1000-10000 times lower than that of the extracellular fluid.[9] This gradient is maintained mainly to the action of a Ca^{2+} sequestering system in the mitochondria and the endoplasmic reticulum and by the Na/Ca-ATPase pump.[9] An increase in cytosol calcium concentration damages intracellular organelles leading to liberation of calcium from intracellular stores causing a further cytosolic increase. The mechanisms of calcium-induced injury are complex and multifactorial and include excess uptake of calcium by mitochondria and calcium-induced activation of enzymes such as phospholipases A_1, A_2 and C and proteases.[10] Phospholipase C mediates hydrolysis of phosphatidylinositol-4,5-biphosphate to inositol-1,4,5-triphosphate which in turn mobilizes intracellular calcium.[11] Degradation of phospholipids (required for the integrity of the cell membrane) also facilitates cellular influx of calcium.[12]

Structural Changes

Ischemia of various durations induces similar structural changes to all cell types. Early changes include reversible mitochondria swelling, followed by loss of mitochondrial matrix and finally disintegration as the ischemia time is prolonged. Expansion and formation of vesicles occurs in the endoplasmic reticulum and cytoplasm. Lysosomal rupture with enzyme release may represent the final step leading to cell death.[13]

Organ Injury: Pathophysiology of Reperfusion Phase

Several damaging effects of ischemia will not become apparent until the ischemic organ is reperfused with oxygenated blood. For example, little injury in the intestinal mucosa is seen during the ischemic period while the majority occurs upon reperfusion.[14] The injury seen after 3 hours of ischemia and 1 hour of reperfusion is more pronounced than that after 4 hours of ischemia alone.[14] Along this concept, anoxic reperfusion of ischemic tissues results in little additional damage.[15] Reperfusion injury therefore, implies some reaction initiated by return of oxygenated blood to the ischemic tissue. It appears that ischemia-reperfusion injury is initiated by the formation of cytotoxic oxidants derived from molecular oxygen. It must be emphasized that reperfusion injury exacerbates ischemic injury that occurs during the limited time of ischemia. Prolonged depravation of oxygen will result in tissue necrosis and cell death and restoration of oxygenated blood flow will not affect the ultimate viability of the tissue in this setting.

Oxygen Free Radicals

Oxygen free radicals have been implicated to be of importance for ischemia- induced cell injury during the reperfusion phase.[16] Within seconds after the onset of ischemia, xanthine dehydrogenase is converted to xanthine oxidase, possibly triggered by calcium inflow.[8,16] When the ischemic tissue is subsequently reperfused and reoxygenated, xanthine oxidase catalyses the conversion of hypoxanthine to xanthine and uric acid and these molecules generate large amounts of OFR.[16] OFR are hyperreactive chemicals that differ from other compounds in that they have unpaired electrons in their outer orbits.[17] OFR are potent oxidizing and reducing agents can significantly damage cellular integrity. The accumulation of OFR is normally prevented by the action of endogenous scavengers, such as catalase and superoxide dismutase and by antioxidants, such as vitamin E.[16,18] Following ischemia, the activity of such endogenous scavengers and antioxidants may be reduced which augments the accumulation of OFR.

When OFR are not removed by natural scavengers, a chain reaction occurs leading to lipid peroxidation of structurally important lipids within the phospholipid layer of the cell membrane.[17] Lipid peroxidation activates phospholipases A_2 (PLA_2), which preferentially removes oxidized fatty acids from phospholipids, protecting the membranes from oxidative injury.[19,20] Reduction of toxic lipid hydroperoxides by glutathione peroxidase requires the action of PLA_2.[21] On the other hand, PLA_2 once activated, hydrolyses membranous phospholipids and releases free fatty acids, which are precursors of biologically active eicosanoids. Free fatty acids are metabolized by cyclooxygenase, via the unstable endoperoxide intermediates to prostanoids including PGE_2 and $F_{2\alpha}$ and the more labile prostacycline PGI_2 and thromboxane TXA_2. The alternative metabolism route for free fatty acids is the lipooxygenase pathway. OFR can also cause tissue injury by degrading hyaluronic acid and collagen in the extracellular matrix and nucleic acids within the nucleus and cytoplasm.[21]

No-Reflow Phenomenon

Another described tissue ischemic effect is the so-called "no-reflow phenomenon" characterized by vascular obstruction at the capillary level. An important precipitant of this phenomenon is thought to be cellular swelling during ischemia compressing the capillary bed with resultant occlusion of the lumen.[22] The lack of adequate blood flow after reperfusion represents a second mechanism by which ischemic injury may occur. Initially viable cells after an ischemic period may lose viability if the circulation is sluggish upon reperfusion.

Following prolonged ischemia, despite restoration of oxygenated blood flow, the ability of the tissue to regenerate ATP may also be critically impaired. During ischemia, ATP, ADP and AMP are degraded by the action of the enzyme 5'-nucleotidase resulting in nucleoside adenosine which diffuses freely through the cell membrane.[4,5] During reperfusion, the purine metabolites are washed out of the cell and it thus loses the needed substrate for ATP regeneration.[5]

Hypothermia

Hypothermia is the mainstay of organ preservation. For every 1°C decrease in temperature, metabolic rate and oxygen requirement decrease by 5%.[7,23] The activity of many enzymes decreases 12 to 13 fold when the temperature is reduced from 37°C to 0°C.[3,5] At 5°C, the oxygen consumption in the kidney is only about 5% of that at normothermia.[7,23] Since most organs can tolerate warm ischemia of 30-60 minutes, theoretically cold ischemia will be tolerable for up to 6-12 hours. Clearly, hypothermia does not stop metabolism, but markedly slows metabolism rates and delays cell death.

Two major methods have been described for preservation of solid organs for transplantation, cold storage and hypothermic perfusion preservation.[4] Cold storage involves initial flushing the organ during the organ procurement with chilled preservation solution and then preserving the organ on ice (0-4°C) in an excess of the preservation solution. Perfusion preservation requires extracorporeal perfusion of the organ using a continuous pump.[4] Perfusion preservation has several advantages over cold storage. It allows continuous delivery of oxygen and nutrients, it removes waste metabolites and it provides for better pH control.[4]

Cold Preservation

The effectiveness of simple cold storage greatly depends on the composition of the preservation solution. An ideal preservation solution should (1) minimize hypothermia-induced cell swelling, (2) prevent tissue edema, (3) prevent intracellular acidosis, (4) provide substrates for regenerating high-energy phosphate compounds during reperfusion and (5) prevent injury from OFR.

Prevention of Cell Swelling

To counteract the intracellular osmotic forces, cellular impermeants of 110-140 mOsm/kg must be added to the preservation solution to prevent the cellular swelling.[4,24] The Collins solution, which was the main preservation solution until 1988, contains glucose as the main impermeant.[24] While glucose as an impermeant works well for kidneys, it is not ideal for other organs. Wahlberg et al, however, found that neither glucose nor mannitol were effective in preventing cell swelling in pancreatic slices.[25] The same authors demonstrated that the trisaccharide raffinose and the acid of lactose, lactobionate, completely prevented pancreatic cell swelling during preservation.[25] Similar results have been shown in liver tissue.[26] A solution based on lactobionate and raffinose was shown to successfully preserve the canine pancreas up 72 hours.[27] These results were superior to all previous studies on preservation of the pancreas. This solution (UW, Viaspan[R]) introduced in clinical practice in 1988, has dramatically improved liver and pancreas transplantation.

It has been postulated that elimination of the concentration gradients for sodium and potassium ions between the inside and outside of the cell would prevent ionic shifts and subsequent cell damage. Collins previously attempted to counteract the deleterious effects of ionic shifts during the hypothermic period by the use of a modified intracellular fluid rich in potassium and magnesium and buffered with phosphate.[24] However, the same author obtained similar results of kidney preservation using solution in which potassium has been replaced entirely by sodium.[28] Hyperkalemic preservation solution also induces vasospasm and possibly endothelial injury. Systemic overload of potassium in the recipient following revasularization might result in cardiac arrythmias. However, UW solutions either with a high Na-low K, or a high K-low Na content were equally effective.[29] Prevention of cellular swelling by adding an effective impermeant to the preservation solution, such as lactobionate, seems to be the main factor for successful preservation. The action of these impermeants may be multifactorial. Lactobionate, for instance, the main impermeant in the Viaspan[R] solution, cannot be replaced by another anionic organic impermeant, gluconate, without loss of viability.[30] Lactobionate, therefore, seems to have additional important effect in preservation.

In fact, it has been suggested that lactobionate is a strong chelator of calcium and iron,[27] features that may be of importance in minimizing cell injury from calcium inflow and OFR.

Prevention of Tissue Edema

Colloids have been extensively used in perfusion preservation for transplantation.[5] The function of the colloids is to balance the perfusion pressure across the capillary endothelial lining, which will prevent the development of interstitial edema and thereby facilitate adequate perfusion.[31,32] The value of colloids in preservation by cold storage is not yet established.

Colloids may improve the quality of preservation by several mechanisms. First, the oncotic pressure exerted by the colloids prevents leakage of fluid into the interstitial space by counteracting the intravascular pressure. This is especially important during continuous perfusion preservation. This will help prevent collapse of the capillary vascular bed as a consequence of increased interstitial swelling that is associated with the impaired washout of red blood cells and a reduced tissue distribution of the preservation medium.[4] Second, certain colloids may reduce erythrocyte aggregation,[33] which improves the removal of red blood cells during the initial washout procedure.[33] Third, a colloid may prevent endothelial cell injury.[34] Finally, it is possible that colloids directly prevent hypothermia-induced cell swelling.

Prevention of Intracellular Acidosis

The prevention of intracellular acidosis during preservation is important for successful preservation.[4,5,31] During preservation, there is an increase in anaerobic glycolysis and glycogenolysis concomitant to lactic acid and hydrogen ions production leading to tissue acidosis. All known preservation solutions contain a buffering system such as phosphate in EC and UW and histidine in HTK (Table 1).

Providing Substrates for ATP Regeneration

The tissue level of ATP at the end of the preservation period predicts the viability of the organ following transplantation.[4,5] Despite the availability of oxygen during reperfusion, the lack of ATP precursors prevents the cells from generating high energy phosphate molecules. Therefore, precursors for ATP such as adenosine and phosphate are also included in preservation solution.[5,8]

Prevention of OFR Injury

Xanthine oxidase inhibitors such as allopurinol or OFR scavengers such as superoxide dismutase (SOD) or catalase can prevent OFR injury during reperfusion. Administration of SOD before revascularization of pancreatic grafts preserved in Eurcollins solution for 12 has hours prevented the increase level of phospholipase A_2 activity, lipoperoxide, thromboxane TXB_2 and 6-keto prostaglandin $F_{1\alpha}$.[23] Viaspan[R] solution contains 1 mM of allopurinol, an inhibitor of the enzyme xanthine oxidase, to prevent OFR production.

Preservation Solutions

Belzer developed a machine perfusion (MP) device in the 1960's that used a blood component to perfuse a kidney at a constant pressure and flow rate.[3] This flow and pressure could be quantified and changed as needed. The device was subsequently converted to allow pulsatile flow and oxygenation of the perfusate.

The cryoprecipitated plasma (cellular swelling inhibitor) initially used in the Belzer's solution was difficult and cumbersome to prepare. The lipoprotein fractions present in plasma had a tendency to precipitate at low temperatures and with changes in pH. The requirements for microfiltration and avoidance of exposure to gas made the preparation of this solution both complex and costly.

Belzer new MPS solution mimicked extracellular fluids in regards to the electrolyte content, containing 140 mmol/L sodium and 10 mmol/L potassium. HEPES was used as the buffer and glutathione and dexamethasone were added as inhibitors of metabolic end-products. Glucose was also added to support the low level of ongoing metabolism during hypothermia.

Table 1. Composition of preservation solutions

	Collins	Euro Collins	UW	HTK	Celsior
K+ (mmol/L)	115	115	120	10	15
Na+ (mmol/L)	10	10	30	15	100
Mg^{2+} (mmol/L)	30	0	5	4	13
Ca^{2+} (mmol/L)	0	0	0	0.015	0.25
PO^4 (mmol/L)	100	100	25	0	0
Cl-(mmol/L)	15	15	0	50	0
HCO$^-$ (mmol/L)	10	10	0	0	0
SO^{2-} (mmol/L)	30	0	5	0	0
Lactobionate (mmol/L)	0	0	100	0	80
Glucose (mmol/L)	126	198	0	0	0
Raffinose (mmol/L)	0	0	30	0	0
Hydroxyethyl starch (g/L)	0	0	50	0	0
Glutathione (mmol/L)	0	0	2	0	3
Adenosine (mmol/L)	0	0	5	0	0
Allopurinol (mmol/L)	0	0	1	0	0
Insulin (U/L)	0	0	40	0	0
Dexamethasone (g/L)	0	0	16	0	0
Penicillin (U/L)	0	0	200,000[d]	0	0
Histidine (mmol/L)	0	0	0	180	30
Tryptophan (mmol/L)	0	0	0	2	0
α-Ketoglutarate (mmol/L)	0	0	0	1	0
Mannitol (%)	0	0	0	30	60
Glutamate (mmol/L)	0	0	0	0	20
PH (at 25°C)	7.0-7.3[c]	7.0-7.3[c]	7.4	7.2	7.3
Osmolality (mOsm/kg)	320	406	320	310	360

Substitution of human serum albumin for cryoprecipitated plasma[35] or the use of silica gel treatment[36] eliminated lipoproteins as well as fibrinogen and other entities. These maneuvers obviated the need for microfiltration and provided a stable perfusate of predictable composition. These modifications allowed for development of newer solutions,[37-41] some of which continue to be used today. The solutions, developed by Collins[24] Marshall[37] and Bretschneider[38] were usually phosphate buffered with additional agents such as dextran, glucose, histidine, tryptophan, HEPES citrate, sucrose and mannitol. They were used to help maintain pH and to stabilize cell membranes. Sacks et al[39] subsequently developed a hyperosmolar solution for kidney preservation. The use of a modified UW perfusate in 72-hour preservation of dog kidneys demonstrated that albumin could effectively be replaced with hydroxyethyl starch (HES), which gave superior results. This solution became widely used with machine perfusion and continues to be used today. Subsequently, machine perfusion became more commonplace because it provided a safer, longer storage time, with consistently good outcomes and results.

Newer solutions have also been developed for use in static cold storage. The success of these solutions allow for longer cold storage times, avoiding the need for expensive machines and for technicians to monitor the machine performance.

Specific Preservation Solutions

Four basic preservation solutions have been widely utilized (Table 1). Collins solution was initially described in 1969[24] and modified in 1976 as the Euro-Collins solution (EC) from which $MgSO_4$ was omitted and glucose was added. The basic composition of EC solution is the intracellular electrolyte environment, i.e., high potassium and low sodium concentrations and a strong phosphate buffer. This solution uses glucose rather than cryoprecipitated plasma or human serum albumin for inhibition of cellular swelling, making it simpler and less expensive to prepare. Its high osmotic pressure is based on the glucose content, which functions as an impermeant in kidneys but not in other organs. For this reason, EC solution worked reasonably well only for kidney preservation. The use of this intracellular electrolyte composition extended the allowable cold preservation time for kidneys.[28,40,41]

The University of Wisconsin solution (UW or Viaspan®) was introduced more recently. With the widespread success in pancreas and liver transplantation in the 1980s, there was a need for a preservation solution with improved efficacy that facilitated longer preservation times. The major alteration of Belzer MPS to create the UW solution was the replacement of sodium gluconate with lactobionic acid. The markedly improved outcomes observed with lactobionic acid containing solutions like UW suggest that it may possess properties beyond the ability to inhibit cellular swelling. Raffinose is also included as an inhibitor of cell swelling. It has colloid as well as other OFR scavengers. UW solution has an unequivocal outcome advantages over previous solutions, especially for liver and pancreas preservation. Canine studies demonstrated excellent results with kidney, pancreas and liver preservation[42,43] Clinically, UW solution has been shown to have advantages over EC in kidney transplant recipients. In the European Multicentre Study Group (54 transplant centers), 695 kidneys were randomly preserved either in the EC solution or the UW solution.[44] There was a significant decrease in delayed graft function (33% vs 25%) and primary nonfunction (6.4% vs 3.7%) and serum creatinine level in UW group vs EC group. One year graft survival was higher in the UW group than in the EC group.[44] Another study compared 185 liver preserved in UW solution with 180 livers preserved in EC solution and demonstrated UW to safely increase preservation time to more than 15 hrs compared with EC solution.[45] These and other studies have resulted in UW solution becoming the gold standard for multiorgan cold storage.

HTK solution was developed as a physiologic cardioplegic solution. It has potent buffering systems and low viscosity. Mannitol, tryptophan, histidine and α-ketoglutarate are large, impermeable molecules and act to prevent cellular swelling. Histidine acts as both a buffer and a free-radical scavenger. It has been used successfully for heart preservation.[46] Recently the efficacy of HTK solution for preservation in other organs was compared to UW. For pancreas preservation with a mean time of 11 hours of cold ischemia, UW and HTK demonstrated similar efficacy.[47] One prospective mutlicenter randomized study compared HTK and UW solution for kidney preservation from heart beating donors,[48] while another compared HTK and EC. It was reported that delayed graft function (DGF) and long term 3-year graft survival were not different between HTK and UW solutions. However, there was a significantly higher risk of DGF using EC as the preservation solution with 29% of the HTK (n = 85) versus 43% of the EC (n = 119) requiring dialysis in the first posttransplant week. The study concluded that HTK is comparable to UW in its preservative abilities, using kidneys from heart-beating kidney-only donors, whereas HTK was superior to EC as a renal preservation solution.[48] A prospective randomized study compared HTK to UW solution in 60 liver transplant patients.[49] There was no significant difference in the outcome between groups and the study concluded that both HTK and UW solutions are appropriate for clinical use in liver transplantation.

Celsior solution is a mixture of the impermeant components of UW solution and the strong buffers from the HTK solution. Celsior has an extracellular composition, with high sodium and low potassium and is a low-viscosity preservation solution already used successfully for heart and lung preservation. A prospective multicenter study of 70 cardiac transplant recipients was performed to evaluate the Celsior solution for cardiac preservation with a minimum follow up of 5 years.[50] Operative mortality was 8.6% with a 5 year actuarial survival of 75%. The authors concluded that the use of Celsior in cardiac transplantation was safe and very effective. Another multicenter, randomized, controlled, open-label trial was conducted to compare Celsior solution to conventional preservation solutions in 131 heart transplant.[51] No significant difference was measured between the Celsior and control groups in 7-day patient survival (97% versus 94%) and the proportion of patients with one or more adverse events (Celsior, 88%; /control 87%) or serious adverse events (Celsior, 38%; control, 46%). Significantly fewer patients in the Celsior group developed at least one cardiac-related serious adverse event (13% versus 25%). They concluded that Celsior solution is as safe and effective as conventional solutions for flush and cold storage of cardiac allografts before transplantation. Recently Cesior solution has been used for other organs. A prospective multicenter randomized study was performed to evaluate the efficacy of the Celsior solution in the clinical preservation of kidneys.[52] Of 187 renal transplants studied, 99 kidneys were stored in Celsior solution and 88 in UW solution. There were no significant differences in DGF, mean serum creatinine levels, mean daily urinary output and two year graft survival between both groups. They concluded that preservation of kidneys in Celsior solution was equivalent to that of UW solution. A single-center, prospective, randomized pilot study compared Celsior solution with UW solution for pancreas transplant recipients.[53] They found that UW and Celsior solutions have similar safety profiles with similar outcomes following pancreas transplantation. A prospective, randomized, multicenter, pilot study compared UW solution to Celsior solution in liver transplantation.[54] The authors concluded that Celsior solution was similar to UW solution as a preservation solution in liver transplantation.

Clinical Implications of Preservation

The goal of preservation is to extend the period of viability of the procured organ until the transplant procedure can be completed. This begins with in situ cooling in the donor. Once the organ has been surgically removed, ex vivo preservation begins. Static cold storage is used in all extra-renal organs and the majority of the kidneys. Following excision, the organ is usually reperfused with the a selected perfusate on the back table and is then packaged, placed on ice in a sterile container and transported to the transplanting center. Kidney preservation includes a different option, namely pulsatile preservation. This is accomplished using a closed circuit with a contained pump that forces fluid through the kidney, simulating physiologic conditions. As a result of the benefits observed in pulsatile preservation of kidneys, several basic research projects have addressed it's use in other solid organs, including liver[55,56] and heart.[57,58]

Delayed graft function (DGF) is a descriptive term and can be viewed as a surrogate marker of cold ischemic injury to the kidney. It is commonly defined as the requirement for dialysis in the first posttransplant week. Clearly cold ischemia is not the only cause of DGF. However, most other factors can be ameliorated by careful attention donor management, procurement techniques, recipient hemodynamic management and minimizing the use of nephrotoxic immunosuppressive agents in the early posttransplant period. DGF has been associated with an increased incidence of acute rejection episodes and poorer long term graft function and survival,[59,60] as well as increased costs due to the requirements for dialysis posttransplant and increased length of hospital stay.[61-63]

The most clinically recognizable positive outcome of pulsatile preservation is a decrease in posttransplant DGF.[64-66] Pulsatile preservation allows for ongoing perfusion of the kidney which maintains the dilated vasculature and dilutes and removes the products of anaerobic metabolism. In addition, it maintains the metabolic integrity of the organ by providing substrates necessary to synthesize high energy phosphate compounds (i.e., ATP), necessary for optimal cellular function following reperfusion.[67] This synthesis does not occur with static cold storage. Pharmacologic agents, in particular vasodilators, can be added to the pump circuit and deliver the medication directly into the renal vasculature during cold perfusion.[66] While cold ischemic injury is a continuum, with an increasing incidence of DGF with increasing times of static cold storage,[68,69] it can be amelio-rated with pulsatile preservation, allowing for excellent outcomes even with considerably longer cold ischemic times.[70] Finally, as described by Henry[71] and subsequently others,[72] the measurable dynamics of this preservation method allows one to predict the likelihood of early posttransplant graft function. This is done by indirectly determining the status of the renal vasculature by mea-suring intrarenal resistance which is calculated as flow/mean pulsate pressure. Since circuit flow is continuously measured during machine pumping and mean pressure can be readily determined from the pulse pressures that are also continuously measured by the machine, resistances can be calculated at any time that is desired.

There has recently been a renewed interest in pulsatile presesrvation as a result of increasing interest in maximum utilization of ECD (extended criteria donors) and DCD (donation after cardiac death) donor kidneys to positively impact the longer waiting lists for deceased donor kid-neys. As noted above, examining dynamic flow characteristics and intrarenal resistance has allowed clinicians the ability to identify those kidneys with poor characteristics (very high resistance) that are quite likely to do poorly posttransplant. This can greatly aid the clinician in the decision to use or discard an organ, or to choose a recipient that might better tolerate prolonged dialysis/renal dysfunction posttransplant. Several studies have shown improvements in outcomes of the kidneys preserved with pulsatile preservation, presumably by improving the metabolic integrity of those organs.[66,73,74] Pulsatile preservation of ECD kidneys is associated with significantly lower discard rates and therefore allows for more transplantable organs.[74] The use of pulsatile preservation is ideally suited to augment aggressive attempts to increase procurement of organs from donor sources not previously utilized.

Conclusion

Preservation solutions and the methods utilized have improved significantly over the last 25 years. Optimal organ preservation has allowed for improved organ function early posttransplant and subsequent improved graft and patient outcomes. Additionally, improved preservation allows the clinician more opportunity to choose and prepare a particular recipient for the transplant op-eration. Broader sharing of organs, particularly non-renal solid organs, can be accomplished with current preservation methods. Pulsatile preservation can be used to improve outcomes in kidney transplantation, especially from donors that yield less than optimal organs (ECD and DCD do-nors). Dynamic flow measurement and calculated vascular resistance during preservation can help the clinician decide which organs are best utilized (or not utililized) in which recipients. Ongoing research may allow the benefits of pulsatile storage to be extended to nonrenal organs.

References

1. Carrel A, Linbergh CA. The culture of whole organs. Science 1935; 21:621.
2. Calne RY, Pegg DE, Pryse-Davies J et al. Renal preservation by ice cooling: an experimental study relating to kidney transplantation from cadavers. Br Med J 1963; 2(5358):651-5.
3. Belzer FO, Ashby BS, Gulyassy PF et al. Successful seventeen-hour preservation and transplantation of human-cadaver kidney. N Engl J Med 1968; 278(:608-610.
4. Belzer FO, Southard JH. Principles of solid-organ preservation by cold storage. Transplantation 1988; 45:673-676.
5. Pegg DE. Organ preservation. Surg Clin N Am 1986; 66:617-632.
6. Wicomb WN, Cooper DKC, Barnard CN. Twenty-four hour preservation of the pig heart by a portable hypothermic perfusion system. Transplantation 1982; 34:246-50.
7. Levy MN. Oxygen consumption and blood flow in the hypothermic, perfused kidney. Am J Physiol 1959; 197:11.
8. Green CJ, Healing G, Simpkin S et al. Increased susceptibility to lipid peroxidation in rabbit kidney: a consequence of warm ischemia and subsequent reperfusion. Comp Bioch Physiol 1986; 83:603-606.
9. Kretsinger RH. The informational role of calcium in the cytosol. Adv Cyclic Nucleotide Res 1979; 11:1-26.
10. Opie LH. Proposed role of calcium in reperfusion injury. Int J Cardiol 1989; 23:159-164.
11. Berridge MJ, Irvine RF. Inositol triphosphate, a novel second messenger in cellular signal transduction. Nature 1984; 312:315-321.

12. Spector AA, Yorek MA. Membrane lipid composition and cellular function. J Lipid Res 1985; 26:1015-1035.
13. Wattiaux R, Wattiaux-De Coninck S. Effect of ischemia on lysosomes. Int Rev Exp Pathol 1984; 26:85-106.
14. Parks DA, Granger DN. Contributions of ischemia and reperfusion to mucosal lesion formation. Am J Physiol 1986; 250:G749-G753.
15. Korthuis RJ, Smith JK, Carden DL. Hypoxic reperfusion attenuates postischemic microvascular injury. Am J Physiol 1989; 256:H315-H319.
16. Fuller BJ, Gower JD, Green CJ. Free radical damage and organ preservation: fact or fiction? Cryobiology 1988; 25:377.
17. Southorn PA. Free radicals in medicine. I. Chemical nature and biological reactions. Mayo Clin Proc 1988; 63:381-389.
18. Southard JH, Marsh DC, McAnulty JF et al. Oxygen-derived free radical damage in organ preservation: activity of superoxide dismutase and xanthine oxidase. Surgery 1987; 101:566-570.
19. Van Kuijk FJGM, Sevanian A, Handelman GJ et al. A new role for phospholipase A_2: protection of membranes from lipid peroxidation damage. Trends in Biochem Sci 1987; 12:31-34.
20. Sevanian A, Muakkassah-Kelly SF, Montestruque S. The influence of phospholipase A_2 and glutathione peroxidase on the elimination of membrane lipid peroxides. Arch Biochem Biophys 1983; 223:441-452.
21. Brawn K, Fridovich I. Superoxide radical and superoxide dismutase: threat and defense. Acta Physiol Scand 1980; 492:9-18.
22. Flores J, Di Bona DR, Frega N et al. Cell volume regulation and ischemic tissue damage. J membr Biol 1972; 10:331-343.
23. Nagano K, Gelman S, Bradley EL et al. Hypothermia, hepatic oxygen supply-demand and ischemia-reperfusion injury in pigs. Am J Physiol 1990; 258:G910-G918.
24. Collins GH, Bravo-Shugarman M, Terasaki PI. Kidney preservation for transplantation. Initial perfusion and 30 hours' ice storage. Lancet 1969; 2:1219-1222.
25. Wahlberg JA, Southard JH, Belzer FO. Development of a cold storage solution for pancreas preservation. Cryobiology 1986; 23:477-482.
26. Sundberg R, Ar'Rajab A, Ahrén B et al. The functional effects of suppression of hypothermia induced cell swelling in liver preservation by cold storage. Cryobiology 1992; 28:150-158.
27. Wahlberg JA, Love R, Landegaard L et al. 72-hour preservation of the canine pancreas. Transplantation 1987; 43:5-12.
28. Collins GM, Hartley LC, Clunie GJ. Kidney preservation for transportation. Experimental analysis of optimal perfusate composition. Br J Surg 1972; 59:187-9.
29. Moen J, Claesson K, Pienaar H et al. Preservation of dog liver, kidney and pancreas using the belzer-UW solution with a high-sodium and low-potassium content. Transplantation 1989; 47:940-945.
30. Jamieson NV, Lindell S, Sundberg R et al. An analysis of the components in UW solution using the isolated perfused rabbit liver. Transplantation 1988; 46:12-16.
31. Belzer FO, Ashby BS, Dunphy JE. 24-hour and 72-hour preservation of canine kidneys. Lancet 1967; 2:536-538.
32. Southard JH, Senzig KA, Hoffmann R et al. Denaturation of albumin: a critical factor in long-term kidney preservation. J Surg Res 1981; 30:80-85.
33. Hitchcock CR, Kiser JC, Telander RL et al. Effect of low molecular weight dextran on organ perfusion and sludging. Surgery 1964; 56:533-539.
34. McKeown CMB, Edwards V, Phillips MJ et al. Sinusoidal lining cell damage: the critical injury in cold preservation of liver allografts in the rat. Transplantation 1988; 46:178-191.
35. Johnson RWG, Anderson M, Flear CTG et al. Evaluation of a new perfusate for kidney preservation. Eur Surg Res 1971; 3:215.
36. Toledo-Pereyra LH, Condie RM, Callender CO et al. Hypothermic pulsatile kidney preservation. Arch Surg 1974; 109:816.
37. Marshall VC, Ross H, Scott DF et al. Preservation of cadaver renal allografts: comparison of ice storage and machine perfusion. Med J Aust 1977; 2:353-356.
38. BretschneiderHJ. Myocardial Protection. J Thorac Cardiovasc Surg 1980; 28:295.
39. Sacks SA, Petritsch PH, Kaufman JJ. Canine kidney preservation using a new perfusate. Lancet 1973; 1:1024-8.
40. Collins GM, Green RD, Halasz NA. Importance of anion content and osmolarity in flush solutions for 48 to 72 hr hypothermic kidney storage. Cryobiology 1979; 16:217.
41. Hardie I, Balderson G, Hamlyn L et al. Extended ice storage of canine kidneys using hyperosmolar Collins' solution. Transplantation 1977; 23:282-3.

42. den Butter G, Saunder A, Marsh DC et al. Cold Storage Solutions for Liver Preservation. Transpl Proc 1993; 25:3218-9.
43. Ploeg RJ, Goossens D, McAnulty JM et al. Successful 72-hour cold storage of dog kidneys with UW solution. Transplantation 1988; 46:191-196.
44. Ploeg RJ, van Bockel JH, Langendijk PT et al. Effect of preservation solution on results of cadaveric kidney transplantation. The european multicentre study group. Lancet 1992; 340:129-37.
45. Todo S, Nery J, Yanaga K et al. Extended preservation of human liver grafts with UW solution. JAMA 1989; 261:711-4.
46. Reichenspurner H, Russ C, Uberfuhr P et al. Myocardial preservation using HTK solution for heart transplantation. A multicenter study. Eur J Cardiothorac Surg 1993; 7:414-9.
47. Fridell JA, Agarwal A, Milgrom ML et al. Comparison of histidine-tryptophan-ketoglutarate solution and university of wisconsin solution for organ preservation in clinical pancreas transplantation. Transplantation 2004; 77:1304-6.
48. de Boer J, De Meester J, Smits JM et al. Eurotransplant randomized multicenter kidney graft preservation study comparing HTK with UW and Euro-Collins. Transpl Int 1999; 12:447-53.
49. Erhard J, Lange R, Scherer R et al. Comparison of histidine-tryptophan-ketoglutarate (HTK) solution versus university of wisconsin (UW) solution for organ preservation in human liver transplantation. A prospective, randomized study. Transpl Int 1994; 7:177-81.
50. Remadi JP, Baron O, Roussel JC et al. Myocardial preservation using celsior solution in cardiac transplantation: early results and 5-year follow-up of a multicenter prospective study of 70 cardiac transplantations. Ann Thorac Surg 2002; 73:1495-9.
51. Vega JD, Ochsner JL, Jeevanandam V et al. A multicenter, randomized, controlled trial of celsior for flush and hypothermic storage of cardiac allografts. Ann Thorac Surg 2001; 71:1442-7.
52. Faenza A, Catena F, Nardo B et al. Kidney preservation with university of Wisconsin and Celsior solution: a prospective multicenter randomized study. Transplantation 2001; 72:1274-7.
53. Boggi U, Vistoli F, Del Chiaro M et al. Pancreas preservation with university of wisconsin and celsior solutions: a single-center, prospective, randomized pilot study. Transplantation 2004; 77:1186-90.
54. Cavallari A, Cillo U, Nardo B et al. A multicenter pilot prospective study comparing celsior and university of wisconsin preserving solutions for use in liver transplantation. Liver Transpl 2003; 9:814-21.
55. van der Plaats A, Maathuis MH, 'T Hart NA et al. The Groningen hypothermic liver perfusion pump: functional evaluation of a new machine perfusion system. Ann Biomed Eng 2006; 34:1924-34.
56. Minor T, Olschewski P, Tolba RH et al. Liver preservation with HTK: salutary effect of hypothermic aerobiosis by either gaseous oxygen or machine perfusion. Clin Transplant 2002; 16:206-11.
57. Ozeki T, Kwon MH, Gu J et al. Heart preservation using continuous ex vivo perfusion improves viability and functional recovery. Circ J 2007; 71:153-9.
58. Poston RS, Gu J, Prastein D et al. Optimizing donor heart outcome after prolonged storage with endothelial function analysis and continuous perfusion. Ann Thorac Surg 2004; 78:1362-70.
59. Ojo AO, Wolfe RA, Held PJ et al. Delayed graft function: risk factors and implications for renal allograft survival. Transplantation 1997; 63:968-74.
60. Gjertson DW. Impact of delayed graft function and acute rejection on kidney graft survival. In: Cecka JM, Teraski P, eds. Transplant Proc. Los Angeles UCLA Immunogenetics Center, 2002; 34:2432.
61. Almond PS, Troppmann C, Escobar F, Frey DJ, Matas AJ. et al. Economic impact of delayed graft function. Transplantation Proc 1991; 23:1304.
62. Rosenthal JT, Danovitch GM, Wilkinson A et al. The high cost of delayed graft function in cadaveric renal transplantation. Transplantation 1991; 51:1115.
63. Light JA, Gage F, Kowalski AE et al. Immediate function and cost comparison between static and pulsatile preservation in kidney recipients. Clin Transplant 1996; 10:233.
64. Burdick JF, Rosendale JD, McBride MA et al. National impact of pulsatile perfusion on cadaveric kidney transplantation. Transplantation 1997; 64:1730.
65. Sellers MT, Gallichio MH, Hudson SL et al. Improved outcomes in cadaveric renal allografts with pulsatile preservation. Clin Transplantation 2000; 14:543-549.
66. Polyak MM, Arrington BO, Stubenbord WT et al. The influence of pulsatile preservation on renal transplantation in the 1990s. Transplantation 2000; 69:249-58.
67. Stubenitsky BM, Ametani M, Danielewicz R et al. Regeneration of ATP in kidney slices after warm ischemia and hypothermic preservation. Transplant Int 1995; 8:293-7.
68. Ojo AO, Wolfe RA, Held PJ et al. Delayed graft function: risk factors and implications for renal allograft survival. Transplantation 1997; 63:968-974.
69. Gjertson G. DGF by cold ischemia time and donor age. In: Cecka JM, Teraski P, eds. Clin Transpl. Los Angeles UCLA Immunogenetics Center, 2000:467-480.
70. Henry ML, Pelletier RP, Elkhammas EA et al. Pulsatile preservation is associated with a low incidence of delayed graft function. American Jour of Transpl 2003; (Suppl 5)3:S1081,428.

71. Henry ML, Sommer BG, Ferguson RM. Renal blood flow and intrarenal resistance predict immediate renal allograft function. Transplant Proc 1986; 18:557-558.
72. Polyak MM, Arrington BO, Kapur S et al. Calcium ion concentration of machine perfusate predicts early graft function in expanded criteria donor kidneys. Transpl Int 1999; 12:378-82.
73. Jacobbi LM, Gage F, Montgomery RA et al. Machine preservation improves functional outcomes in cadaveric renal transplantation. American Jour of Transpl 2003; (Suppl 5)3:S1080,428.
74. Schold JD, Kaplan B, Howard RJ et al. Are we frozen in time? Analysis of the utilization and efficacy of pulsatile perfusion in renal transplantation. Am J Transplant 2005; 5:1681-8.

CHAPTER 4

Organ Freezing

Luis H. Toledo-Pereyra,* Fernando Lopez-Neblina and Alexander H. Toledo

Introduction

Organ freezing remains, so far, an elusive goal of transplant specialists and cryobiologists interested in maintaining long-term preservation as a mean of storing organs with the possibility of establishing effective banking for better matched transplanted organs.

Crystal formation continues to represent the major obstacle for successful organ freezing. Thus, how to prevent the formation of ice crystals is the primary objective for reaching organ viability after freezing, thawing and transplantation. Two other difficulties during organ freezing are cryoprotectant toxicity and recrystallization during rewarming.[1] Chilling injury caused by reduction of temperature, even though less evident, can still be present after freezing and transplantation.

The use of cryoprotectants becomes critical to securing certain degree of protection when freezing is taking place. To reach freezing temperatures at less than $-76\,^\circ$C and maintain viability is a lofty goal under any circumstances. The evaluation behind all freezing attempts in organs for transplantation is the main concern of this work.

Difficulties of Organ Freezing

There is no single method that is consistently acceptable to reach organ freezing as a potential clinical possibility. Organ function and structure difficulties begin at the time of lowering the temperature below $-0\,^\circ$C and obviously these difficulties increase as the temperature reaches $-20\,^\circ$C which is the supercooling state, then when the temperature advances to $-76\,^\circ$C to reach freezing and afterwards to continue to the vitrification state at $-196\,^\circ$C. The principal consideration when evaluating the distant probability of organ freezing is how achieve these states without interfering with cell and tissue function, knowing that crystal formation can affect any attempts at improving function and survival. Accomplishing organ freezing without cryoprotectant toxicity is also vital to maintaining organ viability. In addition, reduction of the organ temperature from room temperature to extremely low temperatures is challenging. Determination of the best speed of freezing and thawing is also essential. These are just a few of the critical questions in better understanding organ freezing viability.

Organ freezing has not been fully understood. Many important considerations remain as part of the study of this critical biological state. Organ freezing is not the same as cell freezing, where concerns for the organ are not frequently present in cells. Another important question has to do with the susceptibility of certain organs to freezing, that is, are livers more difficult to freeze than kidneys or hearts? Are any one of these organs more apt to tolerate the freezing process than the thawing response? Is it possible that the structural and functional organ condition is better preserved with certain cryoprotectants? The questions abound and not enough answers are available.

*Corresponding Author: Luis H. Toledo-Pereyra—Departments of Research and Surgery Michigan State University, Kalamazoo Center for Medical Studies, Kalamazoo, Michigan 49008, USA. Email: toledo@kcms.msu.edu

Organ Preservation for Transplantation, Third Edition, edited by Luis H. Toledo-Pereyra. ©2010 Landes Bioscience.

Kidney Freezing

There have been many attempts at freezing kidneys since the initial observations put through in the 1960's.[2] In 1967, Halasz and his associates[3] took canine kidneys to −50°C and 4 of 38 sustained life. The thawing procedure varied from water bath, diathermy or microwave. Glycerol and propylene glycol were the cryoprotectants used. Six years later, in 1973, Dietzman and his group[4] utilizing DMSO, supercooled dog kidneys at −22°C and thereafter microwave thawed them, but did not achieve normal serum creatinine after transplantation. In 1974, Kubota[5] working in the Dietzman laboratories, used similar conditions as indicated before, but this time the thawing technique included long/short pulses with pause. This change permitted 8 of 21 dog kidneys to produce urine with an average survival of 10 days after contralateral nephrectomy.

In 1977, Guttman and his collaborators[6] froze canine kidneys at −80°C using DMSO 1.0-1.4 M as a cryoprotectant like Dietzman and Kubota before. The big difference had to do with the thawing procedure. Guttman used microwave thawing from −80°C to 0°C in 90 seconds with rotation of the kidneys. This important variation allowed his group to reach survival with transplanted kidneys on 8 of 16 animals. These results indicated that thawing is particularly important in obtaining some success. More studies investigating this aspect need to be evaluated to better understand the way that kidneys respond to freezing and thawing changes.

In 1978, Pegg and his associates[7] did not reach kidney function after freezing at −80°C with various thawing protocols. These protocols were not completely similar to the ones utilized by Guttman and his group, since the freezing and thawing units were different in the two groups and the number of animals was limited in Pegg's group.[7] Thus, it is difficult to compare these studies and to define the importance of thawing within the whole process of organ freezing. Two years later, we introduced a completely different protocol that highlighted the integration of organ perfusion and preservation within the process of organ freezing.[8] We observed limited evidence of kidney function, but the idea of combining some of the successful principles or factors to approach the possibility of acceptable function after kidney transplantation was encouraging.

From 1982, when the first edition of this book of Organ Preservation was published,[2] to the second edition in 1997, there were no significant new works introduced in the organ kidney freezing arena.[9] During this period, attention was given to cell freezing and the basic principles relevant to cryobiology such as, changes in the aqueous solutions due to ice, cell osmotic responses, osmotic responses at low temperatures, optimal cooling rate, cryoinjury and protection, minimization of ice formation and toxicity stresses.[10] In the following period, from 1997 to the present time, we have encountered the development of limited renewed interest.

In 2000, Green[11] moved ahead with the auto-transplantation of kidneys from hibernating squirrels in the summer to nonhibernating squirrels of the same species. The tolerance to preservation times of 24, 48 and 72 hours at 4°C was more evident in the kidneys obtained from hibernating animals. This interesting study points to the importance of certain molecular factors protecting the kidneys obtained from hibernating animals. This is the first and only study using kidneys from hibernating animals for transplantation.

The same year, Kheirabadi and Fahy[12] took rabbit kidneys and perfused them with VS4, a vitrification solution containing dimethyl sulfoxide (DMSO) and formamine in a 1:1 molar ratio and 15% 1, 2-propanediol, with a total concentration of 7.5 M. (Fig. 1) Even though low urinary output was observed for several days, kidney function was completely recovered by day 10 after immediate auto-transplantation in the low temperature group (−2°C). (Fig. 2) The prostacyclin analog, iloprost, was used in all kidneys prior to nephrectomy and before VS4 administration. These worthwhile experiments clearly demonstrated that is possible to use a vitrification solution with prostacyclin to obtain minimal to moderate kidney damage with near normal function. The authors suggested mitigation of this transient kidney injury by increased pharmacological manipulation. After solving this problem, the next challenge would be the use of vitrification in kidneys to prevent ice formation when reaching glass transition temperatures (−125°C). Vitrification and devitrification would be the ultimate goals to achieve.

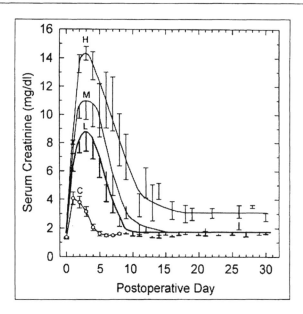

Figure 1. Postoperative serum creatinine levels after autografting kidneys previously perfused with VS4 at three different temperatures. H, M and L refer to exposure to cryoprotectant at higher, medium and lower temperatures, respectively. C (open points) refers to VS4-free controls (perfused with EC solution only. Means ± SEM. From Kheirabadi and Fahy.[12] Used with permission of Lippincott Williams and Wilkins, Inc.

In 2001, Pegg[13] reached similar conclusions as expressed before, that is, the prevention of the formation of ice was at the center of understanding and protecting organs from freezing. Pegg identified cryoprotectant toxicity and recrystallization during rewarming as major problems to be overcome. Another problem he referred to was chilling injury, associated with injury due to temperature decrease. In addition, Pegg accepted that the best possible solution, so far, would be vitrification.

Four years later, Bottomley and his group[14] were unable to vitrify embryonic kidneys for transplantation. Metanephroi isolated from pregnant Lewis rats were vitrified and stored at −135°C for 48 hours in one of the groups and then rewarmed and tested in vitro. The cytoplasm showed little disruption, but mitochondrial and nuclear injury were evident. It was clear then, that more studies were required to make vitrification successful.

In the same year, 2005, Monzen and his associates[15] utilized an electrostatic field in organs obtained from Sprague-Dawley rats subjected to −4°C–supercooling, but not frozen state-in a refrigerated unit for 72 hours for kidneys and 24 hours for heart and liver. An electrode was placed inside the chamber and voltage was conducted through the electrode to delivered 100 V or 500 V. The organs were evaluated by in vitro testing with no transplantation. The results indicated improved findings with the application of 100 V or 500 V at −4°C, when compared to organs subjected to no electrostatic force and at 4°C. It is difficult to determine the applicability of this technique to clinical transplantation when no voltage was given at 4°C and transplantation was lacking as part of the overall testing of this procedure.

In 2006, Pegg[16] summarized some of his ideas on organ freezing for preservation for transplantation. He emphasized, as indicated before, that the formation of ice was fundamental in producing the cryogenic injury. The site of crystallization being of significant consideration. Vitrification represented a path for better protection, if the formation of ice was prevented. The issue here was related to the concentration and amount of cryoprotectant that were necessary

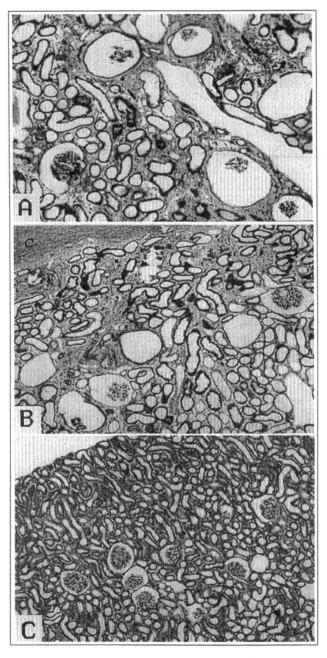

Figure 2. Histologic features ≥3 weeks after perfusion with VS4 under different conditions. The cortex was the most sensitive region of the kidney. A) VS4 perfused at about 2.5°C; middle cortex, showing extensive loss of functional mass, B) VS4 perfused at about 0°C; outer cortex (c = capsule, showing middle cortex. C) VS4 perfused at about –2.5°C; outer and middle cortex (renal surface visible at upper left), showing preservation of all cortical structures. From Kheirabadi and Fahy.[12] Used with permission of Lippincott Williams and Wilkins, Inc.

to reach vitrification. He implied that under this situation, toxicity was a potential consequence when using high cryoprotectant concentrations.

For the last three years, our group[17] has been utilizing a combination of cryoprotectants (DMSO, glycerol and 1,2 butanediol) with HBS, a new preservation solution containing macromolecules, high energy substrates and other anti-inflammatory compounds, to attempt the protection of rat kidneys subjected to very low temperature preservation (−20°C and −76°C) without ice formation. The results are mixed since in vitro data is moderately acceptable (with histology and expansion experiments) but in vivo results after transplantation did not end in successful graft survival for one or more days. This information is critical to enhance our ability to understand the most significant parameters that would allow us to reach protection during organ freezing.

Liver Freezing

The liver is an organ as sensitive, or even more sensitive, than the kidney, regarding cryopreservation tolerance. The results observed with frozen livers resemble those obtained with kidneys, in many ways, which are consistently poor, except for occasional signs of successful response. The first attempts at liver freezing began with the studies of Brown and his group,[18] in 1966. They stored the livers for 5 days at −6°C with 2M glycerol as cryoprotectant. No meaningful survival was attained after orthotopic liver transplantation. In the same year, Moss and his group[19] were unsuccessful in storing livers at −20°C and −60°C with 10-33% glycerol. In 1971, Zimmerman and associates[20] stored rabbit livers at −60°C with 2M glycerol and observed partial metabolic activity in vitro. No attempt at transplantation was even considered.

Many years went by until the next generation of researchers began new studies on liver preservation at lower temperatures than 4°C. In this case, in 1994, Rubinsky and his group[21] stored rat livers at −3°C for 6 hours using a cryoprotective solution based on the proteins existent in freeze tolerant animals. They concluded that some function was evident based on bile production and liver morphology. In 1996, Scotte and associates[22] studied the role of 8% 2,3-Butanediol in UW solution to preserve rat livers at −4°C. In an isolated liver perfusion model, they tested the viability of livers after low temperature preservation. They observed maintenance of good function in vitro after 72 hours storage. In 1999, Rubinsky and associates[23] pursued the same idea put forward before, utilizing antifreeze glycoproteins (0.5 mg/mL) together with 0.5 M glycerol to store rat livers at −2°C. Their results demonstrated that endothelial cell damage was a constant pathological finding in the supercooled livers. A year later, using the same parameters and anti-freezing components[24] in rat livers cooled at −2°C, Ishine and colleagues observed lack of microvascular structure protection, even though some bile was present (Fig. 3).

In 2001, Soltys and his group[25] showed that rat livers perfused with 10% 2,3 butanediol, UW solution and Type I anti-freeze protein at −4°C were effective in preventing ice formation and showing in vitro cell viability and increased adenine nucleotides. Demonstration of function after orthotopic transplantation has not been obtained so far. There are significant barriers to overcome to reach successful liver transplantation postfreezing. Vascular integrity, in general and microcirculation preservation, in particular, are two significant factors that need to be protected to obtain functioning livers following freezing and transplantation. In addition, endothelial cell damage with their sinusoidal detachment is universally present after supercooling temperatures.

Pancreas Freezing

In the last two decades, because of the intense interest in islet cell transplantation and freezing of islet cells as a way of preservation, less enthusiasm has been directed towards whole organ pancreas transplantation. After isolated attempts realized in the 1970s by Lazatin and his group[26] and Zimmerman and his associates[27] who demonstrated partial pancreatic function after storage with 10% DMSO at −50°C of tail pancreases and 2M DMSO, 2M glycerol or ethylene glycol of pancreas and duodenum stored at −60°C respectively. In 1981, Toledo-Pereyra and

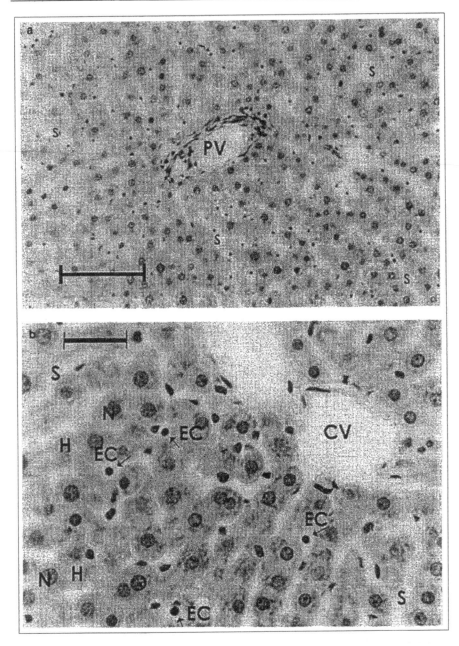

Figure 3. Histological cross-section of liver preserved in a frozen state with glycerol and AFG{, after freezing and thawing (a, × 200; b × 400). The structure of the lobules was well preserved and intercellular connection of the hepatocytes (H) was not destroyed. The portal vein (PV), central vein (CV) and sinusoids (S) are observed to be intact. Hepatocytes appeared a little shrunken, but nuclei (N) of the hepatocytes were intact. Most endothelial cells (EC) are round and detached from the sinusoids (S). The damage to the endothelial cells is most pronounced in the sinusoids, while in the larger blood vessels they are intact. Bar, 100 μm in (a) and 30 μm om (b). From Ishine, et al.[24] Used with permission of Academic Press.

colleagues[28] preserved pancreases at −196°C with 7.5% DMSO without success. Since then, we are not aware of any other studies using freezing of the whole or partial pancreas as a means to preserve the organ for transplantation. In general, it can be surmised that successful whole or partial pancreas freezing remains unresolved and therefore constitutes a possibility for future studies. The freezing of pancreatic islets cells is a different story and although successful in many fronts, is not the focus of this analysis.

Small Bowel Freezing

In 1967, Hamilton and Lehr, were able to obtain some success in two canine small bowel segments frozen at −197°C with 10% DMSO. Two years later, in 1969, Guttman and his group[30] showed some success of small bowel segments stored at −76°C with 10% glycerol and/or DMSO and transplanted subsequently. Other studies of Guttman[31] using 5% inositol and 5-10% DMSO reached greater success of small bowel segments frozen at −76°C and transplanted thereafter. Makita from Guttman's group[32] described several ultrastructural changes in the various cell types of preserved and frozen small bowel segments subjected to a similar protocol used by their group before.

In 1985, Guttman and colleagues[33] froze segments of fetal rat small bowels to −20°C or −40°C, which grew isogenically at 50% and 89% rate, demonstrating the feasibility of cryopreservation. In 2001, Tahara and associates[34] froze rat newborn small intestines to −180°C with 10% DMSO and implanted, thereafter, into the omental pouch. In the syngeneic group, there was revascularization in the omentum of the recipient with normal histology. The allogeneic tissue was rejected and could not sustain revascularization or normal structure. (Figs. 4 and 5)

The possibility of freezing the small intestine appears to be more feasible than other organs. More studies in this field with better immunosuppressant drugs might offer a significant advantage for successful organ freezing.

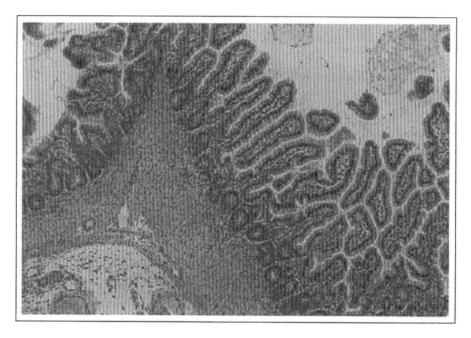

Figure 4. Representative photomicrograph of small intestine graft from group preserved with RPMI 1640 prior to syngeneic transplantation. The villi, the crypts and muscle remain almost intact (H and E, original magnification × 20). From Tahara, et al.[34] Used with permission of WB Saunders Co.

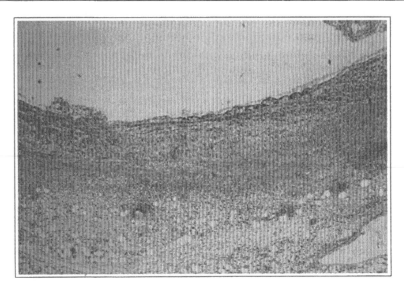

Figure 5. Representative photomicrograph of small intestinal graft from group which was stored in RPMI 1640 before allogeneic transplantation. The mucosal tissue is not found. The deformity is severe (H7E, original magnification × 20). From Tahara, et al.[34] Used with permission of WB Saunders Co.

Heart Freezing

The story of attempts at heart freezing goes back to the 1950s.[35] When the first edition of this book of Organ Preservation appeared in 1982, I indicated in relationship to the heart, "most studies have been disappointing and have yielded minimal success".[2] The field continues about the same with minimal improvement. In 1994, Wang and his group[36] studied the effect of Antarctic fish antifreeze proteins (15 mg/mL) as cryoprotectants for isolated rat hearts maintained at $-1.4°C$. All hearts failed to beat upon thawing and reperfusion. Practically, ten years later, in 2003, Arvin, Rubinsky and others[37] demonstrated that rat hearts tolerated temperatures of $-1.2°C$ average when perfused with UW solution and antifreeze proteins I or III. The hearts upon heterotopic transplantation survived with preserved myocyte and mitochondrial structure. In the same year, Hernandez and associates[38] preserved rabbit hearts at $-1.6°C$ with 5% PEG. In vitro, under the Langendorf System, adequate electrical, mechanical, metabolic and morphological recovery of the heart was seen.

In 2004, Amir, Rubinsky and others[39] subjected rat hearts to preservation of $-1.3°C$ with UW solution plus antifreeze protein III (15 mg/mL). The study similar to the one mentioned above, had the hearts transplanted heterotopically. Transplant viability was improved in the hearts perfused with antifreeze protein III. Better histological and ultrastructural integrity were maintained as well. (Fig. 6) A year afterwards, Monzen and his group[15] utilized a supercooling refrigerator for rat hearts and applied an electrostatic field inside the box maintaining the temperatures from $0°C$ to $-4°C$. Improved organ preservation was obtained in these studies. However, the temperatures are not those observed with supercooling or freezing conditions. Thus, there are limitations associated with these results.

This year, in 2008, Elami and associates[40] proceeded to examine the role of 10% ethylene glycol in UW solution for cryoprotection of rat hearts undergoing storage at $-8°C$. Testing in vitro in the Langendorf preparation indicated functional recovery after low temperature preservation. The possibility of keeping hearts at low temperatures is worthwhile to continue to explore. The observations above mentioned need to be confirmed with orthotopic transplantation of hearts stored at low temperatures. As indicated before, the temperatures studied in the previous works

Figure 6. Longitudinal section of left ventricular muscle after freezing in UW and 10% ethylene glycol, thawing and normothermic reperfusion. This section is showing regular arrays of myofibrils, divided into sarcomeres. Note the presence of numerous intact mitochondria between the myofibrils and the presence of T-tubules (T), which penetrate into the muscle at the level of the Z-bands (× 21,000). From Elami, et al.[40] Used with permission of The American Association for Thoracic Surgery.

were at the supercooling level and not to the freezing level, which starts at temperatures below −76°C. Temperatures of −180°C are classically referred to in freezing preservation as part of the vitrification state.

Freezing of Other Organs

In 1973, Okaniwa and his associates noted moderate to severe perivascular edema of canine lungs stored at subzero temperatures of −25°C and −70°C.[41] It is not clear whether DMSO 15% was used in all experiments maintained at subzero temperatures for 30 min to 6 hours. Evidence of endothelial cell toxicity is reported with DMSO at this concentration. There was no transplantation performed on these lungs which were only tested in vitro. In 1986, Groscurth and his group[42] analyzed the effect of low temperature preservation at −80°C on the morphology, viability and differentiation capacity of human fetal lung, kidney, small intestine, thyroid, liver and spleen obtained from 10 fetuses (Figs. 7-9). DMSO 10% was used as cryoprotectant. Frozen livers and spleens were practically destroyed with necrotic hepatocytes and reticular cells. Fragments from lung, kidney and intestine appeared slightly necrotic on the periphery when tested grew under in

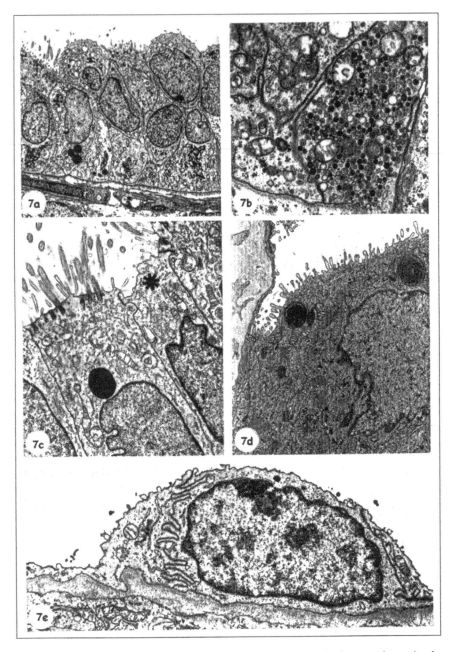

Figure 7a-c. TBM appearance of xenotransplanted lung fragments studied six to eight weeks after implantation. a) Wall of a primitive bronchus covered by an intact pseudostratified epithelium. × 2,500; b) Endocrine cell in the epithelium. × 22,000; Ciliated as well as undifferentiated (asterisk) epithelial cells in the wall of a primitive bronchus. × 8,100; d) Prospective type II pneumocyte characterized by lamellar bodies in the apical cytoplasm. × 9,000; e) Primitive pneumocytes of type I covering the wall of saccules. Note the attenuated cytoplasmic processes. × 12,500 From Groscurth, et al.[42] Used with permission of Springer-Verlag.

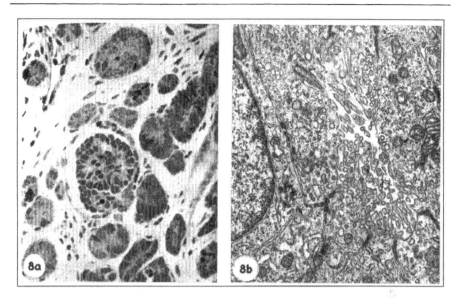

Figure 8. Xenotransplanted kidney of a 14 week old fetus. a) By light microscopy numerous tubules as well as a glomerulus of stage IV are detectable. × 375; b) Detail of a primitive tubule displaying microvilli and cilia in the lumen. Note junctional complexes between adjacent epithelial cells. × 14,000. From Groscurth, et al.[42] Used with permission of Springer-Verlag.

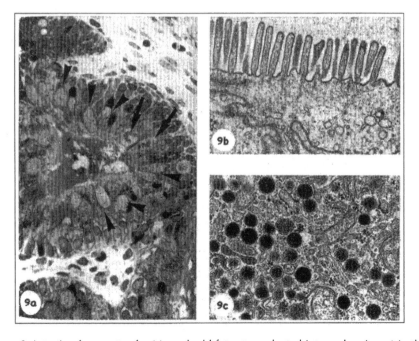

Figure 9. Intestine fragments of a 14 week old fetus transplanted into nude mice. a) In the epithelium numerous goblet cells (*arrowheads*) are detectable. The *arrows* point to mitotic figures of epithelial cells. × 590; b) Apical brush border of a prospective enterocyte. × 21,000; c) Detail of an endocrine cell in the epithelium characterized by several membrane bound granules. 33,000. From Groscurth, et al.[42] Used with permission of Springer-Verlag.

vitro conditions. Following subcutaneous transplantation into nude mice, kidney, small intestine and thyroid were accepted, but not liver and spleen.

In 2002, Wang, Gosden and associates[43] performed successful rat auto-transplantation of ovaries, fallopian tubes and the upper segment of the uterus en bloc after storage in liquid nitrogen. Prior to freezing, the tissue was perfused for 30 min at 0.35 ml/min with M2 medium containing 0.1 M fructose and increasing concentrations of dimethylsulphoxide (0-1,5 M). The treated organs were slowly cooled and ice nucleation was induced at −7°C. After overnight storage, the organs were rapidly thawed and the cryoprotectant was removed by perfusion with a reversed concentration gradient. Ovarian function was compromised to some extent by freezing, however, one animal that had received a frozen-thawed transplant conceived two fetuses. A majority of the animals receiving cryopreserved grafts had follicular ovaries and their copora lutea indicated recent ovulation. Tubal and uterine morphology of controls and frozen tissue were comparable. This work should stimulate further investigation.

Conclusions

In spite of some isolated examples of organ cryopreservation successes, the field continues to suffer lack of consistent results worth pursuing clinically. The search for the right cryoprotectant and the understanding of how an organ is better frozen and subsequently thawed are important factors that have not been clearly defined. The science of cryopreservation is lagging behind the field of organ preservation in general. Efforts at studying the best organ cryopreservation conditions will permit identification of the ideal method and cryopreservation solution for acceptable organ freezing.

References

1. Pegg DE. The current status of tissue cryopreservation. Cryo Letters 2001; 22:105-14.
2. Toledo-Pereyra LH. Organ Freezing. In: Toledo-Pereyra LH ed. Basic concepts of organ procurement, perfusion and preservation for transplantation. New York: Academic Press, 1982:159-180.
3. Halasz NA, Rosenfield HA, Orloff MJ et al. Whole organ preservation. II. Freezing studies; Surgery 1967; 61:417-21.
4. Dietzman RH, Rebelo AE, Graham EF et al. Long-term functional success following freezing of canine kidneys. Surgery 1973; 74:181-9.
5. Kubota S, Graham EF, Crabo BG et al. Influence of DMSO distribution upon renal function following Freezing and thawing. J Surg Res 1974; 16:582-91.
6. Guttman FM, Lizin J, Robitaille P et al. Survival of canine kidneys after treatment with dimethyl-sulfoxide, freezing at −80 degrees C and thawing by microwave illumination. Cryobiology 1977; 14:559-67.
7. Pegg DE, Green CJ, Walter CA. Attempted canine renal cryopreservation using dimethyl sulphoxide helium perfusion and microwave thawing. Cryobiology 1978; 15:618-26.
8. Toledo-Pereyra LH. Factors involved in successful freezing of kidneys for transplantation. Preliminary experimental observations. J Surg Res 1980; 28:563-70.
9. McGann LE, Acker JP. Cryopreservation. In: Toledo-Pereyra LH, ed. Organ Procurement and Preservation for Transplantation. Second Edition. Austin/New York: Landes Bioscience/Chapman and Hall, 1997.
10. Clark P, Fahy GM, Karow AM Jr. Factors influencing renal cryopreservation. Cryobiology 1984; 21:260-273.
11. Green C. Mammalian Hibernation: Lessons for Organ Preservation. Cryo Letters 21, 91-98.
12. Kheirabadi BS, Fahy GM. Permanent life support by kidneys Perfused with a vitrifiable (7.5 molar) cryoprotectant Solution. Transplantation 2000 Jul 15;70(1):51-7.
13. Pegg DE. The current status of tissue cryopreservation. Cryo Letters 2001; 22:105-14.
14. Bottomley MJ, Baicu S, Boggs JM et al. Preservation of embryonic kidneys for transplantation. Transplant Proc 2005; 37:280-4.
15. Monzen K, Hosoda T, Hayaski D et al. The use of a supercooling refrigerator improves the preservation of organ grafts. Biochem Biophys Res Commun 2005; 337:534-9.
16. Pegg DE. The preservation of tissues for transplantation. Cell Tissue Banking 2006; 7:349-358.
17. Lopez-Neblina F, Toledo-Pereyra LH, Toledo AH. Personal Communication 2008.
18. Brown H, Patel J, Blair DW et al. Biochemical studies with preserved transplanted canine liver. JAMA 1966; 196:775-9.

19. Moss GS, Reed P, Riddell AG. Observations on the effects of glycerol on the cold storage of the canine liver. J Surg Res 1966; 6:147-51.
20. Zimmerman G, Tennyson C, Drapanas T. Studies of preservation of liver and pancreas by freezing techniques. Transplant Proc 1971; 3:657-9.
21. Rubinsky B, Arav A, Hong JS et al. Freezing of mammalian livers with glycerol and antifreeze proteins. Biochem Biophys Res Commun 1994; 200:732-41.
22. Scotte M, Eschwege P, Cherruau C et al. Liver Preservation below 0 degrees C with UW Solution and 2, 3-butanediol. Cryobiology 1996; 33:54-61.
23. Ishine N, Rubinsky B, Lee CY. A histological analysis of liver injury in freezing storage. Cryobiology 1999; 39:271-7.
24. Ishine N, Rubinsky B, Lee CY. Transplantation of mammalian livers following freezing: vascular damage and functional recovery. Cryobiology 2000; 40(1):84-9.
25. Soltys KA, Batta AK, Koneru B. Successful nonfreezing, subzero preservation of rat liver with 2,3-butanediol and type 1 antifreeze protein. J Surg Res 2001; 96:30-4.
26. Lazatin LL, Thorn RG, Lehr HB. Dimethyl sulfoxide in the cryopreservation of canine pancreatic tissue. Cryobiology 1971; 8:396.
27. Zimmerman G, Tennyson C, Drapanas T. Studies of preservation of liver and pancreas by freezing techniques. Transplant Proc 1971; 3:657-659.
28. Toledo-Pereyra LH, Gordon D, MacKenzie G. Cryopreservation of whole pancreas versus islet cells. Trans Am Soc Artif Intern Organs 1981; 27:259-62.
29. Hamilton R, Lehr HB. Survival of small intestine after storage for seven days at −197°C. Cryobiology 1967; 3:375.
30. Guttman FM, Khalessi A, Huxley BW et al. Whole organ preservation. 1. A technique for in vivo freezing of canine intestine using intraarterial helium and ambient nitrogen. Cryobiology 1969; 6:32-6.
31. Guttman FM, Khalessi A, Berdnikoff G. Whole organ preservation. II. A study of the protective effect of glycerol, dimethyl sulfoxide and both combined while freezing canine intestine employing an in vivo technique. Cryobiology 1970; 6:339-46.
32. MakitaT, Khalessi A, Guttman FM et al. The ultrastructure of small bowel epithelium during freezing. Cryobiology 1971; 8:25-45.
33. Guttman FM, Nguyen LT, Laberge JM et al. Fetal rat intestinal transplantation: cryopreservation and cyclosporin Pediatr Surg 1985; 20:747-53.
34. Tahara K, Uchika H, Dawarasaki G et al. Experimental small bowel transplantation using newborn intestine in rats:III. Long-Term Cryopreservation of Rat Newborn intestine. J Ped Surg 2001; 36:602-694.
35. Luyet B, Gonzales F. Survival of cells in embryonic heart of chick after freezing. Biodynamica 1951; 7:61-66.
36. Wang T, Xhu Q, Layne JR et al. Antifreeze glycoproteins from Antarctic notothenoid fishes fail to protect the rat cardiac explant during hypothermic and freezing preservation. Cryobiology 1994; 31:185-92.
37. Amir G, Rubinsky B, Kassif Y et al. Preservation of myocyte structure and mitochondrial integrity in subzero cryopreservation of mammalian hearts for transplantation using antifreeze proteins- an electron microscopy study. Eur J Cardiothorac Surg 2003; 24:292-6.
38. Hernandez E, Gutierrez E, Borrego JM et al. Morphologic and metabolic evaluation of the donor heart after an experimental freezing Protocol. Tran Proc 2003; 35:729-731.
39. Amir G, Rubinsky B, Horowitz L et al. Prolonged 24-hour subzero preservation of heterotopically transplanted rat hearts using antifreeze proteins derived from arctic fish. Ann Thorac Surg 2004; 77:1648-55.
40. Elami A, Gavish Z, Korach A et al. Successful restoration of function of frozen and thawed isolated rat hearts. J Thorac Cardiovasc Surg 2008; 135:666-72.
41. Okaniwa G, Nakada T, Kawakami M et al. Studies on the preservation of canine lung at subzero temperatures. J Thorac Cardiovasc Surg 1973; 65:180-186.
42. Groscurth P, Erni M, Balzer M et al. Cryopreservation of human fetal organs. Anat Embriol 1986; 174:105-113.
43. Wang X, Chen H, Yin H et al. Fertility after intact ovary transplantation. Nature 2002; 415:385.

CHAPTER 5

Cellular and Molecular Biology of Ischemia Reperfusion

Shohachi Suzuki*

Introduction

The interruption of organ blood flow followed by its restoration frequently occurs as inevitable events in a wide variety of clinical situations, such as aortic cross-clamping, cardiopulmonary bypass, hemorrhagic shock, stroke, myocardial infarction, liver resection and organ transplantation.[1-15] Prolonged organ ischemia is characterized by a lack of tissue oxygen and a conversion of cellular metabolism to anaerobic pathways, which results in the loss of cellular function and ultimately cell death.[1-4] Prompt blood restoration is needed for functional recovery of the organ lapsing into ischemia, but reperfusion itself initiates a cascade of adverse reactions that paradoxically injures ischemically damaged tissue. This pathophysiological phenomenon has been recognized as ischemia reperfusion injury (IRI). Much attention has been paid to clarify the mechanisms of IRI over the past three decades. Early researches on IRI had focused on the role of reactive oxygen species (ROS) as pivotal factors for its development.[1,2,5,16-20] Subsequently, IRI in various organs was recognized as complex phenomena directly associated with the inflammatory response. Recent researches on the pathophysiological process of IRI have been advanced in cellular and molecular aspects.

In the field of liver surgery, surgeons may encounter poor outcome in patients subjected to a longer period of hepatic ischemia concomitant with liver resection and transplantation.[7,9,21,22] IRI under these conditions induces hepatocellular damage which may result in the impairment of liver regeneration and primary graft dysfunction or failure.[10,21-23] Numerous studies using different animal models of hepatic IRI have demonstrated that the interaction between polymorphonuclear neutrophil (PMN) and endothelial cell through ROS, inflammatory mediators such as cytokines and chemokines and adhesion molecules may be closely associated with the development of the injury.[24-29] Recent studies have attempted to clarify in detail the mechanisms of IRI in the levels of intracellular molecular basis.[30-38]

This chapter will provide a brief overview of the pathophysiology of organ IRI followed by a discussion of the possible cellular and molecular mechanisms, with a central focus on hepatic IRI.

Cellular Mechanisms

Implication of Hepatic Component Cells in IRI of the Liver

The pathophysiology of organ IRI is multifactorial and complex. The inflammatory aspect of this injury includes the cellular and molecular events. Because tissue component cells vary with each organ, mechanisms of IRI may have partly organ-dependent ingredient. Regarding liver tissue, four cell types, such as parenchymal cells (hepatocytes) and nonparenchymal cells (resident

*Shohachi Suzuki—Second Department of Surgery, Hamamatsu University School of Medicine, 1-20-1 Handayama,Higashi-hu Hamamatsu 431-3192, Japan. Email: shohachi88@msn.com

Organ Preservation for Transplantation, Third Edition, edited by Luis H. Toledo-Pereyra. ©2010 Landes Bioscience.

macrophages (Kupffer cells), sinusoidal endothelial cells (SECs) and hepatic stellate cells (HSCs)) are associated with IRI (Fig. 1). The basic mechanisms of liver injury caused by warm and cold ischemia are similar, but there are significant difference in the main victims for ischemic insult. Warm ischemia is associated with greater oxidative stress and mitochondrial dysfunction, whereas cold ischemia is related to reduced oxidative phosphorylation, lower cellular adenosine triphosphate (ATP) levels and increased glycolysis.[39-41] The main sites of injury in warm ischemia are hepatocytes, though nonparenchymal cells (Kupffer cells, SECs and HSCs) are more susceptible to cold IRI than hepatocytes.[42-44]

Hepatocytes tend to be resistant to damage by ROS, because these hepatic parenchymal cells contain high intracellular concentrations of glutathione (GSH), superoxide dismutase (SOD), catalase and lipid soluble antioxidants. However, intracellular formation of ROS can be detected in hepatocytes after extended periods of hypoxia or ischemia of the liver. Hepatocytes are the site of biochemical and functional changes given by the breakdown of ATP products, the oxidative stress based on the reduction of glutathione and the generation of ROS in hepatic IRI.[16,45] Moreover, capillary plugging is formed through PMN-endothelial interaction and results in the derangement of microcirculation with the development of organ IRI.[46,47] Though it has been considered that hepatocytes face two hepatic sinusoids and are less susceptible to capillary plugging, stagnant PMNs contribute to the flow hindrance in the sinusoidal network of hepatic microcirculation which leads to structural and functional liver damages.[48]

Extensive experimental researches by Jaeschke and associates have demonstrated that there are two distinct phases of hepatic IRI composed of Kupffer cell-mediated injury phase and subsequent PMN-induced injury phase.[25,27,28] Kupffer cells are the largest population of fixed tissue macrophages in the body. These cells are mainly associated with the initial phase of IRI occurring in the first 4 hours after reperfusion following hepatic ischemia.[27] Activated complement products during reperfusion are critically involved in the stimulation of primed Kupffer cells.[49] These resident macrophages appear to be relatively resistant to ischemia, yet become activated during reperfusion. At the initial phase of reperfusion, activated Kupffer cells are predominant sources of ROS and implicated in the production of potentially harmful mediators, including tumor necrosis factor (TNF)-α, interleukin (IL)-1, IL-6, platelet-activating factor (PAF) and the other mediators (Fig. 2).[50-52] Involvement of Kupffer cell in hepatic IRI was directly or indirectly verified by studies using the blockade of Kupffer cell function, including gadolinium chloride and methyl palmitate, or an anti-inflammatory immunonutrient.[27,51,53-55] ROS produced by Kupffer cells may be associated with the development of IRI through the activation of redox-sensitive transcription factors, such as nuclear factor (NF)-κB and activating protein (AP)-1 in endothelial cells and hepatocytes, thereby regulating proinflammatory genes.[56-58]

Impaired hepatic microcirculation is a critical feature induced by IRI of the liver. Both of SECs and HSCs are important cell types relevant to microcirculatory disturbance in hepatic IRI through sinusoidal constriction. Endothelin (ET)-1, a potent vasoconstrictor, is released from SECs, endothelial cells and HSCs and this vasoconstrictive substance is associated with microcirculatory derangement in hepatic IRI via ET receptor abundantly present on HSCs.[59,60] Thus, cellular events are complexly intertwined with various molecular events in the development of hepatic IRI.

Polymorphonuclear Neutrophils (PMNs)

Seminal works using circulating PMN-depleted animals have indicated that PMNs exert a harmful influence to ischemically damaged tissue in IRI.[61,62] Subsequent abundant evidences have demonstrated a central role of PMNs and specific adhesion molecules involved in the PMN-vascular endothelium interaction in organ IRI.[28,29,38,63-68] The reperfusion injury is closely linked with the inflammatory response characterized by activation, adhesion and diapedesis of PMNs.[66,69,70] In hepatic IRI, the activation of complement system and the production of proinflammatory cytokines and chemokines are responsible for PMN recruitment into liver tissue.[26,49,52,71] The first observation on the role of PMNs in the microcirculatory disturbance occurring in hepatic IRI was introduced around two decades ago.[24] The involvement of PMNs

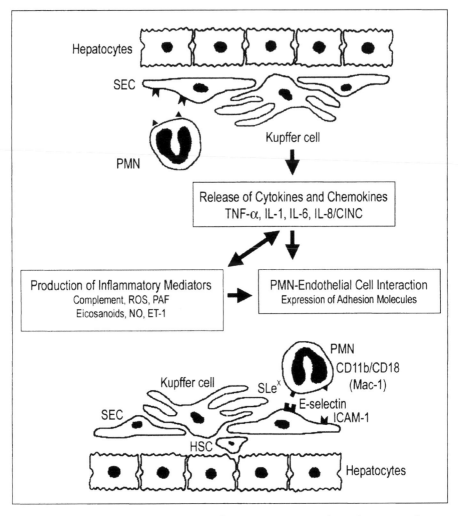

Figure 1. Schematic illustration of hepatic inflammatory response after ischemia-reperfusion. Abbreviations: SEC: sinusoidal endothelial cell; PMN: polymorphonuclear neutrophil; ROS: reactive oxygen species; PAF: platelet-activating factor; NO: nitric oxide; CINC: cytokine-induced neutrophil chemoattractant; ET: endothelin; HSC: hepatic stellate cell; SLe^x: sialylated Lewis-X blood group antigen.

in hepatic IRI has been supported by following evidences; (i) the increase of PMNs accumulation in reperfused tissue,[24-26] (ii) the attenuation of IRI by monoclonal antibodies (Mabs) to PMNs and specific adhesion molecules,[28,29,38] and (iii) the exacerbation of IRI through inflammatory mediators released by activated PMNs.[28,51,72] As in the case of hepatic IRI, the importance of PMN population has been demonstrated in studies on myocardial, pulmonary or intestinal IRI, showing markedly attenuated functional and structural damages.[62,73,74] Complement factors, TNF-α, IL-1 and PAF have a potential to induce priming of PMNs for ROS generation.[49,52,75,76] ROS are generated by membrane-associated nicotinamide adenine dinucleotide phosphate (NADPH) oxidase system and cytotoxic enzymes, such as elastase and myeloperoxidase (MPO), are released from cytoplasmic granules.[77] These inflammatory mediators not only enhance PMN activation

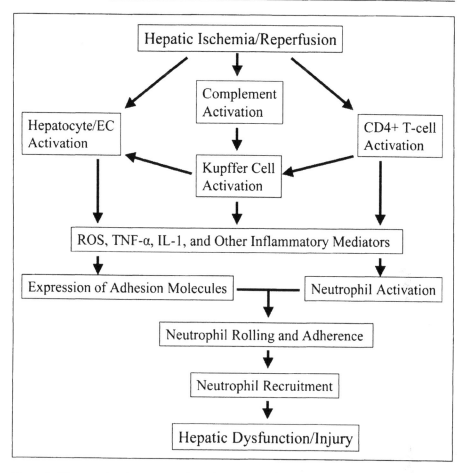

Figure 2. Diagram showing proposed mechanisms on the development of hepatic ischemia-reperfusion injury. Abbreviations: EC: endothelial cell; ROS: reactive oxygen species.

but also stimulate PMN-endothelium adherence by via mechanisms involving the upregulation of cell surface molecule expression on PMN and vascular endothelium.[78,79] Through these processes, PMNs may accumulate in hepatic sinusoids and in postsinusoidal venules.[80] Various families of adhesion molecules are responsible for PMN rolling and firm adherence in postcapillary venules.[81] PMN-endothelium adherence is, in part, a response to ROS as well as cytokines with chemotactic properties.[82-85] This adherence results in microvascular failure, frequently called 'no-reflow phenomenon' and tissue injury.[46,86] This critical phenomenon is evoked by changes in cell metabolism, cell volume regulation, synthesis and release of vasoactive substances, endothelial interaction with circulating cells and the permeability property of the microvessel wall.[86]

Tissue MPO activity is used as a quantitative assessment of PMN infiltration into tissue subjected to IRI and activated PMNs secrete MPO into extracellular space.[51,87] This enzymes may be indirectly involved in tissue injury via production of hypochlorous acid (HOCl), a powerful oxidant.[88] Thus, PMNs certainly play an essential role in perpetuating reperfusion injury after organ ischemia, but it has been considered that their function is less predominant in the early phase of reperfusion.[25,27,73,89] Multiple mechanisms including microvascular occulusion, release of ROS, cytotoxic enzymes and cytokines and increased vascular permeability, have been postulated for

PMN-mediated tissue damage in IRI. The other leukocytes such as macrophages may have a more important role in the early phase of reperfusion.[27,51,53,73,89] Because the apparently delayed response of PMNs is caused by multiple steps in organ IRI, three therapeutic strategies directed toward the inhibition of PMN activation, the inhibition of PMN adhesion molecule synthesis and the inhibition of PMN-endothelial adhesion have been developed to attenuate PMN-mediated IRI. Therapeutic advantage in treating IRI will be strongly supported, provided the beneficial effect in basic research findings is demonstrated in clinical application.

Macrophages

Macrophages play a pivotal role for initiating the inflammatory cascade in IRI of various organs, such as the liver, the lung, the kidney, the intestine and the heart.[27,51,53,89-92] As noted earlier, Kupffer cells, resident macrophages of the liver, release biological active molecules, including ROS, cytokines and other inflammatory mediators, in response to various stimuli, including IRI.[50] These tissue macrophages are a major component of the reticuloendothelial system (RES) and have a critical function in host defense mechanisms.[93] Wam hepatic ischemia up to 60 minutes is tolerable for normal rat liver without a detrimental effect on phagocytic activity, but when the hepatic ischemic time increased to 90 minutes, marked elevation of TNF-α activity as well as a decreased phagocytic activity is seen in animal model.[94] Paradoxical alterations in Kupffer cell functions, such as cytokine release and phagocytosis, may be related to different activation steps in severe hepatic IRI.

Pulmonary macrophages are present in the alveoli, interstitium, pleural space and alveolar capillaries.[95] These macrophages are equipped to orchestrate the manifestation of acute lung damage through the release of various mediators which sequestrate and activate PMNs. In the initiation of lung IRI, alveolar macrophages function as an essential player to produce proinflammatory cytokines and chemokines.[91] A study by Bishop and associates has indicated that a remarkable increase in macrophages is observed in histological sections of the reperfused lung when compared to the contralateral lung.[96] In reperfusion injury after lung transplantation, attention has been paid to a distinction between the role of donor macrophages on the one hand and the role of recipient macrophages on the other.[97] Activation of donor macrophages could be the initial consequence of ischemia and early reperfusion. In reaction to activation, donor macrophages deliver cytokines, chemotactic agents and proteolytic enzymes responsible for early reperfusion injury. Moreover, the implication of macrophages in renal and intestinal IRI has been investigated in experimental studies using macrophage-depleted models.[98,99] These studies have demonstrated that systemic macrophages in renal IRI and resident macrophages in intestinal IRI are important elements in the initiation periods of those insults.[98,99] With regard to the role of macrophages myocardial IRI, current research has indicated that macrophages infiltrating the ischemic and early reperfused myocardium may promote myocardial PMN infiltration and subsequent PMN-induced tissue dysfunctions through the production of TNF-α.[92] Therefore, strategies aimed at interfering with macrophage activation may be promising as a prime target of therapeutic intervention for organ IRI.

T-Cells

T-cells constitute one of primary arms of the adaptive immune response. Current studies from numerous different laboratories have demonstrated the direct contribution of T-cells to IRI of the liver as well as the extrahepatic organs such as the kidney, the intestine, the lung and the heart.[100-104] The fact that immunosuppressive agents such as FK506, cyclosporine and azathioprine attenuate hepatic IRI implies the involvement of T-cells in the pathophysiology of IRI.[26,105] In normothermic hepatic ischemia model in mice, CD4+ T-cells are associated with the aggravation of postischemic liver injury by interacting with SECs and platelets.[106] Moreover, unlike antibody deletion of CD8+ T-cells, CD4+ T-cell-deficient nude mice show less PMN accumulation in the postischemic liver tissue and sustain less injury during the later phase of reperfusion.[100] Thus, CD4+ T-cells appear to be the major subset of T-cells involved in IRI (Fig. 2). The current paradigm is that CD4+ T-cells have two main functional subsets, Th1 and Th2 cells, with specific cytokine profiles and cellular responses.[107] These cells can be differentiated by two distinct factors, signal transducers and activators of transcription (STAT)-4 and STAT-6. STAT-4 pathway that is mainly activated by IL-12 leads to Th1 differentiation and contributes to the development of IRI. The activation of STAT-6 gene by IL-4 is implicated in Th2 differentiation and exerts protective effect for IRI.[108,109] On the basis of these findings, T-cells via the paradigm of Th1/Th2 may play a pivotal role in the pathoprogression of IRI. Elucidation of the mechanism on T-cell activation for PMN recruitment may result in the development of a useful treatment to ameliorate of organ IRI.

Molecular Mechanisms

Complement System

The complement system has been recognized as a central mediator of innate immune defense and inflammation.[110] Though the exact molecular mechanisms of complement activation following oxidative stress have not insufficiently elucidated, experimental and clinical studies have demonstrated that reperfusion after ischemia results in local activation of the complement system in various organs.[49,111,112] Attention has been paid to the role of complement system in hepatic IRI by an evidence that depletion of serum complement by cobra venom factor (CVF) results in a significant attenuation of this insult.[49] Plasma levels of activated complement components are increased in hepatic IRI model using pigs and several complement components become localized in ischemic liver.[113-115] Complement components are activated through either the antibody-dependent classical pathway, the alternative pathway, or the mannose-binding lectin (MBL) pathway.[116] Complement cascade is rapidly activated at the initial phase of reperfusion following organ ischemia, leading to the release of biologically active substances such as the anaphylatoxins, complement factors 3a (C3a) and 5a (C5a) and the cytolytic terminal membrane attack complement complex C5b-9.[49,116] Thus, results in models of examining complement activation after organ ischemia and reperfusion have suggested a potential novel therapeutic approach to reduce harmful injury following organ ischemia and reperfusion.

The soluble form of complement receptor type one (sCR1) as the first complement-specific inhibitor showed the beneficial effect in animal models subjected to myocardial, intestinal or hepatic IRI.[49,115,117,118] In hepatic IRI, treatment with sCR1 provided the evidence that this complement-specific inhibitor was highly effective in attenuating liver necrosis during the later phase of reperfusion through the suppression of Kupffer cell-induced oxidative stress, the priming of Kupffer cells and PMNs for enhanced ROS production and PMN accumulation in liver tissue.[49] Implication of C5 and its cleavage product C5a in IRI has been extensively investigated in experimental models.[119,120] C5a is considered to be one of the most potent phlogistic peptide and binds to the C5a receptor (C5aR) on neutrophils, monocytes and macrophages with high affinity.[121,122] Among various liver cells, the C5aR mRNA in Kupffer cells is strongly expressed than in HSCs and SECs.[123] Moreover, C5a primes and activates neutrophils and Kupffer cells to generate ROS in hepatic IRI.[49] C5a also induces the secretion of lysosomal enzymes from neutrophils and macrophages and proinflammatory cytokines from monocytes and macrophages.[124,125] The observation that C5aR antagonist reduces deleterious effects of C5a in hepatic IRI may support an important role of C5a in this insult.[126] As another ihibitor of complement activation, the plasma-derived protein C1 esterase inhibitor (C1-INH) has also been adapted for organ IRI.[127,128] Increased perfusion after the treatment with C1-INH may result from reduced endothelial swelling and interstitial edema as well as effectively reduced PMN adhesion through blockade of the classical pathway.[127,128] However, it remains unclear which pathway among the classical, the alternative, or the MBL pathways is responsible for the initiation of complement activation, because complement inhibitors used in published studies do not fully discriminate among those pathways.

Blockade of complement activation has been shown the beneficial effect for attenuation of IRI in various organs. Because complement blockade alone may not result in a complete inhibition of PMN adherence, total inhibition of elevation of biochemical markers, or complete protection from liver pathology, various other mediators also would regulate inflammatory processes in hepatic IRI. The novel inhibitors of complement products may find wide clinical application as a pharmaceutical potential for the treatment or prevention of IRI.

Reactive Oxygen Species

A huge number of investigations have suggested that ROS including superoxide anion, hydroxyl radical and hydrogen peroxide are one of the most important components of tissue injury after reperfusion of ischemically damaged organs.[1-5,12-16,129,130] These molecules are produced by endothelial cells, macrophages and PMNs during IRI via mitochondria and microsomal transport system. ROS react with lipid component of the cell membrane and lipid peroxidation occurs during reperfusion after organ ischemia.[17] These serial events cause the destruction of cell membranes, resulting in increased permeability and cell lysis. In hepatic IRI, the prime sources of ROS generation are cytosolic xanthine oxidase, Kupffer cells and PMNs.[16,20,27,45,131]

During organ ischemia, cellular ATP is broken down to hypoxanthine and xanthine dehydrogenase (XD) is converted to xanthine oxidase (XO).[2,3,16] The hypoxic stress triggers the conversion of XD to the oxygen radical-producing XO. When organ subjected to ischemic insult is reoxygenated by restoration of blood flow, XO transforms hypoxanthine to xanthine to produce a burst of superoxide radical (Fig. 3).[2] This event leads to the formation of other ROS.[2,16] Production of ROS from hepatocytes and endothelium may be dependent on this pathway.[16,45,132] Phagocytic cells, such as PMNs and Kupffer cells have also been shown to produce both superoxide anion and hydrogen peroxide.[18,27] Another mechanism leading to ROS production is dependent on NADPH oxidase system. This membrane-associated system is an efficient source of ROS by activated phagocytes, such as PMNs and macrophages.[18,133] Endothelial damage caused by ROS results in the loss of microvascular integrity and the decline in tissue blood flow.

The human body is endowed with many endogenous antioxidants including SOD, catalase and GSH, but this specific defense system succumbs to the attack of ROS production during IRI. Various types of antioxidant agents have been shown to protect functional and structural tissue insult after reperfusion following organ ischemia.[16,17,19,134-138] Antioxidant therapy is a promising therapeutic strategy to ameliorate organ IRI, however, the debate partially stems from limited efficiency of supplementation of antioxidants to protect against tissue damage induced by IRI.

Cytokines

The cytokines, a group of polypeptide mediators, are not only an integral part of normal homeostasis, but also have various biological effects in many pathophysiological processes.[139] TNF-α and IL-1 may act to attract PMNs to the site of inflammation.[140,141] These proinflammatory cytokines might induce PMNs to generate other toxic metabolites as well as ROS.[142,143] The TNF transcription factor, NF-κB is induced in response to oxidative stress,[144] leading to subsequent TNF synthesis in IRI of many organs. In a positive feedback, proinflammatory fashion, binding of TNF to specific TNF membrane receptors can reactivate NF-κB. This provides a mechanism by which TNF can upregulate its own expression as well as facilitate the expression of other genes pivotal to the inflammatory response in organ IRI.[15] During the initial phase of reperfusion after hepatic ischemia, a vast potential for TNF-α production exists in Kupffer cells.[71,94] Furthermore, the endothelial cells exposed to TNF-α produce a variety of inflammatory mediators, including IL-1 and PAF.[145,146] Through these complicated interactions, TNF-α plays a pivotal cytokine to propagate the inflammatory response in IRI. This mediator is also responsible for enhancing the expression of vascular cell adhesion molecules, such as intercellular adhesion molecule-1 (ICAM-1) and P-selectin.[147-149] Consequently, the susceptibility of the vascular endotheliums to the PMN-mediated attack is enhanced through the expression of hepatic vascular adhesion molecules and CXC chemokines as chemotactic factors for PMNs.[35,52,147,150,151] A substantial body of experimental evidence that the blockade of TNF-α abolishes tissue damage caused by ischemia and reperfusion has been reinforced the significance of this cytokine as a central propagating factor in organ IRI.[71,152-154] IL-1, a macrophage-derived cytokine, is the most potent agent known to be capable of causing extravasation of PMNs into inflamed tissues.[141] This proinflammatory cytokine plays an important role in the induction of ROS production, PMN adhesion and tissue injury after organ IRI.[51,143,155,156] Thus, TNF-α and IL-1-dependent inflammatory cascades have long been thought to be the primary initiating event for propagation of inflammatory response in IRI.[71,143,155-158] In contrast, IL-1 receptor knockout mice have failed in the improvement of hepatic

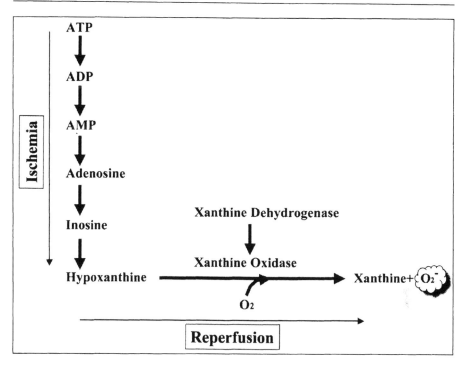

Figure 3. Mechansism of oxygen free radical production during reperfusion of ischemic tissue. During ischemia, cellular energy level is rapidly broken down from adenosine triphosphate (ATP) to hypoxanthine, a substrate for xanthine oxidase. Concomitantly, xanthine dehydrogenase is transformed xanthine oxidase. Reoxygenation by restoration of tissue blood flow leads to xanthine oxidase-mediated formation of superoxide radicals. Abbreviations: ADP: adenosine diphosphate; AMP: adenosine monophosphate. Modified from McCord JM.[2]

tissue injury compared to their wild-type counterparts, irrespective of reduction of PMN accumulation.[159] Controversy exists as to whether IL-1 may play an essential role in the development of hepatic IRI though this cytokine acts to augment PMN recruitment.

IL-6 has remarkably pleiotropic biologic activities that is synthesized by monocytes, PMNs, macrophages, lymphocytes and endothelial cells.[160] This cytokine clearly plays a major role as a mediator of inflammatory reactions. IL-6, along with IL-1 and TNF-α, induces the expression of ICAM-1 in myocytes and endothelial cells, respectively.[161] Both IL-6 and TNF-α production has been observed after a 60-minute period of hepatic ischemia and subsequent reperfusion in rodent animals.[162] Wanner and associates have indicated that early cytokinemia with increased TNF-α, IL-1α and IL-6 may be associated with dysfunction/injury of the liver, lung and kidney as systemic inflammation in hepatic IRI.[158] Thus, IL-6 had recognized as an inflammatory mediator associated with the development of IRI. In contrast, this cytokine has been found to prevent warm hepatic IRI, conceivably through its anti-inflammatory properties, modulation of TNF-α and/or promotion of liver regeneration.[163] Anti-inflammatory effects of IL-6 have been also demonstrated in IRI of extrahepatic organs, such as the intestine and the lung.[164,165] These modulating effects for IRI are induced the reduction of proinflammatory cytokine upregulation and PMN infiltration through the decreased NF-κB and STAT-3.[165] Though NF-κB is known to upregulate IL-6 expression, high IL-6 protein levels in turn may limit NF-κB activity in a negative feedback loop.[165] This may partly explain the reason that IL-6 is at times proinflammatory, while at other times anti-inflammatory in ischemia and reperfusion models. Recently, there is growing evidence that a brief prior ischemia-reperfusion period, termed 'ischemic preconditioning', is hepatoprotective in hepatic IRI.[166-168] This can be mimicked by IL-6 and TNF-α involved in priming hepatocytes to enter the cell cycle.[167] Base on these observations, it has been suggested that treatment with IL-6 might become a promising strategy to modulate organ IRI in clinical situations.

With increasing attention for the role of T-cells in the inflammatory milieu,[100-104] the influence of IL-12, a Th1 cytokine and Th2 cytokines such as IL-10 and IL-13 have been investigated as regulatory factors in IRI.[169,170] Some experimental data support the hypothesis that Th1 inflammatory pattern is pathogenic and Th2 pattern is protective in IRI.[108,109,171] The importance of IL-12 in the initiation of hepatic inflammatory response associated with IRI has been demonstrated by Lentsch and associates.[169] The expression of IL-12 is required for the full expression of TNF-α. When hepatic IL-12 levels reach maximal, TNF-α as a central mediator in IRI begins to increase shortly after reperfusion at a time.[169] An experimental observation using neutralizing antibodies against IL-12 or mice lacking IL-12 p40 suggests that the regulation of IL-12 is important for the prevention of increase in TNF-α expression and subsequent development of PMN-dependent liver injury after warm hepatic ischemia.[169] While the source of IL-12 generation has not yet been identified except for Th1 cells, Kupffer cells and SECs in liver cell population may be responsible for producing this cytokine.[170]

IL-10 is an endogenous, potent anti-inflammatory cytokine. The influence of IL-10 on hepatic IRI has been extensively studied by Le Moine and associates.[172-174] This cytokine has hepatoprotective effects during liver transplantation as well as galactosamine/lipopolysaccharide (LPS)-induced liver injury in rodent models.[175,176] In hepatic IRI model, it has been shown that recombinant murine IL-10 suppresses TNF-α, chemokine and ICAM-1 mRNA expression.[31] IL-10 may act, in part, by inhibiting the maladaptive activation of TNF-α associated with PMN activation and adhesion.[31,177] Deficiency of IL-10 enhances the infiltration of PMNs into tissues after ischemia and reperfusion.[31,177-180] Most of the evidence on the role of anti-inflammatory cytokines in the attenuation of IRI derives from experimental studies, but a clinical study by Pulitano and associates has determined a pathogenic role of proinflammatory Th1 cytokines including TNF-α and anti-inflammatory cytokines such as IL-4 and IL-10 in patients subjected to warm hepatic IRI in elective hepatic resection.[181] From their data, the local expression of anti-inflammatory Th2 cytokines in the early postreperfusion phase (90 minutes after surgery) may contribute to the attenuation of post-ischemic liver damage with reduced serum TNF-α levels in humans.[181] Thus, the Th1/Th2 or pro-inflammatory cytokine/anti-inflammatory cytokine balance may be essential for the regulation of organ IRI.[108,109,171] As an anti-inflammatory cytokine, IL-10 may check the inflammatory response by exerting an inhibitory effect on NF-κB, thereby attenuating organ IRI.[31]

IL-13 is also classified as a Th2 cytokine and this mediator is capable of modulating inflammatory responses through the down-regulation of proinflammatory cytokines, such as TNF-α, IL-6, macrophage inflammatory protein (MIP)-1α and MIP-2.[182,183] IL-13 is known to prevent LPS-induced lethality and IgG immune complex-induced lung injury.[184,185] Exogenously administered IL-13 has a great potential to suppress TNF-α and MIP-2 production and hepatic PMN recruitment by activating STAT-6 as a negative regulator of proinflammatory mediator in hepatic IRI.[32] In addition, prominent protective effects of IL-13 on hepatocytes and hepatic endothelial cells have led to the inhibition of hepatic IRI.[186] These novel insights on IL-13 may provide the basis for a better understanding of the pathophysiological aspects and a potential therapeutic strategy of hepatic IRI, but futher studies would be required to determine whether this cytokine yields the similar effects for IRI of the extrahepatic organs.

Ability of cytokines to act in an autocrine, paracrine and on humoral manner causes the production of various inflammatory mediators, such as other cytokines, chemokines, ROS, eicosanoids and ET-1. Thereby, the cross-talk among these inflammatory mediators promotes the inflammatory response in IRI. The cytokine milieu during reperfusion after organ ischemia depends on timing and the severity of the insult. Abundant experimental approaches have been shown that neutralization of specific cytokines by antiserum, Mabs and receptor antagonists decreases the severity of IRI, as evidenced by attenuated PMN infiltration and parenchymal damage. However, the possibility of the therapeutic intervention targeting cytokines may be limited in the treatment of IRI under clinical settings, because only a minimal increase in plasma TNF-α levels is observed during reperfusion period after orthotopic liver transplantation.[187]

Chemokines

A family of homologous chemotactic cytokines is now known as chemokines. The superfamily of chemokines is subclassified on the basis of the arrangement cysteine residues lacated in the N-terminal region of these molecules. These are designated C, CC, CXC and CXXXC, where C represents the number of N-terminal region cystein residues and X represents the number of intervening amino acids.[188] Among four known receptors for CXC chemokines, CXCR1, CXCR2, CXCR3 and CXCR4, only CXCR1 and CXCR2 are expressed on PMNs to mediate the chemotactic response of PMNs towards their ligands.[189] Accordingly, most CXC chemokines are chemoattractants for PMNs whereas CC chemokines generally attract monocytes, lymphocytes, basophils and eosinophils. Like cytokines, these chemokines are produced through the activation of NF-κB and AP-1, transcription factors regulating inflammatory gene expression.[190,191] Chemokines secreated by tissue cells, cytokine-activated endothelial cells and recruited PMNs elicit their effects on PMNs via receptors expressed on the plasma membrane.

TNF-α is a potent inducer of chemokine synthesis in various cell types, including PMNs, macrophages and endothelial cells.[84,150,188] In hepatic IRI, Kupffer cells and hepatocytes are considered as sources of chemokine production induced by cytokines.[52,84] Hepatic chemokine production forms a chemotactic gradient, which serves to direct PMN recruitment into the injured liver. Chemokines can induce an activated state in the endothelium and PMNs and upregulate expression of adhesion molecules on their surfaces, which enhance PMN-endothelial adhesion.[192,193] Thus, the interaction between cytokines and chemokines appears to be involved in upregulation of adhesion molecules that is critical for PMN-mediated liver inury. Among various chemokines, CXC chemokines, such as MIP-2, KC and cytokine-induced neutrophil chemoattractant-1 (CINC-1), a member of the IL-8 family, are most involved in organ IRI.[52,86,171,194-198] Chemokines are indispensable for the development of organ IRI because of their ability to recruit PMNs into remote organ as well as ischemically damaged tissue after reperfusion.[52] Since antibodies against chemokines have yielded the amelioration of PMN-mediated IRI in extrahepatic organs as well as the liver,[194,195,197] Targeting chemokines or chemokine receptors may provide therapeutic benefit in the management of organ IRI, because PMNs are principal effector cells in IRI.

Cell Adhesion Molecules

The adhesive interaction between PMNs and vascular endotheliums is a critical event in IRI. According to individual structure, adhesion molecules are divided into three major families; the selectins, the integrins and the immunoglobulin superfamily.[199] These adhesion molecules are responsible for PMN rolling and firm adhesion in postcapillary venules.[200] There exists a multistep paradigm of PMN emigration to inflamed tissues.[201,202] To emigrate into inflamed tissues, PMNs initially tether to and roll on the vascular endothelial cells. PMN rolling is dependent on selectin-mediated interaction between PMNs (L-selectin) and endothelial cells (P-selectin and E-selectin).[203] The carbohydrate ligand of these three selectins is a member of a class of sialylated and fucosylated structures related to the sialylated Lewis-X blood group antigen (SLe^x).[204] P-selectin is constitutively stored in secretory granules in endothelial cells and platelets and E-selectin is mainly expressed on endothelial cells activated by IL-1 and TNF-α. Subsequent firm adherence of PMNs occurs when PMNβ2 integrin (CD11b/CD18), Mac-1, binds to an immunoglobulin-like counter-receptor, such as ICAM-1 expressing on endothelial cells (Fig. 1).[205] ICAM-1 is upregulated by various proinflammatory cytokines, such as TNF-α and IL-1 and PMN-endothelial adhesion is enhanced after endothelial exposure to ROS or anoxia-reoxygenation in ICAM-1-dependent manner.[206,207] PMNs transmigrate between vascular endotheliums in the final stage and then travel through the extracellular matrix to the source of tissue injury, in accordance with a concentration gradient of cytokines and chemokines generated at the site of injury. These sequential interactions mediate extravascular PMN migration as well as firm adhesion and are essential for PMN recruitment into tissue subjected to reperfusion after organ ischemia.

The blockade of the critical adhesion step is highly specific and appears to be beneficial approaches to reduce PMN-mediated injury. Since a pioneer work by Hernandez and colleagues on anti-adhesion therapy using a CD18 Mab for IRI,[208] numerous studies have performed to examine whether blocking against various types of adhesion proteins leads to the protection of tissue damage in various organ IRI models.[28,29,63,64,67,68,209-221] Treatment with CD 18 or CD11b Mab has shown to prevent increase in microvascular permeability in intestinal IRI,[208] PMN accumulation and activation in hepatic IRI,[28] and infact size in myocardial IRI.[209] Moreover, anti-ICAM-1 Mab has also yielded the inhibitory effect against myocardial IRI, showing significantly inhibited PMN adherence to ischemic-reperfused coronary artery endothelium.[217] During reperfusion after hepatic ischemia, ICAM-1 is transcriptionally induced on hepatocytes and SECs.[58] Under these circumstances, anti-adhesion therapy using Mab against ICAM-1 has a potential to conducive to the amelioration of hepatic IRI, as shown by markedly reduced liver enzyme levels and hepatic necrosis with less PMN accumulation.[212,218-219] In addition to affected tissue damage, intestinal ischemia and reperfusion causes remote organ injury, in particular to the lung. The activation of CD11/CD18 glycoprotein complex and its ligand, ICAM-1, on pulmonary endothelial surface is required as causal factors of lung injury implicated in intestinal IRI. Salutary effect of Mabs against CD11/CD18 and ICAM-1 for these injuries has been reported[220] and immunoneutralization against these adhesion molecules may prevent remote organ injury relevant to IRI.

With regard to the implication of P-selectin in HIR, this glycoprotein plays a key role on PMN rolling and adherence in hepatic microvasculature in the early phase of reperfusion and mediates reperfusion injury through platelet sequestration as a P-selectin-dependent mechanism.[221] Thus, P-selectin may have a biphasic response after hepatic IRI. Pretreatment with P-selectin Mab has shown to improve IRI of the liver with reduced hepatic MPO levels.[213] Moreover, improved liver function, decreased PMN infiltration and reduced MIP-1α and MIP-2 responses have been observed in an experimental study using P-selectin deficient animals.[216] Another study has suggested that blockade of P-selectin expression results in significantly decreased PMN adhesion and platelet sequestration and improved survival, when compared with wild-type controls.[221] P-selectin glycoprotein ligand-1 (PSGL-1) is a homodimeric mucin expressed on the surface of most leukocytes that binds NH2-terminal domain.[222] This glycoprotein ligand is essential for primary tethering and rolling of PMNs. In ex vivo hepatic cold ischemia and reperfusion model, pretreatment with Mab against PSGL-1 mediating the initial tethering of PMNs to activated platelets and endothelium has yielded ameliorated portal venous flow, increased bile production and decreased hepatocellular damage.[38] In IRI of extrahepatic organ as well, it appears that platelet-endothelial interactions may contribute to the development of IRI. Reperfusion of the ischemic lung is associated with increased pulmonary vascular resistance and reduced alveolar perfusion in conjunction with an inflammatory response. These results indicate that in the intact lung, pulmonary IR causes platelet rolling and adhesion along arteriolar walls and suggest that this process mediated by P-selectin, contributes to vasoconstriction and hypoperfusion.[223] Early modulation of the interaction between P-selectin and its ligand may result in decreased platelet sequestration and PMN infiltration with the regulation of chemokine production and consequently diminish liver tissue damage.

Considerable number of exciting observations due to the inhibition of PMN adherence to endothelium may herald the development of novel anti-adhesion strategies for use in human disease. However, PMN adherence to endothelium is required for the eradication of bacterial infections at extravascular sites as clearly demonstrated by the CD11/CD18-deficient patients.[224] Thus, anti-adhesion therapy could be a double-edged sword to be handled with great caution. If a critical problem in host defense is solved in the future, targeting of selective adhesion molecules might be an elegant method to prevent organ IRI.

ET-1

Endothelin (ET) is a powerful vasoactive peptide that is derived from multiple cell types, including vascular endothelium, smooth muscle cells and hepatic SECs.[225,226] Among three isotypes of ET (ET-1, -2 and -3), ET-1 is the most powerful vasoconstrictive substance. This peptide acts as a potent vasoconstrictor with a predominantly high affinity for ET-A receptors,[227] but the ET-B receptors mediate vasodilatation (ET-B1) and vasoconstriction (ET-B2).[228] It has been suggested that ET-1 is responsible for microcirculatory incompetence as a critical component of IRI. Indeed, high expression of this peptide has been reported in various pathological circumstances, such as sepsis and myocardial, pulmonary, renal, intestinal and hepatic IRI.[59,60,229-234] Hepatic microcirculation is regulated through the affinity of ET-1 for ET receptors on HSCs in consert with nitric oxide (NO).[235,236] ET receptors are detectable on all major hepatic cell types but are most abundant on HSCs.[237] An imbalance between these key vasoactive substances occurs in hepatic IRI, increased ET-1 production and without corresponding increase in NO production.[238] ROS formed within reperfusing postischemic tissue lead to microvascular alterations including endothelial cell swelling and increased capillary permeability.[239] Vasoconstriction induced by ET-1 may aggravate narrowing of the sinusoidal lumen and cause subsequent microcirculatory derangement.[60] The presence of the cross-talk among ET-1, platelet-activating factor (PAF) and TNF-α has been suggested in organ IRI.[240-242] In IRI of various organs, including the liver, the heart, the kidney and the intestine, modulation of ET-1 by its receptor antagonists has resulted in the prevention of organ failure with the improvement of microcirculatory disturbance.[59,60,233,234,243-247] On the basis of these evidences, treatment with ET-1 receptor antagonist or inhibitors of ET synthesis may have a potential as a mode of therapy for organ IRI with incompetent microvascular perfusion.

HO-1

Heme oxygenase (HO) catalyzes the conversion of heme into biliverdin, bilirubin, free iron and carbon monoxide (CO).[248] Three isoforms of HO have been identified: inducible HO-1 known as heat shock protein 32; constitutively expressed HO-2 and HO-3. Activation of HO-1 system provides a cellular defence mechanism during oxidative stress through four major cytoprotective effects: (i) antioxidant function, (ii) anti-inflammatory function, (iii) maintenance of tissue microcirculation and (iv) anti-apoptosis.[249] HO-1 is rapidly upregulated during reperfusion and its overexpression has been associated with the cytoprotection during the cascade of events in IRI in various organs, such as the heart, the liver, the kidney and the intestine.[33,250-254] HO-1 exerts anti-inflammatory effects through the modulation of endothelial adhesion molecules and chemoattractant factors.[255] Bilirubin known as an antioxidant elicits protective effects for isolated perfused rat heart.[256] Ferritin produced by activation of free iron may mediate cytoprotection against IRI.[257] Moreover, inhalation of CO with vasodilatory and antiplatelet properties has rescued HO-1-deficient mice subjected to lethal ischemic lung injury.[258] Thus, pharmacologic or genetic HO-1 induction may have a great ability to restore the structural and functional tissue damage after IRI.[33,250-254] Though CO also a candidate for the treatment of organ IRI,[259] therapeutic window must be carefully considered because of a potential to arrest cellular respiration.

Transcription Factors

A number of molecular inflammatory reactions and signaling pathways are activated as a consequence of organ IRI. One of the most important signaling pathways is induced via NF-κB and AP-1. In organ IRI, proinflammatory mediators produced by reoxygenation are the key stimuli in the regulation of NF-κB. The inhibitor of NF-κB (IκB) kinase (IKK) complex consists of 3 subunits, IKK1, IKK2 and NF-κB essential modulator (NEMO) and is implicated in the activation of NF-κB by various stimuli. Stimulation of cells with cytokines, ROS, or inflammatory mediators results in phosphorylation and ubiquitination of IκB.[260] This process leads to nuclear translocation of NF-κB. Activation of NF-κB is the primary transcriptional mechanism for the production of proinflammatory cytokines, chemokines and cell adhesion molecules(Fig. 4),[261-264] but the specific function of NF-κB is not sufficiently understood in this response. In vivo myocardial infarction, a biphasic NF-κB activation is induced after 15 minutes and 3 hours after reperfusion, possibly corresponding to a primary activation by ROS and a secondary activation by proinflammatory cytokines produced by the first activation.[144] Activation of AP-1 occurs in response to various stimuli, including oxidative stress and proinflammatory cytokines.[190,265,266] Hypothermia induces the suppression of AP-1 activation in hepatic IRI, showing reduced pro-inflammatory cytokines and chemokine production.[271] Although the role of AP-1 has not been minutely investigated in

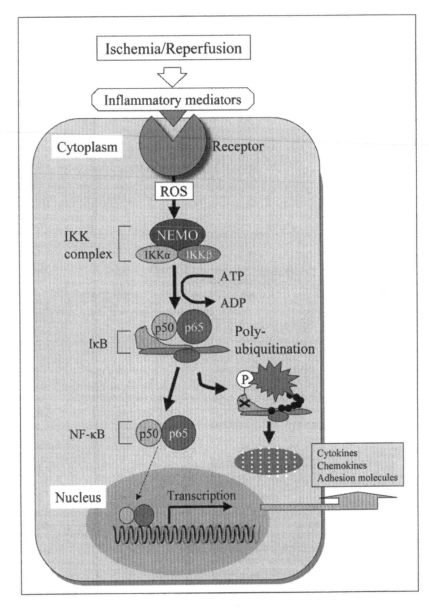

Figure 4. Activation of NF-κB as a step to promote the inflammatory response in ischemia-reperfusion injury. Proinflammatory stimuli activate a specific protein kinase, resulting in the degradation of IκB. NF-κB activated through this process translocates to the nucleus and binds to promoter or enhancer regions of specific genes, initiating transcription.

the inflammatory response of the liver, the activation of c-Jun NH2-terminal kinase (JNK) may be implicated in the activation of AP-1 with subsequent production of pro-inflammatory cytokines and chemokines.[267] A pivotal role of ROS in IRI is to enhance proinflammatory gene expression through the activation of transcription factors, such as NF-κB and AP-1.[268,269] These transcription factors respond effectively to the time and severity of the ischemic insult.[36]

Taking a central role of NF-κB and AP-1 in coordinating the expression of various mediators in the inflammatory response into consideration, it is supposed that selective inhibition of activation of these transcription factors may have a potential as a promising therapeutic strategy for organ IRI. Several studies have evaluated the potential contribution of NF-κB and AP-1 to IRI using genetic and pharmacological strategies. Exogenous IL-10 administration inhibits pro-inflammatory cytokine expression through the suppression of NF-κB activation and consequently protects the development of hepatic IRI.[31] MOL-294, an inhibitor of thioredoxin, is an oxireductase known to regulate the expression of the NF-κB family directly and diminishes PMN influx and TNF-α production in intestinal IRI.[270] Results from this study point out an important role of NF-κB in triggering endogenous proinflammatory networks during intestinal IRI. In contrast, adenoviral expression of mitochondrial SOD in the liver greatly reduces hepatic IRI related to reduced NF-κB and AP-1 activation.[268] Thus, antioxidants can attenuate proinflammatory gene expression through inhibition of NF-κB and/or AP-1 activation. Treatment modalities with inhibitors of NF-κB and/or AP-1 translocation or function would be expected as effective therapeutic strategies for organ IRI.

Conclusion and Perspective

Organ IRI is complex events that create a milieu of inflammation relevant to various clinical settings. Complex networks among different cellular components such as resident macrophages, endothelial cells, PMNs and T-cells and molecules with pleiotropic effects are implicated in the development of IRI. Dogma based on amassed information on IRI has steadily paved the way for a wide variety of interventions to manipulate ischemia-reperfusion-induced tissue damages. In addition to the development of new pharmacological agents, recent attention has been paid to the possibility of ischemic preconditioning to suppress IRI. This promising strategy appears to increase the tolerance of organs for IRI and has been shown to be beneficial in the heart and liver of humans. However, clinical applications to mitigate inflammatory cascades in IRI have often failed to unsuccessful results. These disappointing outcomes may result from the differences in the severity of IRI between animal models and clinical situation and the inherent risk of biological responses among species. An accurate grasp of the biology is essential before a specific intervention is pursued on a therapeutic basis. In future, better understanding of the signal transduction mechanisms and the genetic analysis involved in the regulation of IRI will enable us to develop novel strategies for this insult.

References

1. Parks DA, Granger DN. Ischemia-induced vascular changes: role of xanthine oxidase and hydroxyl radicals. Am J Physiol 1983; 245:G285-9.
2. McCord JM. Oxygen-derived free radicals in postischemic tissue injury. N Engl J Med 1985; 312:159-63.
3. Hasselgren PO. Prevention and treatment of ischemia of the liver. Surg Gynecol Obstet 1987; 164:187-96.
4. Suzuki S, Nakamura S, Koizumi T et al. The beneficial effect of a prostaglandin I2 analog on ischemic rat liver. Transplantation 1991; 52:979-83.
5. Granger DN, McCord JM, Parks DA et al. Xanthine oxidase inhibitors attenuate ischemia-induced vascular permeability changes in the cat intestine. Gastroenterology 1986; 90:80-84.
6. Bjorck M, Troeng T, Bergqvist D. Risk factors for intestinal ischemia after aortoiliac surgery: A combined cohort and case-control study of 2824 operations. Eur J Vasc Endovasc Surg 1997; 13:531-539.
7. Huguet C, Gavelli A, Chieco PA et al. Liver ischemia for hepatic resection: where is the limit? Surgery 1992; 111:251-9.
8. Strasberg SM, Lowell JA, Howard TK. Reducing the shortage of donor livers: what would it take to reliably split livers for transplantation into two adult recipients? Liver Transpl Surg 1999; 5:437-50.
9. Belghiti J, Noun R, Malafosse R et al. Continuous versus intermittent portal triad clamping for liver resection: a controlled study. Ann Surg 1999; 229:369-75.

10. Selzner M, Camargo CA, Clavien PA. Ischemia impairs liver regeneration after major tissue loss in rodents: protective effects of interleukin-6. Hepatology 1999; 30:469-75.
11. Siniscalchi A, Spedicato S, Lauro A et al. Intraoperative coagulation evaluation of ischemia-reperfusion injury in small bowel transplantation: a way to explore. Transplant Proc 2006; 38:820-2.
12. Marczin N, El-Habashi N, Hoare GS et al. Antioxidants in myocardial ischemia-reperfusion injury: therapeutic potential and basic mechanisms. Arch Biochem Biophys 2003; 420:222-36.
13. de Perrot M, Liu M, Waddell TK et al. Ischemia-reperfusion-induced lung injury. Am J Respir Crit Care Med 2003; 167:490-511.
14. Mallick IH, Yang W, Winslet MC et al. Ischemia-reperfusion injury of the intestine and protective strategies against injury. Dig Dis Sci 2004; 49:1359-77.
15. Donnahoo KK, Shames BD, Harken AH et al. Review article: the role of tumor necrosis factor in renal ischemia-reperfusion injury. J Urol 1999; 162:196-203.
16. Atalla SL, Toledo-Pereyra LH, MacKenzie GH et al. Influence of oxygen-derived free radical scavengers on ischemic livers. Transplantation 1985; 40:584-590.
17. Marubayashi S, Dohi Y, Kawasaki T. Role of free radicals in ischemic rat liver cell injury: Prevention of damage by α-tocopherol administration. Surgery 1986; 99:184-192.
18. Metzger J, Dore SP, Lauterburg BH. Oxidant stress during reperfusion of ischemic liver: No evidence for a role of xanthine oxidase. Hepatology 1988; 8:580-584.
19. Toledo-Pereyra LH. The role of allopurinol and oxygen free radical scavengers in liver preservation. Basic Life Sci 1988; 49:1047-52.
20. Jaeschke H. Reactive oxygen and ischemia/reperfusion injury of the liver. Chem Biol Interact 1991; 79:115-36.
21. Toledo-Pereyra LH. Liver transplantation reperfusion injury. Factors in its development and avenues for treatment. Klin Wochenschr 1991; 69:1099-1104.
22. Rosen HR, Martin P, Goss J et al. Significance of early aminotransferase elevation after liver transplantation. Transplantation 1998; 65:68-72.
23. Okamoto K, Suzuki S, Kurachi K et al. Beneficial effect of deletion variant of hepatocyte growth factor for impaired hepatic regeneration in the ischemically damaged liver. World J Surg 2002; 26:1260-6.
24. Koo A, Breit G, Intaglietta M. Leukocyte adherence in hepatic microcirculation in ischemia reperfusion. In: Manabe H, Zweifach W, Messmer K, eds. Microcirculation in Circulatory Disorders. Tokyo: Springer-Verlag, 1988:205-213.
25. Jaeschke H, Farhood A, Smith CW. Neutrophils contribute to ischemia/reperfusion injury in rat liver in vivo. FASEB J 1990; 4:3355-3359.
26. Suzuki S, Toledo-Pereyra LH, Rodriguez FJ et al. Neutrophil infiltration as an important factor in liver ischemia and reperfusion injury. Modulating effects of FK506 and cyclosporine. Transplantation 1993; 55:1265-72.
27. Jaeschke H, Farhood A. Neutrophil and Kupffer cell-induced oxidant stress and ischemia-reperfusion injury in rat liver. Am J Physiol 1991; 260:G355-62.
28. Jaeschke H, Farhood A, Bautista AP et al. Functional inactivation of neutrophils with a Mac-1 (CD11b/CD18) monoclonal antibody protects against ischemia-reperfusion injury in rat liver. Hepatology 1993; 17:915-23.
29. Martinez-Mier G, Toledo-Pereyra LH, McDuffie JE et al. P-selectin and chemokine response after liver ischemia and reperfusion. J Am Coll Surg 2000; 191:395-402.
30. Jaeschke H. Molecular mechanisms of hepatic ischemia-reperfusion injury and preconditioning. Am J Physiol Gastrointest Liver Physiol 2003; 284:G15-26.
31. Yoshidome H, Kato A, Edwards MJ et al. Interleukin-10 suppresses hepatic ischemia/reperfusion injury in mice: implications of a central role for nuclear factor kappa B. Hepatology 1999; 30:203-8.
32. Yoshidome H, Kato A, Miyazaki M et al. IL-13 activates STAT6 and inhibits liver injury induced by ischemia/reperfusion. Am J Pathol 1999; 155:1059-64.
33. Amersi F, Buelow R, Kato H et al. Upregulation of heme oxygenase-1 protects genetically fat zucker rat livers from ischemia/reperfusion injury. J Clin Invest 1999; 104:1631-9.
34. Kato A, Yoshidome H, Edwards MJ et al. Regulation of liver inflammatory injury by signal transducer and activator of transcription-6. Am J Pathol 2000; 157:297-302.
35. Martinez-Mier G, Toledo-Pereyra LH, McDuffie JE et al. Exogenous nitric oxide downregulates MIP-2 and MIP-1alpha chemokines and MAPK p44/42 after ischemia and reperfusion of the rat kidney. J Invest Surg 2002; 15:287-96.
36. Toledo-Pereyra LH, Lopez-Neblina F, Lentsch AB et al. Selectin inhibition modulates NF-kappa B and AP-1 signaling after liver ischemia/reperfusion. J Invest Surg 2006; 19:313-22.
37. Kuboki S, Okaya T, Schuster R et al. Hepatocyte NF-{kappa}B activation is hepatoprotective during ischemia/reperfusion injury and is augmented by ischemic hypothermia. Am J Physiol Gastrointest Liver Physiol 2007;292: 9201-7.

38. Tsuchihashi S, Fondevila C, Shaw GD et al. Molecular characterization of rat leukocyte P-selectin glycoprotein ligand-1 and effect of its blockade: protection from ischemia-reperfusion injury in liver transplantation. J Immunol 2006; 176:616-24.
39. Baumann M, Bender E, Stommer G et al. Effects of warm and cold ischemia on mitochondrial functions in brain, liver and kidney. Mol Cell Biochem 1989; 87:137-45.
40. Churchill TA, Cheetham KM, Fuller BJ. Glycolysis and energy metabolism in rat liver during warm and cold ischemia: evidence of an activation of the regulatory enzyme phosphofructokinase. Cryobiology 1994; 31:441-52.
41. Mochida S, Arai M, Ohno A et al. Oxidative stress in hepatocytes and stimulatory state of Kupffer cells after reperfusion differ between warm and cold ischemia in rats. Liver 1994; 14:234-40.
42. Ikeda T, Yanaga K, Kishikawa K et al. Ischemic injury in liver transplantation: difference in injury sites between warm and cold ischemia in rats. Hepatology 1992; 16:454-61.
43. Clavien PA, Harvey PR, Strasberg SM. Preservation and reperfusion injuries in liver allografts. An overview and synthesis of current studies. Transplantation 1992; 53:957-78.
44. Arii S, Monden K, Adachi Y et al. Pathogenic role of Kupffer cell activation in the reperfusion injury of cold-preserved liver. Transplantation 1994; 58:1072-7.
45. Jaeschke H, Mitchell JR. Mitochondria and xanthine oxidase both generate reactive oxygen species in isolated perfused rat liver after hypoxic injury. Biochem Biophys Res Commun 1989; 160:140-7.
46. Schmid-Schonbein GW. Capillary plugging by granulocytes and the no-reflow phenomenon in the microcirculation. Fed Proc 1987; 46:2397-2401.
47. Ambrosio G, Tritto I. Reperfusion injury: experimental evidence and clinical implications. Am Heart J 1999; 138:S69-75.
48. Wisse E, McCuskey RS. On the interactions of blood cells with the sinusoidal wall as observed by in vivo microscopy of rat liver. In: KIrn A, Knook DL, Wisse E, eds. Cells of the hepatic sinusoid. Rijswijk: Kupffer Cell Foundation, 1986:477-482.
49. Jaeschke H, Farhood A, Bautista AP et al. Complement activates Kupffer cells and neutrophils during reperfusion after hepatic ischemia. Am J Physiol 1993; 264:G801-809.
50. Decker K. Biologically active products of stimulated liver macrophages (Kupffer cells). Eur J Biochem 1990; 192:245-61.
51. Suzuki S, Toledo-Pereyra LH, Rodriguez F et al. Role of Kupffer cells in neutrophil activation and infiltration following total hepatic ischemia and reperfusion. Circ Shock 1994; 42:204-9.
52. Serizawa A, Nakamura S, Suzuki et al. Involvement of platelet-activating factor in cytokine production and neutrophil activation after hepatic ischemia-reperfusion. Hepatology 1996; 23:1656-63.
53. Shiratori Y, Kiriyama H, Fukushi Y et al. Modulation of ischemia-reperfusion-induced hepatic injury by Kupffer cells. Dig Dis Sci 1994; 39:1265-72.
54. Wheeler MD, Ikejema K, Enomoto N et al. Glycine: a new anti-inflammatory immunonutrient. Cell Mol Life Sci 1999; 56:843-56.
55. Mosher B, Dean R, Harkema J et al. Inhibition of Kupffer cells reduced CXC chemokine production and liver injury. J Surg Res 2001; 99:201-10.
56. Baeuerle PA, Henkel T. Function and activation of NF-kappa B in the immune system. Annu Rev Immunol 1994; 12:141-79.
57. Essani NA, McGuire GM, Manning AM et al. Endotoxin-induced activation of the nuclear transcription factor kappa B and expression of E-selectin messenger RNA in hepatocytes, Kupffer cells and endothelial cells in vivo. J Immunol 1996; 156:2956-63.
58. Bell FP, Essani NA, Manning AM et al. Ischemia-reperfusion activates the nuclear transcription factor NF-κB and upregulates messenger RNA synthesis of adhesion molecules in the liver in vivo. Hepatol Res 1997; 8:178-188.
59. Kawamura E, Yamanaka N, Okamoto E et al. Response of plasma and tissue endothelin-1 to liver ischemia and its implication in ischemia-reperfusion injury. Hepatology 1995; 21:1138-43.
60. Mitsuoka H, Suzuki S, Sakaguchi T et al. Contribution of endothelin-1 to microcirculatory impairment in total hepatic ischemia and reperfusion injury. Transplantation 1999; 67:514-20.
61. Romson JL, Hook BG, Kunkel SL et al. Reduction of the extent of ischemic myocardial injury by neutrophil depletion in the dog. Circulation 1983; 67:1016-1023.
62. Litt MR, Jeremy RW, Weisman HF et al. Neutrophil depletion limited to reperfusion reduces myocardial infarct size after 90 minutes of ischemia. Evidence for neutrophil-mediated reperfusion injury. Circulation 1989; 80:1816-27.
63. Ma XL, Weyrich AS, Lefer DJ et al. Monoclonal antibody to L-selectin attenuates neutrophil accumulation and protects ischemic reperfused cat myocardium. Circulation 1993; 88:649-58.
64. Weyrich AS, Ma XY, Lefer DJ et al. In vivo neutralization of P-selectin protects feline heart and endothelium in myocardial ischemia and reperfusion injury. J Clin Invest 1993; 91:2620-9.

65. Naka Y, Toda K, Kayano K et al. Failure to express the P-selectin gene or P-selectin blockade confers early pulmonary protection after lung ischemia or transplantation. Proc Natl Acad Sci USA 1997; 94:757-61.
66. Panes J, Perry M, Granger DN. Leukocyte-endothelial cell adhesion: avenues for therapeutic intervention. Br J Pharmacol 1999; 126:537-50.
67. Martinez-Mier G, Toledo-Pereyra LH, Ward PA. Adhesion molecules in liver ischemia and reperfusion. J Surg Res 2000; 94:185-94.
68. Riaz AA, Wan MX, Schaefer T et al. Fundamental and distinct roles of P-selectin and LFA-1 in ischemia/reperfusion-induced leukocyte-endothelium interactions in the mouse colon. Ann Surg 2002; 236:777-84
69. Smith CW, Kishimoto TK, Abbass O et al. Chemotactic factors regulate lectin adhesion molecule (LECAM-1)-dependent neutrophil adhesion to cytokine stimulated endothelial cells in vitro. J Clin Invest 1991; 87:609-618.
70. Smith CW. Transendothelial migration. In: Harlan JM, Liu DY, eds. Adhesion. Its Role in Inflammatory Disease. New York: W. H. Freeman and Co 1992:83-115.
71. Colletti LM, Remick DG, Burtch GD et al. Role of tumor necrosis factor-alpha in the pathophysiologic alterations after hepatic ischemia/reperfusion injury in the rat. J Clin Invest 1990; 85:1936-43.
72. Hughes H, Farhood A, Jaeschke H. Role of leukotriene B4 in the pathogenesis of hepatic ischemia-reperfusion injury in the rat. Prostaglandins Leukot Essent Fatty Acids 1992; 45:113-9.
73. Eppinger MJ, Jones ML, Deeb GM et al. Pattern of injury and the role of neutrophils in reperfusion injury of rat lung. J Surg Res 1995; 58:713-8.
74. Sisley AC, Desai T, Harig JM et al. Neutrophil depletion attenuates human intestinal reperfusion injury. J Surg Res 1994; 57:192-6.
75. Welbourn CRB, Goldman G, Paterson IS et al. Pathophysiology of ischaemia reperfusion injury: central role of the neutrophil. Br J Surg 1991; 78:651-655.
76. Bautista AP, Spitzer JJ. Platelet activating factor stimulates and primes the liver, Kupffer cells and neutrophils to release superoxide anion. Free Radic Res Commun 1992; 17:195-209.
77. Badwey JA, Karnovsky ML. Active oxygen species and the functions of phagocytic leukocytes. Annu Rev Biochem 1980; 49:695-726.
78. Luscinskas FW, Brock AF, Arnaout MA et al. Endothelial-leukocyte adhesion molecule-1-dependent and leukocyte (CD11/CD18)-dependent mechanisms contribute to polymorphonuclear leukocyte adhesion to cytokine-activated human vascular endothelium. J Immunol 1989; 142:2257-63.
79. Shappell SB, Toman C, Anderson DC et al. Mac-1 (CD11b/CD18) mediates adherence-dependent hydrogen peroxide production by human and canine neutrophils. J Immunol 1990; 144:2702-11.
80. Chosay JG, Essani NA, Dunn CJ et al. Neutrophil margination and extravasation in sinusoids and venules of liver during endotoxin-induced injury. Am J Physiol 1997; 272:G1195-G1200.
81. Granger DN, Kubes P. The microcirculation and inflammation: modulation of leukocyte-endothelial cell adhesion. J Leukoc Biol 1994; 55:662-75.
82. Schlayer HJ, Laaff H, Peters T et al. Involvement of tumor necrosis factor in endotoxin-triggered neutrophil adherence to sinusoidal endothelial cells of mouse liver and its modulation in acute phase. J Hepatol 1988; 7:239-49.
83. Essani NA, Fisher MA, Farhood A et al. Cytokine-induced upregulation of hepatic intercellular adhesion molecule-1 messenger RNA expression and its role in the pathophysiology of murine endotoxin shock and acute liver failure. Hepatology 1995; 21:1632-9.
84. Colletti LM, Kunkel SL, Walz A et al. The role of cytokine networks in the local liver injury following hepatic ischemia/reperfusion in the rat. Hepatology 1996; 23:506-14.
85. Jaeschke H. Mechanisms of Liver Injury. II. Mechanisms of neutrophil-induced liver cell injury during hepatic ischemia-reperfusion and other acute inflammatory conditions. Am J Physiol Gastrointest Liver Physiol 2006; 290:G1083-8.
86. Tucker VL, Victorino GP. Methods for studying microvascular barrier function in ischemia-reperfusion injury. Shock 1997; 8:8-15.
87. Mullane KM, Kraemer R, Smith B. Myeloperoxidase activity as a quantitative assessment of neutrophil infiltration into ischemic myocardium. J Pharmacol Methods 1985; 14:157-67.
88. Weiss SJ. Tissue destruction by neutrophils. N Engl J Med 1989; 320:365-76.
89. Fiser SM, Tribble CG, Long SM et al. Lung transplant reperfusion injury involves pulmonary macrophages and circulating leukocytes in a biphasic response. J Thorac Cardiovasc Surg 2001; 121:1069-1075.
90. Maxey TS, Enelow RI, Gaston B et al. Tumor necrosis factor-alpha from resident lung cells is a key initiating factor in pulmonary ischemia-reperfusion injury. J Thorac Cardiovasc Surg 2004; 127:541-547.
91. Zhao M, Fernandez LG, Doctor A et al. Alveolar macrophage activation is a key initiation signal for acute lung ischemia-reperfusion injury. Am J Physiol Lung Cell Mol Physiol 2006; 291:L1018-1026.

92. Formigli L, Manneschi LI, Nediani C et al. Are macrophages involved in early myocardial reperfusion injury? Ann Thorac Surg 2001; 71:1596-602.

93. Saba TM. Reticuloendothelial systemic host defense after surgery and traumatic shock. Circ Shock 1975; 2:91-108.

94. Suzuki S, Nakamura S, Sakaguchi T et al. Alteration of reticuloendothelial phagocytic function and tumor necrosis factor-alpha production after total hepatic ischemia. Transplantation 1997; 64:821-7.

95. Brain JD. Lung macrophages: how many kinds are there? What do they do? Am Rev Respir Dis 1988; 137:507-9.

96. Bishop MJ, Chi EY, Su M et al. Dimethylthiourea does not ameliorate reperfusion lung injury in dogs or rabbits. J Appl Physiol 1988; 65:2051-6.

97. Fiser SM, Tribble CG, Long SM et al. Pulmonary macrophages are involved in reperfusion injury after lung transplantation. Ann Thorac Surg 2001; 71:1134-8

98. Jo SK, Sung SA, Cho WY et al. Macrophages contribute to the initiation of ischaemic acute renal failure in rats. Nephrol Dial Transplant 2006; 21:1231-9

99. Chen Y, Lui VC, Rooijen NV et al. Depletion of intestinal resident macrophages prevents ischaemia reperfusion injury in gut. Gut 2004; 53:1772-80.

100. Zwacka RM, Zhang Y, Halldorson J et al. CD4(+) T-lymphocytes mediate ischemia/reperfusion-induced inflammatory responses in mouse liver. J Clin Invest 1997; 100:279-89.

101. Rabb H, Daniels F, O'Donnell M et al. Pathophysiological role of T-lymphocytes in renal ischemia-reperfusion injury in mice. Am J Physiol Renal Physiol 2000; 279:F525-31.

102. Shiraishi H, Toyozaki T, Tsukamoto Y et al. Antibody binding to fas ligand attenuates inflammatory cell infiltration and cytokine secretion, leading to reduction of myocardial infarct areas and reperfusion injury. Lab Invest 2002; 82:1121-9.

103. Shigematsu T, Wolf RE, Granger DN. T-lymphocytes modulate the microvascular and inflammatory responses to intestinal ischemia-reperfusion. Microcirculation 2002; 9:99-109.

104. de Perrot M, Young K, Imai Y et al. Recipient T-cells mediate reperfusion injury after lung transplantation in the rat. J Immunol 2003; 171:4995-5002.

105. Kawano K, Kim YI, Ono M et al. Evidence that both cyclosporin and azathioprine prevent warm ischemia reperfusion injury to the rat liver. Transpl Int 1993; 6:330-6.

106. Khandoga A, Hanschen M, Kessler JS et al. CD4+ T-cells contribute to postischemic liver injury in mice by interacting with sinusoidal endothelium and platelets. Hepatology 2006; 43:306-15.

107. Abbas AK, Murphy KM, Sher A. Functional diversity of helper T-lymphocytes. Nature 1996; 383:787-93.

108. Yokota N, Burne-Taney M, Racusen L et al. Contrasting roles for STAT4 and STAT6 signal transduction pathways in murine renal ischemia-reperfusion injury. Am J Physiol Renal Physiol 2003; 285:F319-25.

109. Marques VP, Goncalves GM, Feitoza CQ et al. Influence of TH1/TH2 switched immune response on renal ischemia-reperfusion injury. Nephron Exp Nephrol 2006; 104:48-56.

110. Mastellos D, Morikis D, Isaacs SN et al. Complement: structure, functions, evolution and viral molecular mimicry. Immunol Res 2003; 27:367-86.

111. Dong J, Pratt JR, Smith RA et al. Strategies for targeting complement inhibitors in ischaemia/reperfusion injury. Mol Immunol 1999; 36:957-63.

112. Eppinger MJ, Deeb GM, Bolling SF et al. Mediators of ischemia-reperfusion injury of rat lung. Am J Pathol 1997; 150:1773-84.

113. Bergamaschini L, Gobbo G, Gatti S et al. Endothelial targeting with C1-inhibitor reduces complement activation in vitro and during ex vivo reperfusion of pig liver. Clin Exp Immunol 2001; 126:412-20.

114. Chavez-Cartaya RE, DeSola GP, Wright L et al. Regulation of the complement cascade by soluble complement receptor type 1. Protective effect in experimental liver ischemia and reperfusion. Transplantation 1995; 59:1047-52.

115. Straatsburg IH, Boermeester MA, Wolbink GJ et al. Complement activation induced by ischemia-reperfusion in humans: a study in patients undergoing partial hepatectomy. J Hepatol 2000; 32:783-91.

116. Kirschfink M. Controlling the complement system in inflammation. Immunopharmacology 1997; 38:51-62.

117. Weisman HF, Bartow T, Leppo MK et al. Soluble human complement receptor type 1: in vivo inhibitor of complement suppressing post-ischemic myocardial inflammation and necrosis. Science 1990; 249:146-51.

118. Hill J, Lindsay TF, Ortiz F et al. Soluble complement receptor type 1 ameliorates the local and remote organ injury after intestinal ischemia-reperfusion in the rat. J Immunol 1992; 149:1723-8.

119. Vakeva AP, Agah A, Rollins SA et al. Myocardial infarction and apoptosis after myocardial ischemia and reperfusion: role of the terminal complement components and inhibition by anti-C5 therapy. Circulation 1998; 97:2259-67.

120. Wada K, Montalto MC, Stahl GL. Inhibition of complement C5 reduces local and remote organ injury after intestinal ischemia/reperfusion in the rat. Gastroenterology 2001; 120:126-33.
121. Hugli TE. Biochemistry and biology of anaphylatoxins. Complement 1986; 3:111-27.
122. Huey R, Hugli TE. Characterization of a C5a receptor on human polymorphonuclear leukocytes (PMN). J Immunol 1985; 135:2063-8.
123. Schieferdecker HL, Schlaf G, Jungermann K et al. Functions of anaphylatoxin C5a in rat liver: direct and indirect actions on nonparenchymal and parenchymal cells. Int Immunopharmacol 2001; 1:469-81.
124. Goldstein IM, Weissmann G. Generation of C5-derived lysosomal enzyme-releasing activity (C5a) by lysates of leukocyte lysosomes. J Immunol 1974; 113:1583-8.
125. Haynes DR, Harkin DG, Bignold LP et al. Inhibition of C5a-induced neutrophil chemotaxis and macrophage cytokine production in vitro by a new C5a receptor antagonist. Biochem Pharmacol 2000; 60:729-33.
126. Arumugam TV, Woodruff TM, Stocks SZ et al. Protective effect of a human C5a receptor antagonist against hepatic ischaemia-reperfusion injury in rats. J Hepatol 2004; 40:934-41.
127. Buerke M, Murohara T, Lefer AM. Cardioprotective effects of a C1 esterase inhibitor in myocardial ischemia and reperfusion. Circulation 1995; 91:393-402.
128. Lehmann TG, Heger M, Munch S et al. In vivo microscopy reveals that complement inhibition by C1-esterase inhibitor reduces ischemia/reperfusion injury in the liver. Transpl Int 2000; 13:S547-50.
129. Gross GJ, Farber NE, Hardman HF et al. Beneficial actions of superoxide dismutase and catalase in stunned myocardium of dogs. Am J Physiol 1986; 250:H372-377.
130. Dhalla NS, Elmoselhi AB, Hata T et al. Status of myocardial antioxidants in ischemia-reperfusion injury. Cardiovasc Res 2000; 47:446-56.
131. Fondevila C, Busuttil RW, Kupiec-Weglinski JW. Hepatic ischemia/reperfusion injury—a fresh look. Exp Mol Pathol 2003; 74:86-93.
132. Ratych RE, Chuknyiska RS, Bulkley GB. The primary localization of free radical generation after anoxia/reoxygenation in isolated endothelial cells. Surgery 1987; 102:122-31.
133. Drath DB, Karnovsky ML. Superoxide production by phagocytic leukocytes. J Exp Med 1975; 141:257-62.
134. Cuzzocrea S, Mazzon E, Dugo L et al. Protective effects of a new stable, highly active SOD mimetic, M40401 in splanchnic artery occlusion and reperfusion. Br J Pharmacol 2001; 132:19-29.
135. Gunel E, Caglayan F, Caglayan O et al. Treatment of intestinal reperfusion injury using antioxidative agents. J Pediatr Surg 1998; 33:1536-9.
136. Sisto T, Paajanen H, Metsa-Ketela T et al. Pretreatment with antioxidants and allopurinol diminishes cardiac onset events in coronary artery bypass grafting. Ann Thorac Surg 1995; 59:1519-23.
137. Jones SP, Hoffmeyer MR, Sharp BR et al. Role of intracellular antioxidant enzymes after in vivo myocardial ischemia and reperfusion. Am J Physiol Heart Circ Physiol 2003; 284:H277-282.
138. Toledo-Pereyra LH, Simmons RL, Najarian JS. Protection of the ischemic liver by donor pretreatment before transplantation. Am J Surg 1975; 129:513-7.
139. Fong Y, Moldawer LL, Shires GT et al. The biologic characteristics of cytokines and their implication in surgical injury. Surg Gynecol Obstet 1990; 170:363-78.
140. Ming WJ, Bersani L, Montovani A. Tumor necrosis factor is chemotactic for monocytes and polymorphonuclear leukocytes. J Immunol 1987; 138:1469-1474.
141. Dinarello CA. Biology of interleukin 1. FASEB J 1988; 2:108-115.
142. Klebanoff SJ, Vadas JM, Harlan JM et al. Stimulation of neutrophils by tumor necrosis factor. J Immunol 1986; 136:4220-4225.
143. Shirasugi N, Wakabayashi G, Shimazu M et al. Up-regulation of oxygen-derived free radicals by interleukin-1 in hepatic ischemia/reperfusion injury. Transplantation 1997; 64:1398-403.
144. Chandrasekar B, Freeman GL. Induction of nuclear factor kappa B and activation protein 1 in post-ischemic myocardium. FEBS Lett 1997; 401:30-4.
145. Nawroth PP, Bank I, Hadley D et al. Tumor necrosis factor/cachectin interacts with endothelial cell receptors to induce release of interleukin-1. J Exp Med 1987; 166:1390-1394.
146. Camussi G, Bussolino F, Salvidio G et al. Tumor necrosis factor/cachectin stimulates peritoneal macrophages, polymorphonuclear leukocytes and vascular endothelial cells to synthesize and release platelet-activating factor. J Exp Med 1987; 166:1390-1394.
147. Colletti LM, Cortis A, Lukacs N et al. Tumor necrosis factor up-regulates intercellular adhesion molecule 1, which is important in the neutrophil-dependent lung and liver injury associated with hepatic ischemia and reperfusion in the rat. Shock 1998; 10:182-91.
148. Farhood A, McGuire GM, Manning AM et al. Intercellular adhesion molecule 1 (ICAM-1) expression and its role in neutrophil-induced ischemia-reperfusion injury in rat liver. J Leukoc Biol 1995; 57:368-74.

149. Peralta C, Fernandez L, Panes J et al. Preconditioning protects against systemic disorders associated with hepatic ischemia-reperfusion through blockade of tumor necrosis factor-induced P-selectin up-regulation in the rat. Hepatology 2001; 33:100-13.
150. Colletti LM, Kunkel SL, Walz A et al. Chemokine expression during hepatic ischemia/reperfusion-induced lung injury in the rat. The role of epithelial neutrophil activating protein. J Clin Invest 1995; 95:134-41.
151. Lentsch AB, Yoshidome H, Cheadle WG et al. Chemokine involvement in hepatic ischemia/reperfusion injury in mice: roles for macrophage inflammatory protein-2 and Kupffer cells. Hepatology 1998; 27:507-12.
152. Zhang C, Xu X, Potter BJ et al. TNF-alpha contributes to endothelial dysfunction in ischemia/reperfusion injury. Arterioscler Thromb Vasc Biol 2006; 26:475-80.
153. Souza DG, Cassali GD, Poole S et al. Effects of inhibition of PDE4 and TNF-alpha on local and remote injuries following ischaemia and reperfusion injury. Br J Pharmacol 2001; 134:985-94.
154. Khimenko PL, Bagby GJ, Fuseler J et al. Tumor necrosis factor-alpha in ischemia and reperfusion injury in rat lungs. J Appl Physiol 1998; 85:2005-11.
155. Shito M, Wakabayashi G, Ueda M et al. Interleukin 1 receptor blockade reduces tumor necrosis factor production, tissue injury and mortality after hepatic ischemia-reperfusion in the rat. Transplantation 1997; 63:143-8.
156. Furuichi K, Wada T, Iwata Y et al. Interleukin-1-dependent sequential chemokine expression and inflammatory cell infiltration in ischemia-reperfusion injury. Crit Care Med 2006; 34:2447-55.
157. Suzuki S, Toledo-Pereyra LH. Interleukin 1 and tumor necrosis factor production as the initial stimulants of liver ischemia and reperfusion injury. J Surg Res 1994; 57:253-8.
158. Wanner GA, Ertel W, Muller P et al. Liver ischemia and reperfusion induces a systemic inflammatory response through Kupffer cell activation. Shock 1996; 5:34-40.
159. Kato A, Gabay C, Okaya T et al. Specific role of interleukin-1 in hepatic neutrophil recruitment after ischemia/reperfusion. Am J Pathol 2002; 161:1797-803.
160. Greene WC. The interleukins. In: Gallin JI, Goldstein IM, Snyderman R, eds. Inflammation. Basic principles and clinical correlates, 2nd ed. New York: Raven Press 1992:233-246.
161. Nose PS. Cytokines and reperfusion injury. J Card Surg 1993; 8:305-8.
162. Sakr MF, McClain CJ, Gavaler JS et al. FK 506 pretreatment is associated with reduced levels of tumor necrosis factor and interleukin 6 following hepatic ischemia/reperfusion. J Hepatol 1993; 17:301-7.
163. Camargo CA Jr, Madden JF, Gao W et al. Interleukin-6 protects liver against warm ischemia/reperfusion injury and promotes hepatocyte proliferation in the rodent. Hepatology 1997; 26:1513-20.
164. Kimizuka K, Nakao A, Nalesnik MA et al. Exogenous IL-6 inhibits acute inflammatory responses and prevents ischemia/reperfusion injury after intestinal transplantation. Am J Transplant 2004; 4:482-94.
165. Farivar AS, Merry HE, Fica-Delgado MJ et al. Interleukin-6 regulation of direct lung ischemia reperfusion injury. Ann Thorac Surg 2006; 82:472-8.
166. Lloris-Carsi JM, Cejalvo D, Toledo-Pereyra LH et al. Preconditioning: effect upon lesion modulation in warm liver ischemia. Transplant Proc 1993; 25:3303-4.
167. Teoh NC, Farrell GC. Hepatic ischemia reperfusion injury: Pathogenic mechanisms and basis for hepatoprotection. J Gastroenterol Hepatol 2003; 18:891-902.
168. Matsumoto T, O'Malley K, Efron PA et al. Interleukin-6 and STAT3 protect the liver from hepatic ischemia and reperfusion injury during ischemic preconditioning. Surgery 2006; 140:793-802.
169. Lentsch AB, Yoshidome H, Kato A et al. Requirement for interleukin-12 in the pathogenesis of warm hepatic ischemia/reperfusion injury in mice. Hepatology 1999; 30:1448-53.
170. Kato A, Graul-Layman A, Edwards MJ et al. Promotion of hepatic ischemia/reperfusion injury by IL-12 is independent of STAT4. Transplantation 2002; 73:1142-5.
171. Kojima Y, Suzuki S, Tsuchiya Y et al. Regulation of pro-inflammatory and anti-inflammatory cytokine responses by Kupffer cells in endotoxin-enhanced reperfusion injury after total hepatic ischemia. Transpl Int 2003; 16:231-40.
172. Le Moine O, Louis H, Stordeur P et al. Role of reactive oxygen intermediates in interleukin 10 release after cold liver ischemia and reperfusion in mice. Gastroenterology 1997; 113:1701-6.
173. Le Moine O, Louis H, Demols A et al. Cold liver ischemia-reperfusion injury critically depends on liver T-cells and is improved by donor pretreatment with interleukin 10 in mice. Hepatology 2000; 31:1266-74.
174. Donckier V, Loi P, Closset J et al. Preconditioning of donors with interleukin-10 reduces hepatic ischemia-reperfusion injury after liver transplantation in pigs. Transplantation 2003; 75:902-4.
175. Zou XM, Yagihashi A, Hirata K et al. Downregulation of cytokine-induced neutrophil chemoattractant and prolongation of rat liver allograft survival by interleukin-10. Surg Today 1998; 28:184-91.
176. Santucci L, Fiorucci S, Chiorean M et al. Interleukin 10 reduces lethality and hepatic injury induced by lipopolysaccharide in galactosamine-sensitized mice. Gastroenterology 1996; 111:736-44.

177. Deng J, Kohda Y, Chiao H et al. Interleukin-10 inhibits ischemic and cisplatin-induced acute renal injury. Kidney Int 2001; 60:2118-28.
178. Jones SP, Trocha SD, Lefer DJ. Cardioprotective actions of endogenous IL-10 are independent of iNOS. Am J Physiol Heart Circ Physiol 2001; 281:H48-52.
179. Kozower BD, Kanaan SA, Tagawa T et al. Intramuscular gene transfer of interleukin-10 reduces neutrophil recruitment and ameliorates lung graft ischemia-reperfusion injury. Am J Transplant 2002; 2:837-42.
180. Souza DG, Guabiraba R, Pinho V et al. IL-1-driven endogenous IL-10 production protects against the systemic and local acute inflammatory response following intestinal reperfusion injury. J Immunol 2003; 170:4759-66.
181. Pulitano C, Sitia G, Aldrighetti L et al. Reduced severity of liver ischemia/reperfusion injury following hepatic resection in humans is associated with enhanced intrahepatic expression of Th2 cytokines. Hepatol Res 2006; 36:20-6.
182. Minty A, Chalon P, Derocq JM et al. Interleukin-13 is a new human lymphokine regulating inflammatory and immune responses. Nature 1993; 362:248-50.
183. Berkman N, John M, Roesems G et al. Interleukin 13 inhibits macrophage inflammatory protein-1 alpha production from human alveolar macrophages and monocytes. Am J Respir Cell Mol Biol 1996; 15:382-9.
184. Muchamuel T, Menon S, Pisacane P et al. IL-13 protects mice from lipopolysaccharide-induced lethal endotoxemia: correlation with down-modulation of TNF-alpha, IFN-gamma and IL-12 production. J Immunol 1997; 158:2898-903.
185. Mulligan MS, Warner RL, Foreback JL et al. Protective effects of IL-4, IL-10, IL-12 and IL-13 in IgG immune complex-induced lung injury: role of endogenous IL-12. J Immunol 1997; 159:3483-9.
186. Kato A, Okaya T, Lentsch AB. Endogenous IL-13 protects hepatocytes and vascular endothelial cells during ischemia/reperfusion injury. Hepatology 2003; 37:304-12.
187. Chazouilleres O, Guechot J, Balladur P et al. Tumor necrosis factor-alpha in liver transplantation and resection. No evidence for a key role in ischemia-reperfusion injury. J Hepatol 1992; 16:376-9.
188. Chensue SW. Molecular machinations: chemokine signals in host-pathogen interactions. Clin Microbiol Rev 2001; 14:821-35
189. Murphy PM. International Union of Pharmacology. XXX. Update on chemokine receptor nomenclature. Pharmacol Rev 2002; 54:227-9.
190. Karin M, Takahashi T, Kapahi P et al. Oxidative stress and gene expression: the AP-1 and NF-kappa B connections. Biofactors 2001; 15:87-9.
191. Onai Y, Suzuki J, Kakuta T et al. Inhibition of Ikappa B phosphorylation in cardiomyocytes attenuates myocardial ischemia/reperfusion injury. Cardiovasc Res 2004; 63:51-9.
192. Luscinskas FW, Cybulsky MI, Kiely JM et al. Cytokine-activated human endothelial monolayers support enhanced neutrophil transmigration via a mechanism involving both endothelial-leukocyte adhesion molecule-1 and intercellular adhesion molecule-1. J Immunol 1991; 146:1617-1625.
193. Onuffer JJ, Horuk R. Chemokines, chemokine receptors and small-molecule antagonists: recent developments. Trends Pharmacol Sci 2002; 23:459-67.
194. Sekido N, Mukaida N, Harada A et al. Prevention of lung reperfusion injury in rabbits by a monoclonal antibody against interleukin-8. Nature 1993; 365:654-7.
195. Boyle EM Jr, Kovacich JC, Hebert CA et al. Inhibition of interleukin-8 blocks myocardial ischemia-reperfusion injury. J Thorac Cardiovasc Surg 1998; 116:114-21.
196. Ishii H, Ishibashi M, Takayama M et al. The role of cytokine-induced neutrophil chemoattractant-1 in neutrophil-mediated remote lung injury after intestinal ischaemia/reperfusion in rats. Respirology 2000; 5:325-31.
197. Kohtani T, Abe Y, Sato M et al. Protective effects of anti-neutrophil antibody against myocardial ischemia/reperfusion injury in rats. Eur Surg Res 2002; 34:313-20.
198. Kim YI, Song KE, Ryeon HK et al. Enhanced inflammatory cytokine production at ischemia/reperfusion in human liver resection. Hepatogastroenterology 2002; 49:1077-82.
199. Patarroyo M. Leukocyte adhesion in host defense and tissue injury. Clin Immunol Immunopathol 1990; 60:333-348.
200. Vollmar B, Glasz J, Menger MD et al. Leukocytes contribute to hepatic ischemia/reperfusion injury via intercellular adhesion molecule-1-mediated venular adherence. Surgery 1995; 117:195-200.
201. Butcher EC. Leukocyte-endothelial cell recognition: three (or more) steps to specificity and diversity. Cell 1991; 67:1033-6.
202. Schall TJ, Bacon KB. Chemokines, leukocyte trafficking and inflammation. Curr Opin Immunol 1994; 6:865-73.
203. Lasky LA. Selectins: interpreters of cell-specific carbohydrate information during inflammation. Science 1992; 258:964-9.

204. Foxall C, Watson SR, Dowbenko D et al. The three members of the selectin receptor family recognize a common carbohydrate epitope, the sialyl Lewis(x) oligosaccharide. J Cell Biol 1992; 117:895-902.
205. Diamond MS, Staunton DE, de Fougerolles AR et al. ICAM-1 (CD54): a counter-receptor for Mac-1 (CD11b/CD18). J Cell Biol 1990; 111:3129-39.
206. Gasic AC, McGuire G, Krater S et al. Hydrogen peroxide pretreatment of perfused canine vessels induces ICAM-1 and CD18-dependent neutrophil adherence. Circulation 1991; 84:2154-66.
207. Yoshida N, Granger DN, Anderson DC et al. Anoxia/reoxygenation-induced neutrophil adherence to cultured endothelial cells. Am J Physiol 1992; 262:H1891-8.
208. Hernandez LA, Grisham MB, Twohig B et al. Role of neutrophils in ischemia-reperfusion-induced microvascular injury. Am J Physiol 1987; 253:H699-703.
209. Simpson PJ, Todd RF 3rd, Mickelson JK et al. Sustained limitation of myocardial reperfusion injury by a monoclonal antibody that alters leukocyte function. Circulation 1990; 81:226-37.
210. Moore TM, Khimenko P, Adkins WK et al. Adhesion molecules contribute to ischemia and reperfusion-induced injury in the isolated rat lung. J Appl Physiol 1995; 78:2245-52.
211. Rabb H, Mendiola CC, Saba SR et al. Antibodies to ICAM-1 protect kidneys in severe ischemic reperfusion injury. Biochem Biophys Res Commun 1995; 211:67-73.
212. Suzuki S, Toledo-Pereyra LH. Monoclonal antibody to intercellular adhesion molecule 1 as an effective protection for liver ischemia and reperfusion injury. Transplant Proc 1993; 25:3325-7.
213. Garcia-Criado FJ, Toledo-Pereyra LH, Lopez-Neblina F et al. Role of P-selectin in total hepatic ischemia and reperfusion. J Am Coll Surg 1995; 181:327-34.
214. Misawa K, Toledo-Pereyra LH, Phillips ML et al. Role of sialyl lewis(x) in total hepatic ischemia and reperfusion. J Am Coll Surg 1996; 182:251-6.
215. Palma-Vargas JM, Toledo-Pereyra L, Dean RE et al. Small-molecule selectin inhibitor protects against liver inflammatory response after ischemia and reperfusion. J Am Coll Surg 1997; 185:365-72.
216. Martinez-Mier G, Toledo-Pereyra LH, McDuffie E et al. L-Selectin and chemokine response after liver ischemia and reperfusion. J Surg Res 2000; 93:156-62.
217. Ma X, Lefer DJ, Lefer AM et al. Coronary endothelial and cardiac protective effects of a monoclonal antibody to intercellular adhesion molecule-1 in myocardial ischemia and reperfusion. Circulation 1992; 86:937-946.
218. Nakano H, Kuzume M, Namatame K et al. Efficacy of intraportal injection of anti-ICAM-1 monoclonal antibody against liver cell injury following warm ischemia in the rat. Am J Surg 1995; 170:64-6.
219. Kuzume M, Nakano H, Yamaguchi M et al. A monoclonal antibody against ICAM-1 suppresses hepatic ischemia-reperfusion injury in rats. Eur Surg Res 1997; 29:93-100.
220. Kuzu MA, Koksoy C, Kuzu I et al. Role of integrins and intracellular adhesion molecule-1 in lung injury after intestinal ischemia-reperfusion. Am J Surg 2002; 183:70-4.
221. Yadav SS, Howell DN, Steeber DA et al. P-Selectin mediates reperfusion injury through neutrophil and platelet sequestration in the warm ischemic mouse liver. Hepatology 1999; 29:1494-502.
222. Vachino G, Chang XJ, Veldman GM et al. P-selectin glycoprotein ligand-1 is the major counter-receptor for P-selectin on stimulated T-cells and is widely distributed in nonfunctional form on many lymphocytic cells. J Biol Chem 1995; 270:21966-74.
223. Roberts AM, Ovechkin AV, Mowbray JG et al. Effects of pulmonary ischemia-reperfusion on platelet adhesion in subpleural arterioles in rabbits. Microvasc Res 2004; 67:29-37.
224. Harlan JM. Leukocyte adhesion deficiency syndrome: insights into the molecular basis of leukocyte emigration. Clin Immunol Immunopathol 1993; 67:S16-24.
225. Yanagisawa M, Kurihara H, Kimura S et al. A novel potent vasoconstrictor peptide produced by vascular endothelial cells. Nature 1988; 332:411-5.
226. Rockey DC, Fouassier L, Chung JJ et al. Cellular localization of endothelin-1 and increased production in liver injury in the rat: potential for autocrine and paracrine effects on stellate cells. Hepatology 1998; 27:472-80.
227. Pollock DM, Keith TL, Highsmith RF. Endothelin receptors and calcium signaling. FASEB J 1995; 9:1196-204.
228. Clozel M, Gray GA, Breu V et al. The endothelin ETB receptor mediates both vasodilation and vasoconstriction in vivo. Biochem Biophys Res Commun 1992; 186:867-73.
229. Tschaikowsky K, Sagner S, Lehnert N et al. Endothelin in septic patients: effects on cardiovascular and renal function and its relationship to proinflammatory cytokines. Crit Care Med 2000; 28:1854-60.
230. Margulies KB, Hildebrand FL Jr, Lerman A et al. Increased endothelin in experimental heart failure. Circulation 1990; 82:2226-30.
231. Kawashima M, Nakamura T, Schneider S et al. Iloprost ameliorates post-ischemic lung reperfusion injury and maintains an appropriate pulmonary ET-1 balance. J Heart Lung Transplant 2003; 22:794-801.
232. Ajis A, Bagnall NM, Collis MG et al. Effect of endothelin antagonists on the renal haemodynamic and tubular responses to ischaemia-reperfusion injury in anaesthetised rats. Exp Physiol 2003; 88:483-90.

233. Oktar BK, Gulpinar MA, Bozkurt A et al. Endothelin receptor blockers reduce I/R-induced intestinal mucosal injury: role of blood flow. Am J Physiol Gastrointest Liver Physiol 2002; 282:G647-55.
234. Nakamura S, Nishiyama R, Serizawa A et al. Hepatic release of endothelin-1 after warm ischemia. Reperfusion injury and its hemodynamic effect. Transplantation 1995; 59:679-84.
235. Kawada N, Tran-Thi TA, Klein H et al. The contraction of hepatic stellate (Ito) cells stimulated with vasoactive substances. Possible involvement of endothelin 1 and nitric oxide in the regulation of the sinusoidal tonus. Eur J Biochem 1993; 213:815-23.
236. Housset CN, Rockey DC, Friedman SL et al. Hepatic lipocytes: a major target for endothelin-1. J Hepatol 1995; 22:55-60.
237. Housset C, Rockey DC, Bissell DM. Endothelin receptors in rat liver: lipocytes as a contractile target for endothelin 1. Proc Natl Acad Sci USA 1993; 90:9266-70.
238. Scommotau S, Uhlmann D, Loffler BM et al. Involvement of endothelin/nitric oxide balance in hepatic ischemia/reperfusion injury. Langenbecks Arch Surg 1999; 384:65-70.
239. Kerrigan CL, Stotland MA. Ischemia reperfusion injury: a review. Microsurgery 1993; 14:165-175.
240. Ruetten H, Thiemermann C. Endothelin-1 stimulates the biosynthesis of tumour necrosis factor in macrophages: ET-receptors, signal transduction and inhibition by dexamethasone. J Physiol Pharmacol 1997; 48:675-88.
241. Suzuki S, Serizawa A, Sakaguchi T et al. The roles of platelet-activating factor and endothelin-1 in renal damage after total hepatic ischemia and reperfusion. Transplantation 2000; 69:2267-73.
242. Yang TL, Chen MF, Jiang JL et al. The endothelin receptor antagonist decreases ischemia/reperfusion-induced tumor necrosis factor production in isolated rat hearts. Int J Cardiol 2005; 100:495-8.
243. Wolfard A, Vangel R, Szalay L et al. Endothelin-A receptor antagonism improves small bowel graft perfusion and structure after ischemia and reperfusion. Transplantation 1999; 68:1231-8.
244. Behrend M. The endothelin receptor antagonist TAK-044 in the treatment of reperfusion injury in organ transplantation. Expert Opin Investig Drugs 1999; 8:1079-91.
245. Huang C, Huang C, Hestin D et al. The effect of endothelin antagonists on renal ischaemia-reperfusion injury and the development of acute renal failure in the rat. Nephrol Dial Transplant 2002; 17:1578-85.
246. Ozdemir R, Parlakpinar H, Polat A et al. Selective endothelin a (ETA) receptor antagonist (BQ-123) reduces both myocardial infarct size and oxidant injury. Toxicology 2006; 219:142-9.
247. Erdogan H, Fadillioglu E, Emre MH. Protection from renal ischemia reperfusion injury by an endothelin-A receptor antagonist BQ-123 in relation to nitric oxide production. Toxicology 2006; 228:219-28.
248. Choi AM, Alam J. Heme oxygenase-1: function, regulation and implication of a novel stress-inducible protein in oxidant-induced lung injury. Am J Respir Cell Mol Biol 1996; 15:9-19.
249. Katori M, Busuttil RW, Kupiec-Weglinski JW. Heme oxygenase-1 system in organ transplantation. Transplantation 2002; 74:905-12.
250. Yet SF, Tian R, Layne MD et al. Cardiac-specific expression of heme oxygenase-1 protects against ischemia and reperfusion injury in transgenic mice. Circ Res 2001; 89:168-73.
251. Kato H, Amersi F, Buelow R et al. Heme oxygenase-1 overexpression protects rat livers from ischemia/reperfusion injury with extended cold preservation. Am J Transplant 2001; 1:121-8.
252. Shimizu H, Takahashi T, Suzuki T et al. Protective effect of heme oxygenase induction in ischemic acute renal failure. Crit Care Med 2000; 28:809-17.
253. Ito K, Ozasa H, Kojima N et al. Pharmacological preconditioning protects lung injury induced by intestinal ischemia/reperfusion in rat. Shock 2003; 19:462-8.
254. Attuwaybi BO, Kozar RA, Moore-Olufemi SD et al. Heme oxygenase-1 induction by hemin protects against gut ischemia/reperfusion injury. J Surg Res 2004; 118:53-7.
255. Willis D, Moore AR, Frederick R et al. Heme oxygenase: a novel target for the modulation of the inflammatory response. Nat Med 1996; 2:87-90.
256. Clark JE, Foresti R, Sarathchandra P et al. Heme oxygenase-1-derived bilirubin ameliorates postischemic myocardial dysfunction. Am J Physiol Heart Circ Physiol 2000; 278:H643-51.
257. Balla G, Jacob HS, Balla J et al. Ferritin: a cytoprotective antioxidant strategem of endothelium. J Biol Chem 1992; 267:18148-53.
258. Fujita T, Toda K, Karimova A et al. Paradoxical rescue from ischemic lung injury by inhaled carbon monoxide driven by derepression of fibrinolysis. Nat Med 2001; 7:598-604.
259. Nakao A, Kimizuka K, Stolz DB et al. Carbon monoxide inhalation protects rat intestinal grafts from ischemia/reperfusion injury. Am J Pathol 2003; 163:1587-98.
260. Siebenlist U, Franzoso G, Brown K. Structure, regulation and function of NF-kappa B. Annu Rev Cell Biol 1994; 10:405-55.

261. Collart MA, Baeuerle P, Vassalli P. Regulation of tumor necrosis factor alpha transcription in macrophages: involvement of four kappa B-like motifs and of constitutive and inducible forms of NF-kappa B. Mol Cell Biol 1990; 10:1498-506.
262. Hiscott J, Marois J, Garoufalis J et al. Characterization of a functional NF-kappa B site in the human interleukin 1 beta promoter: evidence for a positive autoregulatory loop. Mol Cell Biol 1993; 13:6231-40.
263. Stein B, Baldwin AS Jr. Distinct mechanisms for regulation of the interleukin-8 gene involve synergism and cooperativity between C/EBP and NF-kappa B. Mol Cell Biol 1993; 13:7191-8.
264. Collins T, Read MA, Neish AS et al. Transcriptional regulation of endothelial cell adhesion molecules: NF-kappa B and cytokine-inducible enhancers. FASEB J 1995; 9:899-909.
265. Karin M, Liu Z, Zandi E. AP-1 function and regulation. Curr Opin Cell Biol 1997; 9:240-6.
266. Funakoshi M, Sonoda Y, Tago K et al. Differential involvement of p38 mitogen-activated protein kinase and phosphatidyl inositol 3-kinase in the IL-1-mediated NF-kappa B and AP-1 activation. Int Immunopharmacol 2001; 1:595-604.
267. Kato A, Singh S, McLeish KR et al. Mechanisms of hypothermic protection against ischemic liver injury in mice. Am J Physiol Gastrointest Liver Physiol 2002; 282:G608-16.
268. Zwacka RM, Zhou W, Zhang Y et al. Redox gene therapy for ischemia/reperfusion injury of the liver reduces AP1 and NF-kappa B activation. Nat Med 1998; 4:698-704.
269. Zhou W, Zhang Y, Hosch MS et al. Subcellular site of superoxide dismutase expression differentially controls AP-1 activity and injury in mouse liver following ischemia/reperfusion. Hepatology 2001; 33: 902-14.
270. Souza DG, Vieira AT, Pinho V et al. NF-kappa B plays a major role during the systemic and local acute inflammatory response following intestinal reperfusion injury. Br J Pharmacol 2005; 145: 246-54.

Chapter 6

Metabolic Management

Sufan Chien*

The normal human body and the bodies of other mammalian animals have complex neurologic and hormonal regulatory systems. Adequate substrates and oxygen supplies allow these systems to adjust the body's function and its metabolism to a wide variety of external and internal environmental changes. Every organ has a vital role in the control of one or more fluid constituents and the efficient exchange of energy and substrates between organs is crucial for inducing this steady state. Once an organ is separated from the body, the regulatory ability becomes very limited or lost and the survival of the organ becomes dependent upon the correct and careful regulation of the new artificial environment.

Metabolic regulation is involved in nearly all aspects of organ preservation, a subject that is overlapped in other chapters of this book. The scope of this particular chapter varies according to the methods used. If a single-flush preservation is used, the major pathophysiology is reduced energy supply and accumulation of metabolic wastes. Thus, either inhibition of tissue metabolism or supplement of energy substrates along with removal of metabolic wastes become the main strategy for extended organ survival. On the other hand, if perfusion is used, whether it is continuous or intermittent, the above problems become less important and the management concerns shift to overcoming the damage caused by perfusion itself. Our understanding of the physiology of isolated organs is superficial and fragmented because our knowledge of normal organ metabolism is still relatively limited. Moreover, due to the rapid advancement of molecular biology, the metabolic pathway is not currently a favorable research topic,[1] leaving many questions unanswered. In this chapter, we will briefly review previous work in these areas plus some findings in the author's laboratory describing his research team's effort to extend organ and tissue survival time during ischemia. Because many topics are also covered in other chapters, these topics will be mentioned only briefly. Metabolic management during reperfusion will not be described here because a separate chapter is dedicated to reperfusion damage.

Ischemic Effects on Tissues and Organs

Solid organs, such as the heart, liver, kidney and brain, require large amounts of oxygen and fuel to support their various specialized functions. Eighty percent of the oxygen consumed by the human body is used for energy production, which is produced when large, complex molecules (such as carbohydrates, lipids and proteins) are broken down into smaller, simpler molecules. The resultant decrease in chemical order causes the liberation of free energy. It is estimated that in humans, the rate of adenosine triphosphate (ATP) synthesis varies daily from 80 moles (about 40 kg) at rest to 1800 moles (~ 910 kg) during strenuous exercise.[2,3] In general, glucose metabolism supplies the largest portion of cellular energy through its various steps (i.e., conversion of substrates to acetyl-CoA in the cytoplasm, the oxidation of acetyl-CoA in the Krebs citric acid cycle and electron transfer activities coupled to oxidative rephosphorylation of ADP to ATP

*Sufan Chien—Department of Surgery, University of Louisville, Louisville, KY 40202, USA.
Email: sufanc@netscape.net

Organ Preservation for Transplantation, Third Edition, edited by Luis H. Toledo-Pereyra.
©2010 Landes Bioscience.

in the respiratory chain of the mitochondria, which is responsible for 90-95% of the total ATP requirement in eukaryotic cells).[2]

A great number of events occur when tissues and organs are removed from the body. As blood circulation stops, there is a combined effect of anoxia, lack of substrate and absence of metabolic waste removal. The tissues use stored glycogen and the leftover oxygen for continuous aerobic metabolism. However, this process soon stops due to depleted oxygen supply. In an effort to maintain tissue energy supply, glycolysis takes over, but this is not an efficient way to produce energy and the process also results in intracellular acidosis. While the pathophysiology of ischemia is complex,[4] some of these events have been clearly delineated, such as depletion of high-energy phosphate, inhibition of cellular energy-dependent processes, generation of oxygen-free radicals, loss of osmotic balance across the membrane, stimulation of glycolysis, accumulation of metabolic products, intracellular acidosis and release of lysosomal enzymes.[5,6] Short episodes of ischemia cause mild damage from which a cell can recover, while longer periods of ischemia cause irreversible cell damage, leading to cell death. An organ's tolerance to ischemia depends on the following three factors: (1) the size of the available stored energy pool; (2) the amount of the energy-demand per time unit; and (3) the efficiency of the anaerobically obtained energy.[7] At 37°C, the maximum tolerable ischemic time is about 30 to 45 minutes for the heart,[8,9] 30 to 60 minutes for the liver[10,11] and less than 5 minutes for the brain.[12,13] Beyond these time limits, irreversible damage results. Even if blood circulation is reestablished, total functional recovery is impossible.[12,14] Furthermore, loss of function always precedes cell death. Some ischemic myocardial tissue, for example, can survive for 30 to 60 minutes, but cardiac muscle contractility is lost within less than 1 minute. Similarly, although neurons survive for some minutes, consciousness is lost almost immediately after a sudden interruption of blood supply to the brain.[15]

Decline of High-Energy Phosphates

The biochemical processes of high-energy phosphate decline are complex and some are still not fully understood. However, it is generally agreed that the following events occur as the result of ischemia:

1. The depletion of oxygen stores occurs within a few seconds in many organs.

   Our bodies require approximately 500 g of molecular oxygen per day.[16] The oxygen is continuously transported to the tissue mainly by oxygen-binding protein hemoglobin. Removal of an organ from the body induces complete tissue ischemia, resulting in shutdown of oxygen and nutrients into the cells. Except for the lungs, residual oxygen in the tissue is very limited and its supply lasts only a few seconds when blood circulation stops. After this interval, oxidative metabolism ceases. For example, in the brain, tissue pO_2 falls to 0 in less than 30 seconds.[12,17] The myocardial oxygen reserve is depleted within 8 seconds following aortic cross-clamping, when the pO_2 falls below 5 mmHg.[18] The blood oxygen content in the liver or kidney at the moment of separation from blood supply is about 1.5 μmol/g fresh weight of tissue. This would provide less than a 1 minute supply of oxygen for the liver and much less for the kidney, which consumes about 5 μmol of O_2/min per gram of tissue.[19]

2. Stored energy in the form of creatine phosphate (CP) and adenosine triphosphate (ATP) is reduced and depleted.

   Although there is a progressive decrease of ATP levels after ischemia, the early decrease of creatine phosphate is more striking, especially in the myocardium. When the ATP/ADP ratio declines, ATP is regenerated rapidly by the reversible creatine kinase (CK) reaction to replenish the ATP pool. CP is utilized while the ATP concentration is only slightly reduced from its resting level.[20,21]

   The exact times at which the changes of high-energy phosphates occur are difficult to determine because of the extreme lability of CP and ATP. Data accumulated from decades of research in many laboratories, including ours, have shown that the level of myocardial CP is reduced by 50% within 1 minute after inducing ischemia. After 5 minutes, CP is nearly depleted.[22-24] ATP levels do not decline as rapidly as CP levels; they decrease to about 50% after 5 to 10 minutes into ischemia.[25,26]

As oxygen levels continue to fall and oxidative phosphorylation ceases, ATP levels are depleted. Although 40 to 910 kg of ATP is synthesized in the adult human body every day, the storage of ATP in tissues is minimal. For example, the human brain can generate 12 kg of ATP per 24 hours, but has only about 5 mmoles of reserve for all species of adenine nucleotides (ATP+ADP+AMP). No large reservoirs of any energy-rich substances such as ATP or CP exist in any tissue.[27] Furthermore, the average ATP molecule has a lifespan of only 1 to 5 minutes and most cells will be depleted of ATP very soon after the cessation of oxidative phosphorylation.[15] Five minutes of complete ischemia is sufficient to lower brain ATP practically to zero. The breakdown of ATP increases levels of ADP, AMP and Pi, thereby increasing the levels of adenosine, inosine, xanthine and uric acid. The continued nucleotide degradation to the membrane-diffusible nucleosides results in the inability to rephosphorylate adenosine monophosphate (AMP) when circulation is reestablished and this slow repletion is especially troublesome for some organs that need immediate function such as the heart.[28]

3. Anaerobic glycolysis increases in an attempt to maintain normal cellular function.

When the oxygen supply to the tissue is normal, the rate of glycolysis is inhibited by high levels of citrate and ATP formed by oxidative metabolism in the citrate cycle (TCA cycle). When there is anoxia or severe hypoxia, oxidative metabolism ceases, citrate and ATP levels fall and glycolysis is stimulated. This feedback mechanism is the Pasteur effect.[29] Glycolysis is fueled by stored glycogen, whose breakdown also is stimulated by a low energy charge. However, this has low efficiency for energy production. One mole of glucose, when completely oxidized, yields 36 to 38 moles of ATP (depending on the shuttle mechanism by which the NADH enters the mitochondria), whereas the same amount of glucose yields only 2 moles of ATP when metabolized to lactate under anaerobic conditions, approximately 18 times less useful energy is produced as compared to aerobic production.[30,31] Normally, anaerobic glycolysis is suitable only for cells with low energy requirements and as a temporary device to survive episodes of hypoxia or to cope with increased energy demands. In most tissues (except red blood cells, the cornea, etc.), glycolysis is only a preparatory pathway for subsequent oxygen-requiring metabolism. In the presence of oxygen, the end-product of glycolysis is not lactate, but pyruvate, which then enters the mitochondria and is oxidized to acetyl CoA by pyruvate dehydrogenase. Acetyl CoA is oxidized to CO_2 by the TCA cycle.

Despite its low efficiency, ATP produced from glycolysis and oxidative phosphorylation is not of equal importance. First, carbohydrates are the only metabolic substrates that can produce ATP under anaerobic conditions.[15] Because glycolysis can provide energy even in the absence of oxygen, inhibition of the Pasteur effect by severe ischemia limits the capacity of the tissue to survive the ischemic insult.[29] Second, glycolysis has a special role in maintaining ion homeostasis. For example, in the heart, aerobically produced ATP supplies energy for contraction, whereas glycolytically produced ATP that contributes only 5 to 10% of the overall ATP supply in normal aerobic state is restricted to processes involved in membrane functions such as ionic pumps and the maintenance of the phosphatidic acid cycle.[32,33] Only a very limited portion of the aerobically produced ATP is available for similar functions.[34,35]

However, glycolysis cannot maintain tissue survival for a longer period of time. When there is not only extreme hypoxia but poor blood flow (ischemia), products of glycolysis accumulate, the protons and lactate inhibit glycolysis and glucose use falls. In the heart, it has been observed that the initial phase of rapid glycolysis lasts only 1 to 2 minutes, followed by a slower rate which lasts about 90 minutes. The terminal slowing and final cessation of anaerobic glycolysis occurs when the ATP of the tissue decreases to less than 1.0 µmol/g dry weight.[36]

The final cessation of glycolysis is most likely due to the combination of inhibition of glycolytic enzymes, depletion of substrates and cofactors and decrease of tissue pH.[36] Lactate is produced by glycolysis under anaerobic conditions. The onset of the maximal rate of lactate formation occurs within 1 minute of severance of blood supply to organs.[19] Being an acid, lactic acid tends to acidify the cell in which it is formed. An increased acidity dampens glycolytic activity when pyruvic and lactic acid, the end products of glycolysis, accumulate to dangerous levels and glycolysis is eventually inhibited.[15]

4. All energy-dependent functions start to cease.

Many living cells must transport molecules against their concentration gradients. This is an active transport and energy provided by ATP-hydrolysis is used to drive the uphill transport of molecules in the energetically unfavorable direction. The ion pumps responsible for maintaining gradients of ions across the plasma membrane provide important examples of active transport driven directly by ATP hydrolysis. The concentration of Na^+ is approximately 10 times higher outside cells than inside cells, whereas the concentration of K^+ is higher inside than out. These ion gradients are maintained by the Na^+/K^+ pump (Na^+/K^+-ATPase), which uses energy derived from ATP hydrolysis to transport Na^+ and K^+ against their electrochemical gradients. The Na^+/K^+ pump is estimated to consume nearly 25% of the ATP utilized by many animal cells and one of the major functions of this pump is to maintain osmotic balance and cell volume.[37] The active transport of Ca^{++} across the plasma membrane is driven by a Ca^{++} pump that is structurally related to the Na^+/K^+ pump and is similarly powered by ATP hydrolysis. The Ca^{++} pump transports Ca^{++} out of the cell, so that intracellular Ca^{++} is maintained to approximately 0.1 μM, in comparison to the extracellular concentration of about 1 mM. In some cells, similar ion pumps are responsible for the active transport of H^+ out of the cell.[37] Together, these pumps function to maintain membrane gradient. Glycolytically derived ATP is preferentially used for maintaining the ion gradients and the loss of K^+ from the cell was shown to be more clearly marked during inhibition of glycolysis.[33,38-40] Similar results were also seen in sodium influx[41] and enzyme release despite similar ATP levels.[42,43] With the declined pump function, membrane gradient disappears and cells swell. If the process is not stopped, death follows quickly.

5. Metabolic waste accumulation and decrease of cellular pH levels.

The acidity of the blood is tightly regulated at a pH level of 7.35 to 7.40. A decrease of the blood pH below this range is called acidosis. Lactic acidosis is the most common type of metabolic acidosis and an impairment of oxidative metabolism is the most common cause of this event. The cellular energy charge decreases, PFK becomes activated and a large amount of pyruvate is formed at a time when its mitochondrial oxidation is impaired and the excess pyruvate is converted to lactate. Continued glycolysis depends on the availability of NAD^+ for the oxidation of glyceraldehyde-3-phosphate. This is achieved by lactate dehydrogenase, which oxidizes NADH to NAD^+ as it reduces pyruvate to lactate. However, pyruvate becomes scarce in ischemic tissues because glycolysis is inhibited not only by NADH but also by excess protons and lactate. Reduced NAD^+ availability has been observed in various tissues and organs during ischemia.[44,45] As oxygen availability inside a cell is decreased, the oxidation of NADH by molecular oxygen, through the electron carrier system, decreases, resulting in an inhibition of the Krebs cycle of mitochondria.[46,47] If glycolysis reaction does not generate enough NAD^+, the glycolytic pathway arrests at glyceraldehyde-3-phosphate and no more ATP is generated. These changes cause mitochondria to cease functioning after a few seconds of severe anoxia.

The importance of metabolic waste can be illustrated from another viewpoint. In animal experiments, cells have survived "pure" anoxia without interruption of the blood flow longer than ischemic anoxia. This fact points to the importance of accumulating metabolic products, such as lactic acid, in ischemic cell injury.[15]

6. Rapid depletion of glycogen levels occurs under anaerobic conditions.

Glycogen is a polymer of approximately 10,000 to 40,000 glucose residues, held together by alpha-1,4 glycosidic bonds. With a molecular weight between 10^6 and 10^7 daltons, it is as big as a complete human ribosome (4.2×10^6 Da). Anaerobic glycolysis from glycogen produces three rather than two molecules of ATP for each glucose residue: the ATP-consuming hexokinase reaction is not required because glucose 6-phosphate is generated by glycogen phosphorylase and phosphoglucomutase, which do not consume ATP.[15,48] However, glycogen is mainly stored in the liver (up to 6%) and muscle (0.7%) and consumed quickly. Very little is consumed in other organs.[49]

7. Mitochondrial ATP generation is blocked because of a lack of molecular oxygen, and insufficient supply of NAD^+ also causes the TCA cycle to cease function.
   A decreased energy charge and decreased $NAD^+/NADH$ ratio are the initial results of oxygen deficiency. The mitochondrial oxidative pathways are arrested for lack of NAD^+. The regulated enzymes of glycolysis and glycogen degradation are stimulated by a low energy charge. The accumulation of lactic acid acidifies the cell. Both the plasma membrane and the organellar membranes are damaged by the increased acidity and by osmotic imbalances that result from the failure of ATP-dependent ion pumps.[50] Reactive oxygen radicals also cause mitochondrial damage.[51,52]

Intracellular Acidosis

The precise mechanism that causes tissue acidosis is not completely understood and is controversial. The cause is believed to be the result of several metabolic derangements during severe ischemia, especially the accumulation of protons, NADH and lactate. The role of metabolic products in the development of ischemic damage has been appreciated for many years. In 1935, Tennant and Wiggers[53] demonstrated that a reduction in coronary flow had a more profound effect on myocardial contractility than did hypoxia. The early decrease in contractile force of ischemic hearts was associated with increased tissue lactate and H^+ and little change in tissue ATP.[54,55] In an animal study, incubation of thin slices of dog myocardium with 50 mM lactate resulted in mitochondrial changes after 10 minutes, a finding similar to those after 1 hour in ischemic myocardium.[56] Thus, high-tissue lactate has been implicated as a factor directly or indirectly causing cellular damage during ischemia,[57] although new evidence has indicated a more complex function of lactate.[58] The formation of ATP by glycolysis rather than by mitochondria has been found to produce more protons that can cause acidosis.[59] Another major source of protons is the continued breakdown of ATP formed by anaerobic glycolysis and, to a lesser extent, from ATP-turnover cycles, such as the triglyceride-free fatty acid (FFA) cycle.[60] Intramitochondrial acidosis can also result from the production of CO_2 during respiration; CO_2 produced in the cytosol from bicarbonate can also penetrate the mitochondrial membrane.[61] Protons can be produced from glycolytic ATP turnover and net ATP breakdown.[59]

There is disagreement regarding the relationship between intracellular acidosis and irreversible cell damage. Conventional textbooks of physiology and medicine state that there is close relationship between these two events. This conclusion has been challenged because there has not been strong enough evidence to support it and some evidence has shown that acidosis may be protective during ischemia.[62] In organ preservation-transplantation, metabolic acidosis is generally believed to play an important role in replantation toxemia.[63]

Osmolar Load and Cell Swelling

With only a few exceptions, the cellular plasma membrane is highly permeable to water molecules, urea molecules and chloride ions because they are considerably smaller than the membrane pores. The volume of a living cell is not constant but changes continuously and the rate per second of diffusion of water in each direction through the pores of a cell is about 100 times as great as the volume of the cell itself.[64,65] The maintenance of cell volume within a physiological limit requires a precise volume-regulatory mechanism. The energy-requiring transport processes present in the plasma membranes of all animal cells play an important role in the regulation of cell volume.[66] All of the ions dissolved in the extracellular and intracellular fluids and other substances, such as glucose, amino acids, free fatty acids and electrolytes, can cause osmotic pressure at the cell membrane. This pressure can be very high, reaching approximately 5400 mmHg (a 230 foot water column).[65] The plasma membrane of the mammalian cell cannot withstand significant hydrostatic pressure gradients. If the osmolarities of extracellular and intracellular fluids are different, water moves across the plasma membrane and volume change occurs (Gibbs-Donnan equilibrium).[67]

Cell swelling is an important factor in tissue damage and is the primary cause of poor function after transplantation.[68] The mechanisms for the modulation of the transport system at the cell membrane are still poorly understood. At least two factors contribute to cell swelling during ischemia:

1. A substantial proportion of cell energy, provided by hydrolysis of ATP, is used to support cell membrane active transport mechanisms. As stated above, the enzyme system Na^+/K^+-ATPase uses energy to expel sodium from the cell and accumulate K^+. Each ion tends to diffuse back down its concentration gradient and the outward extrusion of Na^+ takes place against both chemical and electrical gradients. Ischemia depresses the activity of the Na^+/K^+-ATPase in tissues, so that sodium, chloride and water enter the cell and K^+ moves out.[69,70] The degradation of ATP generates more protons and enhances glycolysis. The intracellular osmotic concentration increases, more water streams into the cell and the cell swells.[71,72] The process creates a vicious cycle, leading to progressive cellular destruction. After the organ is transplanted, this cell swelling prevents normal blood circulation in capillaries and minor blood vessels, resulting in the so-called "no reflow" phenomenon.[73,74] The ATPase of vascular endothelium is particularly sensitive,[75] increasing the no- reflow phenomenon.

2. During ischemia, the accumulation of metabolic intermediates occurs because of the absence of arterial flow. Lactate, H^+, glycolytic intermediates, inorganic phosphate and creatinine are the principal components of this load and, when their levels are significantly increased, are associated with mild cellular edema.[76] Mitochondria swell, mitochondrial reactions are severely limited and finally completely inhibited. The influx of calcium causes precipitation and probably alters cellular reactions. Intracellular pH continues to fall, causing a quantitative reduction of negative charges on proteins by hydrogen ions. This changes the structure and function of regulatory and enzyme proteins. As these irreversible changes progress, the cell dies.[77]

The resultant damage to various functions may be different from organ to organ. Even within an organ, various tissues have different sensitivities to ischemia and low temperature. For example, in the kidney, ATPase activity in the cortex is well maintained at $10°C$, but its activity in the vascular endothelium is completely inhibited at this temperature.[78]

Generation and Accumulation of Oxygen Free Radicals

Oxygen free radicals are generated during ischemia and play a major role in tissue damage. The topic of oxygen free radicals is covered in chapter by Suzuki et al of this book.

Stimulation of Intracellular Lysosomal Enzymes

As the membranes of the cell become swollen and ineffective because of osmotic stress and increased acidity, leaking of enzymes and other critical elements occurs. The decrease in intracellular pH stimulates the activity of lysosomal membranes to release proteolyses, lipases and other enzymes, which can digest structural and functional components of the cells necessary for organ viability.[72] The release of lysosomal enzymes into the cytoplasm can be detected within 30 minutes after ischemia and cellular enzymes leak into the interstitial space and finally appear in the blood.[79] When lysosomal enzymes are released in their acidified environment, they attack cellular proteins, glycoproteins, glycolipids, phosphate esters and other substrates. Some of these hydrolytic cleavages, especially those of carboxylic esters and phosphate esters, further acidify the environment.[15] The results of the released enzyme initiates the process of self-digestion within the cell, although the consequences of this process on organ viability is still controversial.[79,80]

Damage to Mitochondria

Mitochondria play a central role in the generation of energy in aerobic tissue. With the loss of oxygen and energy, mitochondria swell and the matrix becomes electron-lucent. The space between the cristase, which is continuous with the intramembrane space, becomes enlarged and breaks and the mitochondrial reactions are severely limited, or completely inhibited, causing a rapid loss of mitochondrial activity.[72,81] The formation of mitochondrial transition pores also allows the release of proapoptotic mitochondrial intermembrane proteins, such as cytochrome c and apoptosis-inducing factors, into the cytosol.[82] The swelling of mitochondria in the anoxic condition probably has several causes, such as a failure of oxidative phosphorylation, a rise in inorganic phosphate, a rise in free calcium secondary to failure of the calcium pump in the reticulum and the appearance of free fatty acids caused by breakdown of membrane lipoproteins and phospholipids.[25,79] The rate of activity loss is dependent on the organ and has been suggested as an important factor leading to loss of organ viability.[72,83]

Although the biochemical sequence of ischemia, cellular impairment and death has been well established, the exact mechanisms of mitochondrial damage are not totally clear,[83] and it is also unknown at which point the cell reaches the irreversible stage. Short episodes of ischemia cause mild damage from which a cell can recover. Modern techniques have no way of establishing when that critical point is reached in a particular cell. The irreversible stage probably varies markedly between cells and especially between tissues.[77]

Metabolic Inhibition by Hypothermia

The development of hypothermia is based on the principle that reduced temperature decreases metabolism and oxygen consumption, so that organs can survive longer without nutrient supplements.[84] Cooling can hinder the rapid deterioration that occurs in an organ at 37°C when deprived of its blood supply. The precise mechanism by which hypothermia protects the viability of organs is very complex and not entirely understood. It is well known that hypothermia changes the reaction rate of all biochemical processes, especially the enzymatic reaction. This temperature dependence of reaction rates can be expressed by the Arrhenius equation:

$$\text{Rate} = Ae^{-E/RT}$$

where A is a constant known as the pre-exponential factor, T is the absolute temperature, R is the gas constant (8.31 J/mol/K), E is the activation energy (in J/mol) and e is a mathematical number of 2.71828.

An alternative method, the van't Hoff coefficient (Q_{10}), is derived from the Arrhenius equation. It is the decrease in the rate of reaction expressed as a percentage of the initial rate for a 10°C decrease in temperature. For every 10°C drop in temperature, the rate of a chemical reaction is decreased by approximately 50%. The decrease in the metabolic rate during hypothermia varies for the different metabolic processes, depending on the activation energy of the individual reaction, which can give Q_{10} values ranging from 1.5 to 4. The Q_{10} values for enzyme-catalyzed reactions usually fall between 1.5 to 2.5, whereas those for non-enzymatic reactions are usually in the range of 2.0 to 4.0.[85,86] Experimental findings have indicated that the above estimations are suitable for many tissues and organs. Direct measurements of oxygen consumption by dogs showed a 50% reduction of oxygen consumption at 28°C and a 75% reduction at 20°C.[87] The mean value of oxygen consumption for the kidney is 6.26 ± 2.49 ml/min/100 g kidney weight at 39°C. At 30°C oxygen consumption is about 43% of the value at 39°C; at 20°C, it is only 16% of the control rate; and at 5°C, it is less than 5% (Fig. 1).[85,88] However, hypothermia does not produce uniform inhibition of the cellular processes and the inhibition can vary by as much as 400% from one metabolic process to another after a decrease in temperature from 37°C to 0°C.[86] Bretschneider[89] suggested the following points of orientation for pure ischemia and for myocardial protection with nonbuffered solutions:

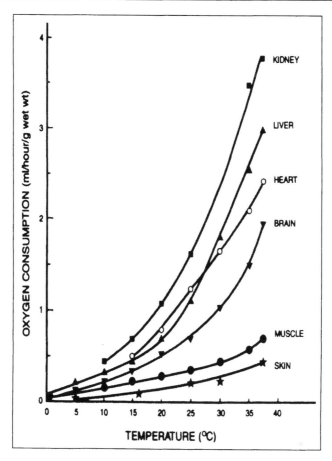

Figure 1. Oxygen consumption of rat tissues at different temperatures. Reprinted with permission from Fuhrman FA, In: Dripps RD, ed. The Physiology of Induced Hypothermia. Washington, DC: National Academy of Sciences, 1956:50-51.

Q_{10} for 35° to 25°C = 2.2
Q_{10} for 25° to 15°C = 1.9
Q_{10} for 15° to 5°C = 1.6
Q_{10} for 35° to 5°C ≈ 7

Some biochemical processes, especially those localized to cell membranes, show a step change in reaction rates at certain critical temperatures. These changes have been termed phase transitions and are thought to be the result of a change in the consistency of the cell membrane from fluid to gel. In mammalian tissues, phase transitions often occur at about 25°C to 28°C and may cause disturbed cell homeostasis. Biophysical processes, such as osmosis and water diffusion, are affected by temperature to a lesser extent, resulting in a linear change of about 3% per 10°C reduction in temperature.[90,91] However, osmolar effects become pronounced as the freezing point of water is approached. Ice formation in tissue concentrates solutes in the residual nonfrozen cytosol, causing marked fluid shifts and membrane disruption. For this reason, the freezing point of water is the physical limit to the beneficial effects of hypothermia, unless protective chemical substances are present. Mammalian tissue will not regain function upon thawing from a frozen state, with the exception of some very simple systems, such as red blood cells.[92]

The use of hypothermia in various medical applications began centuries ago.[87] The protection of hypothermia alone was demonstrated long before various preservation solutions were developed. Hypothermia and freezing techniques were used to store arteries for 1 to 35 days for transplantation early in the 20th century.[93-95] From the 1920s to the 1940s, kidneys were transplanted experimentally after several hours or even within days of simple hypothermia preservation, but without long-term survival.[96-98] Observations in the early 1950s confirmed that hypothermia significantly reduced heart and renal ischemic damage,[99,100] and considerable effort was directed toward total protection of organ function with various types of regional hypothermia. In the late 1950s and early 1960s, it was established that reducing the metabolism of the kidney by cooling enabled it to withstand relatively prolonged periods of ischemia and long-term survival of experimental kidney transplants was reported after 4 to 24 hours of simple cold immersion or refrigeration.[101-103] The simplicity of hypothermic storage stimulated researchers to use this technique as an alternative to perfusion and simple cold storage quickly gained popularity in the 1960s for the preservation of the heart,[104-106] lungs[107,108] and other organs.[92,109]

Limitations of Hypothermia and Single Flush Preservation

Single-flush hypothermia is relatively simple and effective for short periods of preservation. However, hypothermia only delays but does not stop organ deterioration.[110,111] Several limitations make it difficult to use hypothermia for long-term organ preservation. First, although metabolism is reduced, it is not completely suppressed by temperatures above 0°C, even in the presence of chemical metabolic inhibitors. Approximately 5% of the normal activity at 37°C still remains at 0°C.[86] Significant active transport and oxygen consumption persist even at 0°C.[106,112-114] Metabolism continues at a temperature of −60°C, so that even when an organ is completely frozen, metabolism still occurs.[92] Second, not all reaction rates are affected to the same extent by hypothermia. The effect of cooling is complex and can uncouple reaction pathways, producing harmful consequences.[115,116] Third, hypothermia itself also has adverse effects on cell physiology.[72,117-119] Hypothermia suppresses the rate of translocation of adenine nucleotides across the mitochondrial inner membrane, thus decreasing ATP synthesis and causing reduced oxidative phosphorylation.[72,120] The net result is similar to the damage caused by hypoxia. Rapid cooling and profound hypothermia could be deleterious to cell membranes and their function and rewarming the cells to physiologic temperature after exposure to hypothermia also aggravates the cell injury.[82,121]

For single-flush hypothermia preservation, the following additional limitations are well-recognized: 1) substrates for metabolism are not provided after flushing and a collapsed, blocked microcirculation may remain so until the organ is transplanted, which could result in tissue damage;[122] and 2) even though some substrates are added to the preservation solution, metabolic wastes can not be disposed of. Accumulation of toxic wastes in tissue will eventually damage cells, causing irreversible organ malfunction after transplantation.[123]

Despite these limitations, hypothermia is the only universally accepted method for tissue protection through metabolic inhibition. The use of hypothermia is not only limited to organ preservation, but also in other areas such as cardiopulmonary bypass, head injury, aortic surgery and storage of blood and other cells. For solid organs, hypothermic preservation is used most successfully for kidney preservation. For other organs, longer preservation times are still actively sought. Liver preservation times have been improved by using University of Wisconsin solution, but various types of histologic, functional and metabolic damage have been reported at preservation times ranging from 8 to 24 hours. This damage includes loss of adenine nucleotides,[124,125] damage to sinusoidal endothelial cells[126,127] and rapid increase in the levels of liver-related enzymes, such as AST and ALT, after transplantation.[128] Greater posttransplant problems, including the need for retransplantation, have also been reported for livers preserved for more than 20 hours.

We can speculate that current hypothermic storage times will likely be increased further and that the number of preservable organs will be expanded when more effective preservation solutions are developed. Because of the limitations mentioned above, it is unlikely that single-flush hypothermic storage will provide long-term preservation.

Inhibition of Tissue Metabolism by Hibernation

Metabolic rate depression is an important survival strategy for many animal species and a common element of hibernation, torpor, aestivation, anaerobiosis, diapause and anhydrobiosis.[129-132] Hibernation in mammals is a unique circannual adaptation allowing certain species, such as the ground squirrel, woodchuck, brown cave bat, European hedgehog and black bear, to survive extended periods of food deprivation when ambient temperatures may be well- below freezing. For example, hibernating ground squirrels conserve up to 90% of the energy that would be required if they remain active during the winter.[133] Moreover, it has also been suggested that hibernating animals age at a slower rate than those of the same species that are prevented from undergoing hibernation.[134] Profound metabolic changes accompany hibernation, including respiratory depression, a decline in body temperature and a cessation of feeding and renal function. These changes may be of great survival benefit to animals that can subsist without food and water for up to 5 months (up to 8 months in the arctic ground squirrel). In most winter hibernators, except the black bear, body temperatures decline to 4° to 6°C and even 1° or 2°C below freezing in the arctic ground squirrel.[135] It is notable that low temperature is not always required to save energy in some animals: hibernation also occurs in summer (aestivation), during periods of excessive drought and heat. Cactus mice, snails, Texas tortoise and some fish and crabs can hibernate in the summer time.[136-138] These animals can save energy at very high temperatures. However, we know too little about the mechanisms related to aestivation.

Blood or its components from winter hibernating animals have been shown in studies to induce behavioral and physiological depression, including hypothermia, bradycardia, long-term hypophagia without significant weight loss, an anesthetized state and decreased renal function.[139,140] Kidneys from hibernating animals were kept viable for up to 10 days.[141] Studies have also shown that the erythrocytes in hibernating animals have an increased resistance to hemolysis and higher levels of unsaturated fatty acids in the cold blood taken during hibernation.[142] Neither human cells nor ground squirrel cells would agglutinate in ground squirrel serum at low temperatures. During hibernation, stored fat becomes the primary metabolic fuel. Carbohydrate needs are met by gluconeogenesis from amino acids. Urea, the primary nitrogen-containing waste product of protein catabolism, is recycled rather than excreted, thereby negating the need for urination and hence arousal.[143,144] Actively dividing cells, such as those of the intestinal epithelium, become relatively quiescent.[145]

Despite the profound metabolic, biochemical and cellular changes noted in hibernating animals, relatively little information is available concerning the mechanism(s) that induce and maintain these changes or those involved in reversing them. One major reason for this problem is that most hibernation studies rely on in vivo systems, making experiential manipulation both difficult and expensive. For isolated organ preservation, it may not be necessary to induce whole-body hibernation. In the past, we have attempted to use the hibernation principle for the extension of isolated organs with some success.[146-149] Subsequent experiments have also indicated protective effects using similar approaches.[150,151] Plasma from hibernating animals and opioid agonists have both been used to induce a hibernation-like status in nonhibernating animals, but so far no single chemical or chemical combination has been able to induce true hibernation in either whole animals or in an organ. Thus, the theoretical advantages of inducing hibernation in organ preservation have not been successfully translated into the practice. Further characterization of the factors that induce both winter and summer hibernation may prove to be valuable toward extending organ preservation.

Metabolic Substrate Supplement

Supplementation of metabolic substrates to enhance high-energy phosphates represents a more active approach and efforts in this area have been ongoing for decades. The following are several major chemical groups that have shown protective effects in various scenarios. Our research group has focused on the use of fructose-1,6-bisphosphate (FBP, or fructose-1,6-diphosphate, FDP) and direct intracellular delivery of ATP. A more detailed report will be presented here.

Adenosine

During preservation, energy turnover influences the degradation process that affects structural integrity. Adenosine has been shown to enhance the preservation of the heart,[152] liver,[153] and pancreas.[154] Pharmacologic effects of adenosine are complex because many of the effects are related to adenosine receptors[155,156] and/or preconditioning,[152] rather than simply as a precursor of adenine nucleotides. Other protective effects such as vasodilation, anti-oxidants and inhibition of platelet aggregation have also been proposed.[157]

Amino Acids

Amino acids may be involved in many aspects of intermediary metabolism, such as protein synthesis, production of substrates for the tricitric acid cycle, provision of buffering properties and promotion of anaerobic ATP production.[158] Several amino acids, such as histidine, glutamate, aspartate, arginine and glycine, have been shown to improve heart,[159] liver,[160] lung,[161] intestine[162] and kidney[163] preservation.

Glucose and Glycogen

Adding glucose or glycogen may enhance organ preservation by maintaining the small rate of metabolism that exists at low temperatures.[164] Metabolic substrates are particularly important if a temperature higher than 4°C is the optimum temperature, such as for lung storage. Both glucose and glycogen have been used with some success.[162,165] Low-potassium dextran solutions enhanced by substrate or glucose-enriched blood were reported to yielded excellent results in some lung preservation studies.[166-168]

NAD and NAD Precursors

Because the decreased NAD$^+$/NADH ratio is the main restraint of anaerobic glycolysis, adding NAD or its precursors may help restore the NAD$^+$ content. In fact, NAD$^+$, nicotinamide, nicotinic acid and quinolinic acid, have been reported to improve myocardial contractility or skin flap survival.[169,170]

Phosphoenolpyruvate (PEP)

Phosphoenolpyruvate is a high-energy metabolite in the final step of glycolysis. PEP is converted into pyruvate by pyruvate kinase and one molecule of ATP is generated. The intact plasma membrane is considered to be impermeable to PEP under normal conditions, but it has been suggested that a change in membrane integrity may occur during ischemia.[171,172] Furthermore, a small amount of exogenous ATP may transitorily alter the plasma membrane permeability, making it permeable to otherwise nonpermeating substances.[173,174] The combined use of PEP and ATP during ischemia and in the reperfusion phase was reported to have beneficial effects,[175,176] along with higher content of pyruvate at the end of ischemia, indicating that PEP was indeed metabolized to pyruvate and to some extent, also to lactate.[177,178]

However, the protective effects have not been duplicated in other studies.[177,179] It is possible that the altered plasma membrane integrity that enables PEP to enter the cell may also lead to transitory leakage of intracellular constituents such as nucleotides. Our laboratory tested PEP in both cell culture and isolated heart preservation, but we did not find any signs that PEP could penetrate the cell membrane. We did not see a distinctive protective effect in our isolated heart preservation study.

Pyruvate

Pyruvate becomes scarce in ischemic cardiac muscle because glycolysis is inhibited not only by NADH but also by excess protons and lactate. Thus, exogenous pyruvate may help to regenerate NAD$^+$. Infusion of pyruvate was reported to reduce NADH levels and exert a protective effect in heart and liver preservation.[180,181]

AMP Precursors

With the rapid cessation of oxidative phosphorylation during ischemia, not only ATP but also ADP and AMP levels decrease quickly.[18,182] This decrease is the result of the further breakdown of AMP to adenosine, inosine, hypoxanthine, adenine and purines. Hypoxia in the heart results in a marked increase in the above nucleosides and purines and the amounts of these released products are inversely proportional to the oxygen tension of the perfusion fluid.[183,184] The major difference between AMP and the nucleosides and purines is that these latter substances can easily be transported through the cell membrane and lost in the preservation solution or perfusate,[183,185] thereby being unavailable for the nucleotide salvage pathways.[186,187] The cell is now dependent on very slow and energy-consuming de novo synthesis.[188]

Adding AMP precursors such as adenine, ribose, hypoxanthine and inosine has been reported to improve organ function.[183,184] Ribose, a pentose sugar, can be phosphorylated directly to ribose -5-phosphate and then to phosphoribosyl pyrophosphate (PRPP). It can augment both the de novo and salvage pathways.[189,190] The combined use of ribose and adenine has been shown to shorten ATP resynthesis time dramatically in several reports.[190,191] Using temporary LAD ligation, Zimmer and Ibel[192] showed that continuous intravenous infusion of ribose during recovery from a 15-minute period of myocardial ischemia in rats led to restoration of the cardiac ATP pool within 12 hours, whereas 72 hours were needed for ATP normalization without any intervention. Using a chronic dog model, St. Cyr et al[189,191] reported that ATP levels continued to fall during the first 4 hours of reperfusion and the eventual recovery from global ischemic insult was prolonged. The mean recovery rate for ATP levels was about 0.04 nmoles/mg wet wt/day and the projected complete recovery time was 9.9 days (9-14 days). Infusion of adenine and ribose enhanced ATP recovery rate to 2.44 nmoles/mg wet wt/day and recovery was virtually complete (96%) by 24 hours. Similar results were also shown by Mauser et al[183] and Boldwin et al[193] However, mixed results were also reported by others,[190,192] and limited experience in our laboratory has not confirmed any significant protective effect. This may be because the synthesis of the purine ring, in addition to ribose, is also partly rate-determining for restoration of adenine nucleotide.[194]

Insulin and Dichloroacetate (DCA)

During tissue ischemia, the final product of glycolysis is not pyruvate, but lactate. Accumulation of a higher level of lactate has detrimental effects on ionic homeostasis. It inhibits glycolysis itself,[195,196] increases free radical production, induces intracellular acidosis and stimulates calcium release from intracellular stores.[197] This can partly explain the supposition that glycolysis is deleterious to the ischemic heart[57,198] and is another reason why continuous organ perfusion offers better preservation.[72,199]

Insulin and DCA have been used to reduce lactate accumulation because of their ability to stimulate pyruvate dehydrogenase (PDH). The PDH system is a multi-enzyme complex located within the mitochondrion. Three enzymes in the complex catalyze the stepwise conversion of pyruvate to acetyl-CoA and carbon dioxide, with the first step being the irreversible decarboxylation of pyruvate. PDH is regulated by substrate activation, end product inhibition and reversible phosphorylation. In the presence of Mg^{++} and ATP, pyruvate dehydrogenase kinase phosphorylates and inactivates the dehydrogenase, while PDH phosphatase reverses this inhibition. Insulin stimulates PDH indirectly by activating the phosphatase, thus increasing the dephosphorylated, catalytically active form of PDH.[200,201] Insulin treatment decreases extracellular lactate release and improves the preservation of ATP.[202] DCA, however, exerts similar effects by inhibiting the kinase.[203,204] The protective results have been reported in several organs,[205,206] but negative effects have also been reported.[207]

Fructose-1,6-Bisphosphate (FBP)

Of all the glycolytic intermediates, fructose-1,6-bisphosphate (FBP) has received substantial attention for organ preservation and other ischemic conditions. The glycolytic process (Embden-Meyerhof pathway) can be divided into two important phases. In the first phase, glucose is converted to FBP. Two of the enzymes catalyzing the reactions in this phase, hexokinase and phosphofructokinase (PFK), consume 2 ATP molecules. In the second phase, two molecules of glyceraldehyde 3-phosphate are ultimately metabolized to pyruvate and lactate with the production of 4 ATPs per glucose. Thus, the net production of one glucose molecule is 2 ATPs from glycolysis. If FBP were to enter the cell and be used as a substitute for glucose, it enters the glycolytic pathway beyond the two ATP-consuming steps and the two ATP molecules are saved.[208] This results in double the ATP production in glycolysis. FBP has been used in various ischemic conditions, including hypothermic preservation of the liver,[209] intestine,[210] skin flap,[211] isolated heart perfusion,[212,213] neuroprotection,[214] circulatory arrest[215] and sepsis.[216] Increased ATP production has also been reported by many investigators after using FBP.[217,218]

The conventional view portrayed in biochemistry textbooks is that phosphorylated sugars such as FBP cannot cross the cell membrane. Although there have been some studies suggesting that FBP can be metabolized,[218,219] there is no direct evidence that FBP can actually cross the cell membrane.

Our laboratory has studied the potential protective effect of FBP in cardiomyocytes and isolated heart preservation and beneficial effects were observed.[220-222] We were especially interested in its possibility of enhancing glycolytic energy production. The following study was carried out to determine, in the absence of glucose, whether FBP could improve myocardial high-energy phosphate metabolism during hypothermic heart preservation. Our results indicated that adding FBP to St. Thomas solution did, in fact, improve high-energy phosphate reserves during hypothermic rabbit heart preservation.

Forty-two adult New Zealand White rabbits were used. Ten were used for normal controls (normal group), in which the hearts were biopsied fresh (without preservation) to obtain normal values of high-energy phosphate. Thirty-two rabbits were divided into two preservation groups for high-energy phosphate study. The animals were anesthetized with an intramuscular injection of ketamine (50 mg/kg), xylazine (5 mg/kg) and chlorpromazine (0.5 mg/kg). The hearts were removed and preserved in cooled St. Thomas solution. In the study group (n = 15), FBP (5 mM) was added to St. Thomas solution. In the control group (n = 17), fructose (5 mM) was added instead. The hearts were preserved at approximately 4°C in a temperature-controlled refrigerator (Model 252, Sanyo Electric Co., Japan) for 18 hours.

After preservation, myocardial samples were taken from the left ventricle (LV), right ventricle (RV), left atrium (LA) and right atrium (RA) separately and high-energy phosphate and its metabolites were measured using HPLC on a Waters 2690 HPLC system with a Waters 996 photodiode array detector and Waters Millenium 32 manager software.

After 18 hours of hypothermic preservation, LV myocardial ATP concentrations declined in both the study and control groups as compared with normal hearts. The LV ATP content in the study group (1.45 ± 0.11 μmol/g wet weight) was about one third of the LV ATP in the normal group (4.42 ± 0.36 μmol/g WW, $p < 0.05$). However, LV ATP content in the control group (0.65 ± 0.12 μmol/g WW) was only 14% of the LV ATP in the normal group ($p < 0.05$, Fig. 2). ATP contents in the RV had a similar pattern ($p < 0.05$ study vs control). ATP contents in the atria also had similar patterns, but the difference between the study and control groups was not as statistically significant as in the ventricles. ADP content was also decreased, but the reduction was not as dramatic as the ATP. LV ADP content was still about 80% of normal in the study group, whereas it was only about 54% of normal in the control group ($p < 0.05$, study vs control, Fig. 3). ADP content in the RV had a similar change. AMP increased in the hearts after 18-hour preservation. In the control group, AMP levels in the LV increased to 156% of normal. It was also higher than that found in the study group ($p < 0.05$, Fig. 4). AMP concentration in the RV of the control group was nearly twice as high as that in the normal and study groups ($p < 0.05$). Total energy (TE = ATP + ADP + AMP) decreased in both preservation groups. In the study group, total energy (TE) in the LV was 58% of that in the normal group, but was still higher than that in the control group, despite the fact that the control group had a very high AMP level ($p < 0.05$, Fig. 5). The change in energy charge [EC = (ATP + 1/2ADP)/(ATP + ADP + AMP)] exhibited similar patterns as that of ATP. The energy charge (EC) in the LV (0.58 ± 0.03) and RV (0.65 ± 0.03) in the study group

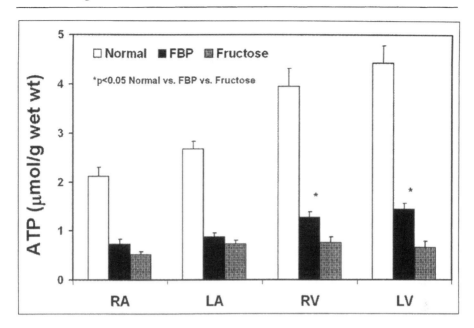

Figure 2. Tissue adenosine triphosphate (ATP) contents after 18-hour preservation in the study (FBP) and control (fructose) groups compared with the normal hearts. In the study group, approximately 33% of normal ATP remained after preservation in each heart chamber, whereas only 15% of ATP remained in the control group.

was higher than those in the control group (0.37 ± 0.04 for LV and 0.43 ± 0.03 for RV, $p < 0.05$, Fig. 6). The ratio of ATP/ADP in the study group was 40% of that in the normal group, but the ATP/ADP ratio in the control group was about 25% of that in the normal group ($p < 0.05$).

When the profile of high-energy phosphates was studied, adenine nucleotides in the four heart chambers in the study group maintained a similar pattern as in the normal hearts. ATP contents increased stepwise from RA, LA, RV, to LV. ATP was higher than ADP, which, in turn, was higher than AMP. However, in the control group, the high-energy phosphate profile was different: the content of ATP was similar in all chambers, but the ratio of ATP:ADP:AMP was reversed (Fig. 7).

In both preservation groups, the trivial amount of adenosine, xanthine and uric acid was nearly immeasurable in our HPLC analyses. However, IMP, inosine and hypoxanthine levels were increased in all four heart chambers. This increase was especially prominent in the control group.

This study clearly indicated that adding FBP to St. Thomas solution attenuated the decrease of high-energy phosphate reserves during hypothermic rabbit heart preservation.

Our results showed that after 18 hours of preservation, the LV still maintained about 30% of normal ATP levels and about 80% of normal EC in hearts treated with FBP. What do these values tell us? Although no causal relationship has been demonstrated, studies of myocardial ischemia have shown that when there is 80% depletion of myocardial ATP during the ischemic period, irreversible myocardial injury does not occur and adequate reperfusion should result in the eventual return of normal structure and function.[55,223] When ATP levels are depleted more than 80% (between 20% and 10% of control), disturbances in cellular biochemical function and homeostatic mechanisms occur. Greater than 90% depletion of ATP levels (below 10% of control) is strongly correlated with irreversible myocardial injury.[55,224] In vivo and in vitro heart studies have also indicated that an EC above 0.60 is generally associated with reversible myocardial injury, but an EC below 0.30 is associated with irreversible injury.[55] If these assumptions hold true in rabbit hearts, it may imply that when FBP is added for hypothermic rabbit heart preservation for 18 hours

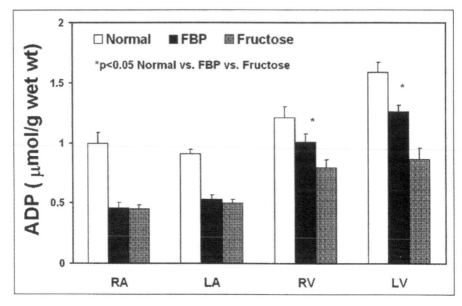

Figure 3. Comparison of tissue adenosine diphosphate (ADP) contents in the study (FBP) and control (fructose) groups compared with normal hearts. There was approximately 80% of normal ADP content in the ventricle in the study group and approximately 60% of normal ADP content in the control group.

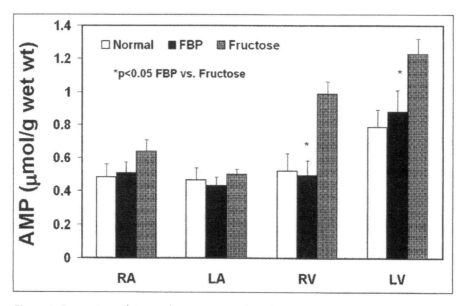

Figure 4. Comparison of tissue adenosine monophosphate (AMP) contents in heart chambers among the 3 groups. Note the higher tissue contents of AMP in the ventricle in the control (fructose) group.

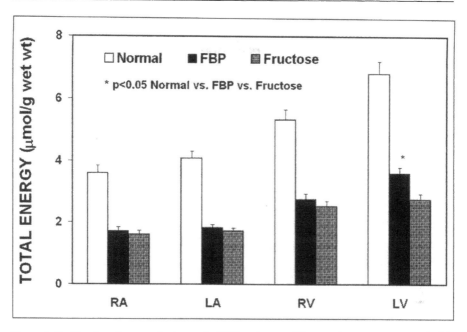

Figure 5. Profile of total energy in the study (FBP) and control (fructose) groups compared with the normal hearts. About 50% of total energy is present in the study group.

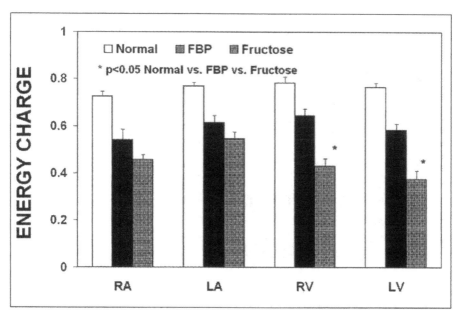

Figure 6. Profile of energy charge among the 3 groups. Approximately 80% of energy charge is still present in the study (FBP) group.

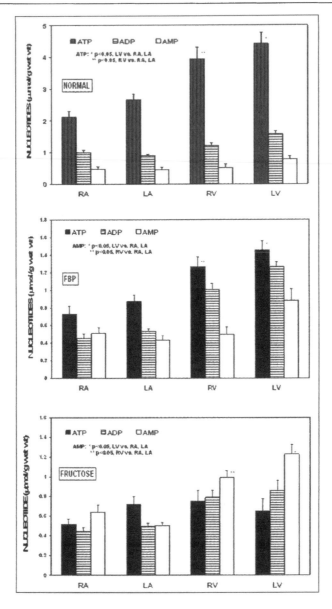

Figure 7. Comparison of adenine nucleotide profile among the 3 groups. After 18-hour preservation, the profile of adenine nucleotides in each heart chamber in the study (FBP) group
is similar to that in normal hearts, but this profile is different or even reversed in the fructose
group.

under hypothermia, although stunned, the myocardium may recover after proper reperfusion.
This assumption is overly simplified, however. Long-term ischemia studies have shown that ATP
levels continue to fall during the first 4 hours of reperfusion after global ischemia and a complete
recovery may take as long as 10 days.[189]

FBP is believed to have multifaceted beneficial effects, including augmentation of high-energy phosphate production during hypoxia,[225,226] prevention of ischemia/reperfusion-induced leukocyte adherence,[227] reduction of lipid peroxidation and oxygen free radical production,[228] stimulation of nitric oxide synthase,[229] increased red blood cell deformability and decreased blood viscosity,[230] chelation of extracellular calcium[231] and prevention of platelet activation.[232] Many of the proposed beneficial effects of FBP can only be achieved in vivo, in the presence of blood, or during re-oxygenation. Our results were obtained in isolated hearts without the presence of blood or reperfusion. Thus, many of these effects can be excluded from consideration in our rabbit study. Augmentation of glycolytic energy production by FBP has been the most elusive and potentially significant mechanism; it remains the most logical explanation for our results.

Although the true mechanism by which FBP enhances glycolytic energy production is not clear, at least four possibilities have been proposed: (1) direct participation of FBP in glycolysis,[218] (2) indirect participation in glycolysis,[233] (3) enhancement of glycolysis by activating glycolytic enzymes directly[234] and (4) activation of glycolytic enzymes indirectly.[235] Except for the first mechanism, the three other possible mechanisms can occur only in a solution containing glucose such as Euro-Collins solution or GIK solution. The St. Thomas solution used in this experiment did not contain glucose and any residual extracellular glucose would have been consumed shortly after the heart was removed. The enhancement of glycolysis induced by FBP would have to occur by the direct participation of FBP in glycolysis and this would require FBP to cross the cell membrane. However, the conventional view is that phosphorylated sugars, such as FBP, do not cross cellular membranes easily. In the past, a few studies have examined FBP movement through the cell membrane, but these studies provided little direct evidence to support this hypothesis. Results from our laboratories have provided some direct evidence that FBP does cross cell membrane.[236,237] Several possible mechanisms have been proposed recently to explain this special feature of FBP. One mechanism is that FBP affects membrane stability, which allows normally nonpermeant FBP to passively diffuse across the membrane bilayer.[236-238] This possibility was not explored in the past, but results from our laboratories have indicated that FBP can passively diffuse through artificial membrane bilayers and cellular membranes.[236] Although the amount of FBP that can passively move through the membrane bilayers is small, this amount is still higher than normal FBP in the cytosol. Since the concentration of free FBP in the cytosol is only 1.2 μmol/L, an infusion of FBP in the millimolar concentration range results in a more than 1000-fold gradient across the plasma membrane.[219] Another mechanism is that FBP might be transported by a dicarboxylate transport system, because its utilization continues to increase at the highest concentration and its transport is inhibited by the dicarboxylate fumarate.[239] Our more recent studies have also indicated that in isolated cardiomyocytes at 21°C, FBP uptake exceeded the uptake of L-glucose by several folds. This movement appeared to be via at least two distinct protein-dependent processes.[222] Because of its stability, ease of use and low cost, FBP may be a valuable metabolic intermediate for use as an adjacent chemical to improve organ function. More research should clarify some of the possible mechanisms.

Direct Supplement of ATP

Providing high-energy phosphate directly to the ischemic tissue is certainly an attractive idea. In fact, this has been an active project for more than a half-century. Direct infusion of free ATP or other high-energy phosphates would be a simple solution if it could be done. ATP entry into muscle cells was suggested in the 1944 study of Buchthal et al,[240] but later work showed that ATP did not enter these cells.[241] Because the idea is so attractive, this finding did not discourage others from trying again. Indeed, administration of either free ATP or CP has been reported to improve tissue protection during ischemia in various models and species, such as in the heart,[242] kidney,[243] liver,[244] brain[245] and also in various forms of shock.[246,247] The theory is that since a significant amount of ATP is released into the extracellular space by several types of cells (myocytes, cardiomyocytes, brain cells), ATP can cross cell membranes.[248,249]

However, this theory has not gained wide acceptance in the scientific community; experimental and clinical studies have not duplicated the many promising results. As described below, strongly charged molecules like ATP normally cannot pass the cell membrane without a transporting mechanism.[242,250] Besides, the half-life of ATP in the blood is less than 40 seconds, which poses technical difficulties for maintaining the ATP supply.[251]

All cells are surrounded by a plasma membrane that is composed of lipids, proteins and carbohydrates. The basic structure of the plasma membrane is the phospholipid bilayer, which is impermeable to most water-soluble molecules and is responsible for the basic function of membranes as barriers between two aqueous compartments. The plasma membranes of animal cells contain four major phospholipids (phosphatidylcholine, phosphatidylethanolamine, phosphatidylserine, sphingomyelin). These phospholipids are asymmetrically distributed between the halves of the membrane bilayer. The internal composition of the cell is maintained because the plasma membrane is selectively permeable to small molecules. Larger, especially charged molecules, for which no specific transport mechanisms exist, cannot cross cell membrane under normal conditions. Specific transport proteins (carrier proteins and channel proteins) are needed for the selective passage of these molecules across the membrane (Fig. 8). Unfortunately, ATP and many energy-rich glycolytic intermediates belong to this category.[252] Although the presence of equimolar $MgCl_2$ reduces the negative charge of ATP from 3 to 1, this does not alter the permeability of ATP because no known transporting mechanism exists in the cell membrane. Results from our laboratory have indicated that a small amount of free ATP can be taken up by the cells, most likely through some pores (about 0.8 nm in diameter) in the cell membrane.[65] The question is whether this small amount can support tissue metabolism.

For the past half century, using various carriers to deliver impermeant substances into the cytoplasm of cells has been a major research interest for cell biologists and clinicians. The ideal system

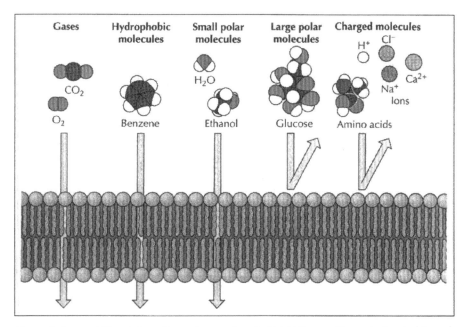

Figure 8. Permeability of the cell membrane phospholipid bilayers. Gases, hydrophobic molecules and small polar uncharged molecules can diffuse through phospholipid bilayers. Larger polar molecules and charged molecules such as ATP cannot. Reprinted with permission from Cooper GM, ed. The Cell. A Molecular Approach. Washington, DC: ASM Press, 1997:478.

for such delivery would couple both specificity and the ability to easily target multiple cells. Five primary approaches have been used to bypass the plasma membrane barrier: permeabilization, microinjection, polymer delivery, liposomal encapsulation and microemulsion. Permeabilization, which allows entry of materials found in the external medium, has been attained by electroporation and by pore-forming proteins or detergents.[253,254] However, the lack of specificity and difficulty in controlling this technique limit its application predominantly to research. Furthermore, as soon as nonspecific pores are formed in the cell membranes by permeabilization, many unwanted molecules can also flood in, loading the cells with toxic chemicals and losing important membrane ion gradients. While micro-injection allows a more targeted delivery than the permeabilization technique,[255,256] this is a time-consuming and technically challenging procedure. This procedure, therefore, has limited clinical application. The recent development of biodegradable polymers may offer a new approach for this purpose,[257,258] but at this time, the main interest is focused on the sustained release of some drugs because polymers are too large to penetrate the cell membrane.[259,260] Microemulsion can produce much smaller oil droplets, making efficient oil-in-water delivery of drugs.[261,262] However, ATP is highly hydrophilic and does not dissolve in oil.

One promising approach is the loading of medication into liposomes, or multilamellar vesicles, which are microscopic sacs made of the very phospholipids that constitute cell membranes. When a high enough concentration of phospholipids is mixed with water, the hydrophobic tails spontaneously herd together to exclude water. In contrast, the hydrophilic heads bind to water. The result is a bilayer in which the fatty acid tails point into the membrane's interior and the polar head groups point outward facing the water. As a liposome forms, any water-soluble molecules that have been added to the water are incorporated into the aqueous spaces in the interior of the spheres, whereas any lipid-soluble molecules added to the solvent during vesicle formation are incorporated into the lipid bilayer.[263] Liposomes can be filled with a variety of medications and, because of their similarity to cell membranes, are not toxic. They also protect their loads from being diluted or degraded in the blood. As a result, when the liposomes reach diseased tissues, they deliver concentrated doses of medication. Individual liposomes were shown to be capable of carrying tens of thousands of drug molecules, making them an efficient carrier for delivery of certain drugs.[264] However, three major problems have limited the widespread use of liposomes for cytosol drug delivery: (1) liposomes are not readily fusogenic, mainly because the stored energy of the vesicles radius of curvature is minimal and the internal layers may inhibit fusion.[265,266] (2) Because of their size, most intravenously infused liposomes are unable to leave the general circulation, except in areas where vessels become leaky (such as inflammation).[264] (3) The body's immune system recognizes the liposomes and removes them from circulation, regardless of the vesicles' composition and size. The liver, spleen and bone marrow take up nearly all liposomes given intravenously, preventing liposomes from circulating long enough to reach many targeted cells and tissues efficiently.[264] To circumvent these problems, various techniques have been developed such as coating liposomes with polyethylene glycol,[267] polyvinyl alcohol,[268] or other polymers[269] to make them "stealth" and combining liposomes with hemagglutinating virus of Japan (HVJ) to enhance fusion.[270]

Our laboratory has focused on specially formulated, highly fusogenic and very small unilamellar lipid vesicles (with diameters around 120-160 nm) to encapsulate Mg-ATP. Lipid vesicle fusion (and the delivery of ATP) occurs primarily by one of four methods: (1) increasing electrostatic interactions; (2) destabilizing the membrane bilayer; (3) increasing nonbilayer phases; or (4) creating dissimilar lipid phases. By changing the phospholipid composition through the use of one or all of these methods, a highly fusogenic lipid vesicle can be created: altering the charge of the phospholipid head group, increasing mean molecular area of the lipids, creating dissimilar regions of lipids and increasing the kinetic energy of the lipid vesicles. The fusogenicity of vesicles is strongly influenced by the dynamic and structural properties of the lipid membrane. The existence of a heterogeneous lipid-bilayer structure, which is composed of fluctuating gel and fluid domains that are prevailing in the temperature range of the main phase lipid membrane transition, leads to a strong increase in the fusion rate. Additionally, membrane curvature stress and the incorporation of various nonlamellar forming agents into the lipid membrane has a dramatic

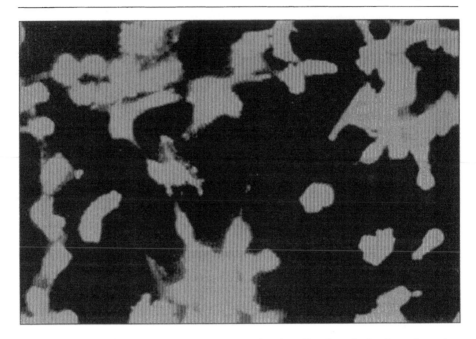

Figure 9. Endothelial uptake of fluorescein-encapsulated small unilamellar lipid vesicles within 5-10 minutes.

effect on the fusion properties.[271] Several formulations have been tested and the chemicals used to make ATP-vesicles included L-α-phosphatidylcholine (Soy PC), Mg-ATP, polyethylene glycol, chloroform and trehalose. Our unilamellar vesicles were found to be highly fusogenic. In our study of endothelial culture, it was clearly shown that these vesicles fused with the cells in 5-10 minutes, delivering water-soluble carboxyfluorescein (Fig. 9). In one rat study, primary cardiomyocytes were incubated in either (1) ATP-vesicles, (2) lipid vesicles only, (3) Mg-ATP (5 mM), or (4) culture medium. After equilibration, potassium cyanide (KCN, 5 mM) was added to the culture media for 30 minutes. Results indicated that cardiomyocytes after 30 minutes of chemical hypoxia had the highest ATP and TE contents in the group treated by ATP-vesicles. Neither vesicles alone nor ATP alone had a similar effect ($p < 0.05$, Fig. 10). The redox status of cardiomyocytes was decreased by the use of KCN. However, in the ATP-vesicles group, this decrease of redox status was significantly different from the other groups. The Alamar-Blue fluorescent intensity reading with ATP-vesicles was higher than with vesicles only, free Mg-ATP only, or with M199 (Fig. 11). Cardiomyocyte contractility was higher (Fig. 12).

During the process, we also have resolved several problems that were expected when highly fusogenic lipid vesicles are used. One problem was the stability of these vesicles. Although it has been proven that small unilamellar lipid vesicles are more stable in blood circulation,[272] it is possible that these vesicles may fuse with each other when left alone for a long period of time.[273] We have used a freeze-dry technique to keep these vesicles stable without particle size change during storage. Another problem was the temperature effect. All lipids have a phase transition temperature that will affect the fusion speed.[271,274] Our current vesicles are created well above the liquid-crystalline-to-gel phase transition, circumventing the problem of lipid phase separation. Efforts are being made to further improve the delivery efficiency. Further research work may prove to be fruitful.

Other Pharmacologic Interventions

Drugs that can provide a physical environment to enhance the hypothermic effect or provide chemical protection, either directly or indirectly, have also been used in organ preservation with

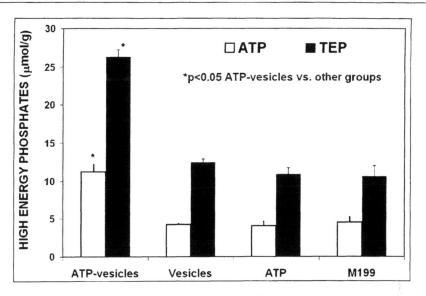

Figure 10. Comparison of ATP and total energy phosphate (TEP) in cardiomyocytes under chemical hypoxia indicates ATP-vesicles increase ATP and TEP by ~2 fold (p < 0.05). Rat cardiomyocytes were incubated with KCN (5 mM) for a period of 30 minutes. The washed cells were then analyzed by reverse-phase HPLC for nucleotide phosphates.

some success. So far, none of the drugs is dramatically effective in itself, but they have been reported as beneficial when used as adjuncts to enhance the quality of storage. Some of these drugs are listed below. Agents for specific organs, such as high potassium in cardioplegia and surfactant for lung preservation,[275] will not be discussed here.

Oxygen Free Radical Scavengers
These chemicals are covered in chapter by Suzuki et al.

Prostaglandins
Prostaglandin E_1 (PGE$_1$), prostacyclin (PGI$_2$) and their analogues have been used in hypothermic preservation. Recognized to have a pulmonary vasodilatory effect to relieve vasoconstriction, they allow uniform distribution of perfusate throughout the lungs. They may also be associated with a number of other beneficial effects, such as inhibition of platelet aggregation, thrombus formation, neutrophil sequestration and lysosomal release. They are used before pulmonary artery flushing and improved preservation in lung and heart-lung transplantation.[276,277] They may also have immunosuppressive and cytoprotective effects and may reduce vascular permeability.[164] Protective effect for hypothermic liver storage has also been reported.[278] However, the value and mechanism of prostaglandins in lung preservation are still controversial and its beneficial effect has been questioned.[166,279]

Calcium-Channel Blockers
Normally, the extracellular free calcium concentration is approximately 10^{-3} M, whereas the cytosolic free calcium concentration is about 10^{-7} M.[280] Calcium influx into cells after reperfusion has been implicated in ischemic damage in a variety of organs and ischemic myocardium accumulates large amounts of calcium during reperfusion.[280] Calcium also accumulates in hepatocytes and renal tubular cells that have been damaged by hypoxia and toxins.[82] Agents blocking calcium flux may prove to be beneficial in preventing lung injury after reperfusion.[166] Calcium-channel blockers like verapamil, nifedipine and nisoldipine have shown protective effect in hypothermic

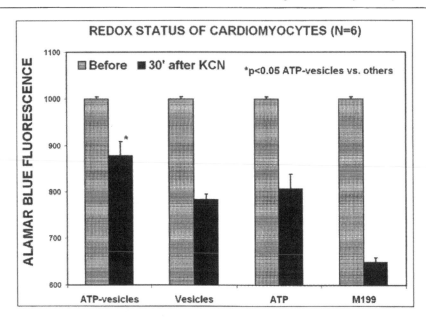

Figure 11. Redox status of cardiomyocytes under chemical hypoxia is increased by ATP delivery. Rat cardiomyocytes were incubated in either media or media + KCN (5 mM). After 30 minutes, the redox status of the cells was determined using Alamar Blue. ATP-vesicles maintained a higher redox status than lipid vesicles, Mg-ATP, or culture media. *Significantly different from vesicles, ATP, or M199 by ANOVA (p < 0.05), N = 6.

organ preservation.[281,282] They are also believed to protect the heart from oxygen and substrate deprivation by decreasing the contractile activity and the energy demand.[283,284]

Na⁺/H⁺ Exchanger Inhibitors

Na⁺/H⁺ exchanger (NHE) inhibitors have received much attention in recent years. As stated above, intracellular Ca^{++} overload during ischemia and reperfusion is an important pathophysiological factor contributing to reduced post-ischemic recovery of cellular function. This is especially important in the myocardium. Although NHE activation is essential for the restoration of physiological pH, hyperactivation of NHE leads to a dramatic increase in intracellular Na⁺ concentration, which subsequently causes a marked increase in intracellular Ca^{++} concentration through the Na⁺/Ca⁺⁺ exchanger.[285] To date, eight isoforms of NHE have been identified. NHE-1 is the most predominant isoform expressed in the heart. The use of NHE-1 inhibitors such as amiloride, eniporide, cariporide, HOE-694, EMD-85131 and many bicyclic guanidines, has been shown to provide significant protection in models of myocardial ischemia and reperfusion, with consistent improvement in functional recovery, metabolic status, attenuation of arrhythmias, preservation of cellular ultrastructure and inhibition of apoptosis.[286-288] However, the final role of these drugs has not been confirmed both in experimental studies or clinical trials.[287,289]

Platelet-Activating Factor (PAF) Antagonists

PAF binds to a specific site on platelets and is a potent inducer of platelet activation and aggregation. Activated platelets are capable of releasing potent mediators of inflammation.[290-292] Platelet-activating factor antagonists, such as WEB2056, BN52021, E-5880 and CV-3988, have been reported to have beneficial effects during tissue ischemia, lung injury and liver preservation.[293-295]

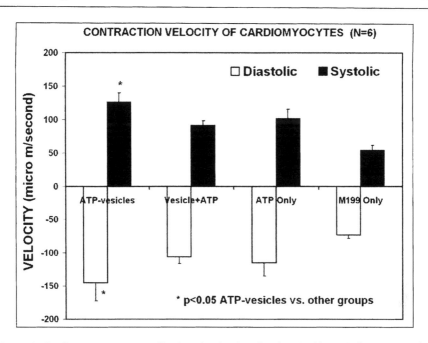

Figure 12. Cardiomyocyte contractility is maintained under chemical hypoxia by ATP-vesicles. The cells were stimulated with 0.5-4 Hz, 8-volt electric stimulator (MyoPacer™) and contractile velocity and duration of contraction were recorded after a 30-minute incubation in KCN (5 mM). Both diastolic and systolic contraction velocities were significantly increased by the presence of ATP-vesicles. In contrast, neither vesicles + non-encapsulated ATP, ATP only, or M199 media could maintain the same velocity of contraction compared to the ATP-vesicles group. *Significantly different from vesicles + ATP, ATP only, or M199 as measured by ANOVA (p < 0.05), N = 6.

Inhibition of Complement Activation

Complement activation has been implicated in the process of ischemia-reperfusion damage. Inhibition of this process may prove beneficial in organ preservation. Preliminary data using the complement inhibitors have shown promising results in lung and liver preservation,[296,297] but controversy still exists.[298]

Nitric Oxide (NO)

One of the most extraordinary discoveries in modern medicine has been the finding that nitric oxide (NO) release, which accounts for the biological activity of the so-called endothelium-derived relaxing factor.[299] With its widespread distribution in tissues and its ability to react with a range of molecules in the organism, a review appears incomplete without mentioning it. NO was Science magazine's "molecule of the year" in 1992 and its research has entered into almost all areas of biology and medicine, including organ preservation. NO donors such as nitrite, nitroglycerin, S-nitrosoglutathione, nitroprusside, or inhaled NO have been reported to be protective in hypothermic organ preservation, especially in lungs and liver.[300-302] However, NO appears to be a double-edged sword that can be protective or destructive on any tissue or organ under different conditions. It is not surprising to see the opposite effect of NO in tissues and organs.[303,304] It has become clear that, in some conditions, there may be too little NO, whereas in others there may be too much. "No one thing does only one thing" is probably more suitable for NO than for any other chemical. It is generally agreed that constitutively generated NO maintains microcirculation and endothelial integrity in many organs, while inducible NO synthase (iNOS)-governed NO production can be either beneficial or detrimental.[305] The result of NO is affected by many other factors, such as the type of isoforms, doses, administering routes, existence of other mediators and, especially, other oxidants. There are many excellent reviews of this mediator[306,307] and the use and controversy surrounding it will certainly continue.

Polyethylene Glycol (PEG)

The use of PEG in preservation solutions has been shown to be cytoprotective with improved organ function. It has also been shown to reduce rejection in clinical and experimental transplantations. These results have been shown in kidney,[308] heart[309] and liver.[310] Several possible mechanisms of PEG have been proposed such as immunosuppression, prevention of osmotic swelling, reduction of lipid peroxidation and improvement of tissue ATP content.[311,312] PEG is frequently incorporated into solutions for organ preservation.

Trimetazidine

Trimetazidine may be related to the restoration of ATP and CP, improvement of mitochondrial function, improved cellular resistance to hypoxic stress and antioxidant activity.[313,314] The use of trimetazidine has shown to be protective in kidney and heart preservations.[315,316]

Vasoactive Drugs

Propranolol can depress myocardial contractility and heart rate, thus exerting an energy-sparing effect on ischemic hearts. Propranolol can also suppress calcium accumulation during reperfusion.[317] Studies of heart preservation have shown positive protective results.[318] Other vasoactive drugs such as vasodilators[319,320] and furosemide[321,322] have also shown some effect.

Prevention of Cell Swelling

A. Mannitol

Mannitol is one of the most popularly used chemicals for reducing cell swelling. Mannitol has been shown to improve renal function when given before warm ischemia.[323] Elevation of osmolality by mannitol increases collateral blood flow to the ischemic region through at least two mechanisms: (1) increasing osmolality results in the dilation of large arterial conductance vessels; and (2) producing an effect on the coronary circulation at a microvascular level, which increases collateral blood flow to ischemic regions.[324]

B. High-Molecular Impermeables

Due to their high molecular weight, these impermeables are used to maintain fluid osmolarity. Both lactobionate and raffinose are important compounds in UW solution because of their large molecular mass.[185,325]

Membrane-Stabilizing Agents

Trehalose has the ability to stabilize the cell membrane structure by binding to phospholipid molecules in the bilayer. Trehalose is incorporated between the polar head groups of the phospholipids, thereby maintaining a specific distance between the molecules, which inhibits gelatinization and subsequent dysfunction of the bilayer under stressful conditions. It is one composition of a solution developed by Kyoto University in Japan.[326-328]

Steroids

Steroids are widely used as an adjunct to preservation, whether in the perfusate or by pretreatment of the donor and treatment of the recipient before reperfusion. The true mechanism of steroids is probably more complex, but their general anti-inflammatory and membrane-stabilizing properties provide a protective role during organ preservation.[115,329]

Inhibition of Apoptosis (Programmed Cell Death)

Cells are eliminated in a variety of physiological settings by apoptosis, a genetically encoded process of cellular suicide. Apoptosis occurs as a consequence of global organ ischemia during isolation and storage before transplantation and various anti-apoptotic compounds have shown beneficial effects.[330] Our group also tested a compound comprised mainly of phospholipids extracted from soybean in rabbit heart and rat liver preservations and obtained some very promising results.[331,332]

Other Metabolic Inhibitors

A. Chlorpromazine

The effect of chlorpromazine in protecting cells against ischemia is believed to be related to 2 mechanisms: (1) its pharmacological α-receptor blocking effect and its consequent vasodilating properties; [333] and (2) its membrane stabilization effect caused by inhibition of endogenous phospholipases, thus inhibiting the breakdown of the phospholipids.[86,121]

B. Butanedione monoxime (BDM)

Contracture of the myocardium is a serious event during ischemia and is thought to be related to depletion of ATP supplies.[334] Butanedione monoxime (BDM) causes a marked depression of myocardial contractility with few side effects.[335] It is also a vasodilator and can depress mobilization of sarcoplasmic reticular Ca^{++}. It has been shown to enhance hypothermic heart preservation.[334,336]

Preconditioning

The earliest study of preconditioning was probably performed more than 4 decades ago by Dahl and Balfour.[337] They found that the ischemic survival of rats was extended by placing them in nitrogen for a short period of time, letting them recover and then inducing anoxia. They also found that concentrations of lactate, pyruvate and ATP in the brain were higher during anoxia in the group subjected to pre-anoxia. When glycolysis was inhibited by iodoacetate, the pre-exposed advantage disappeared. Dahl and Balfour believed that the preconditioning was related to a faster production of ATP during anoxia because the first anoxia resulted in higher pyruvate and lactate reserves.

The phenomenon of preconditioning was described in 1986 by Murry et al,[338] who discovered the protective effect of a brief ischemic period with regard to the detrimental consequences of subsequent prolonged ischemia. Since then, preconditioning has been shown to limit infarct size and to reduce ventricular arrhythmias during sustained ischemia and reperfusion. Brief repetitive periods of ischemia have been shown to retard cardiac energy metabolism during sustained ischemia in various animal experiments. This inhibition leads to sparing of high-energy phosphates and improves the time-averaged energy state during ischemia.[339,340]

Other Important Factors in Organ Preservation

Equilibration

To obtain optimal protective effects, the vascular and extracellular spaces have to be equilibrated with preservation fluid. This process varies for different organs and both flushing time and the volume of flushing fluid are important. For the kidney, fluid should also fill the tubular space. Only by a complete equilibration of the whole extracellular space of the kidney will the desired protection principles of buffering the "intracellular-like" composition be realized. Blood should be cleared from the preserved organs because (1) the trapped erythrocytes become less flexible as their ATP reserves are consumed and they tend to block the microcirculation after revascularization, causing ischemic damage;[341] (2) the residual platelets can aggregate when temperature decreases;[342,343] and (3) blood remaining in the transplanted organ may have undesirable immunological consequences.[344] Cleaning the organs totally is almost impossible because washing out these blood cells can take as long as 10 hours.[121] The equilibration of the different compartments of the kidney is nearly accomplished after 10 to 12 minutes of perfusion with a hydrostatic pressure of 120 cm H_2O. The perfusion volume should be sufficient and that required by the kidney is at least 10 times the organ's weight, according to measurements of fluid substrates of the venous and tubular outflow.[345,346] The heart needs 2 to 3 times its weight and the liver, 6 times its weight.[346]

Buffering

Buffering of the compartments of an organ is important because H+ ions and lactate can pass the cell membrane in anaerobiosis, a process that is pH-dependent. Since anaerobic glycolysis is the only energy source during anaerobiosis, this process is important for structural integrity. This issue has not been studied sufficiently. Using a rat model, Shiraishi et al[347] demonstrated that lungs treated at pH 7.75 showed a significantly lower pulmonary artery pressure and wet-to-dry weight ratio than those treated at pH 7.26 or 7.96. Experimental data and clinical studies show that a pH value of 6.2 at 35°C or 6.6 at 5°C is marginally tolerated. Intrarenal acidosis beyond these values appears to be the first limiting factor for graft function in kidneys.[348,349]

Nature Still Provides the Best Preservation

While various attempts have been made to further extend cold preservation time for solid organs, it is disappointing to realize that decades of research studies have not extended preservation times for the heart and lungs, even for a few hours. It has been shown repeatedly and is also reflected in this book, that using perfusion can extend survival time. This is obvious because perfusion can supply oxygen and nutrients continuously and remove metabolic wastes—all are critical to organ survival. There are complicated issues regarding using the perfusion technique. For one thing, unlike single-flush hypothermic preservation that needs little attention during the preservation period, mechanical perfusion is labor intensive and expensive. It is difficult to gain acceptance in the managed care era. Perfusion itself can pose additional risks to the organs that are not present in hypothermic storage, such as damage to vascular endothelial cells, the possibility of mechanical failure and particle formation in the perfusates.[92,350] To overcome these problems, a simple heart-lung autoperfusion was developed that did provide longer preservation time for the heart, but not the lungs.[351-355] We tried this technique and found one possible reason for lung damage: the circulating aggregates from blood-air (and possible artificial circuit) interface.[356] We eliminated the blood-air interface and expanded the organ block to include the heart, lungs, liver, pancreas, duodenum and both kidneys. In that preparation, the heart pumps blood, the lungs oxygenate the blood, the liver processes normal chemical reactions and metabolic wastes are removed by the kidneys. The organs are self-contained and a respirator is the only artificial equipment needed for its survival. The organs could be preserved for more than 2 days with good function.[146,147,357] In one study, using 6 pairs of adult mongrel dogs, the organs were preserved from 24 to 33 hours (mean 26.7 ± 1.4 hours) in the multiple organ block along with intermittent infusion of plasma from hibernating animals. Orthotopic lung transplantation was performed and the dogs were observed for 24 hours with good lung function.[149] Figure 13 is an example of the lungs at 30 hours of preservation. These lungs show the best quality of any preservation technique we have seen.

Metabolic management during perfusion is not the main focus in this chapter because it requires totally different strategies. However, the above example indicates that physiologically, using an animal's own heart as a pump and its own blood as a perfusion medium, provides the simplest and possibly the best environment for long-term organ survival. The normothermic autoperfusion technique was used early in the history of organ preservation. Due to reasons similar to mechanical perfusion, it is probably difficult to gain acceptance in clinical practice. Because of the technical difficulties in the setup and the monitoring processes, even the simplest autoperfusion was used only sporadically.[292,358,359] We do not know the other possible mechanisms of the plasma from hibernating animals in this preparation, except its obvious role in reducing liver congestion.[146,147,357] The above example appears to show that using natural circulation is still the best option for long-term organ preservation.

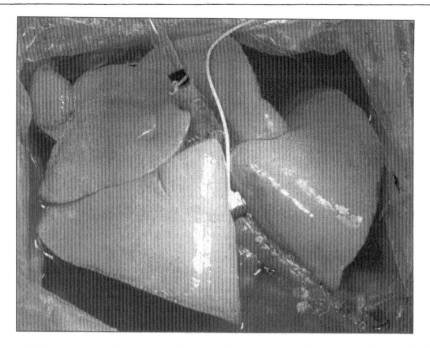

Figure 13. An example of the lungs at 30 hours of preservation. Not only do the lungs look normal, tissue wet/dry ratio was also normal.

Future Perspective

Over the past 40 years, slow progress has been made in organ preservation for transplantation. Despite the extension of preservation times for the kidneys and livers, safe preservation times for the heart and lungs are still very short.[360] In the past decade, there has been a gradual expansion of marginal donor organs used as a result of a more careful selection of solutions for individual organs,[361,362] but there has not been a dramatic extension in preservation times.[363] Longer ischemic times are often associated with early heart failure and chronic fibrosis after transplantation.[364-366] Many of the compounds mentioned in this chapter, albeit with great theoretical advantages and promising experimental results, have not made their way into clinical practice. This is mainly because sustained protective results are lacking in a larger number of animal experiments and/or clinical trials. Among the reasons for this is the period of organ ischemia outside the body, which is subject to a number of biochemical stress factors that cause a combined-damage effect that cannot be prevented or treated by a single drug. Another reason is that our knowledge of known critical components of tissue damage is still incomplete. For example, although the decline of energy production plays a major role in tissue ischemic damage and cell death is well recognized, the critical point at which cells die is far from settled. Evidence shows that it is very unlikely that there is a single critical level of ATP for all tissues and organs. Even in the same organ, this critical level may be different due to the existence of other factors. As such, it is not surprising to realize the controversy toward the relationship between high-energy phosphate concentration and organ function and viability. In the literature, both positive[54,367,368] and negative[72,81] relationships have been reported. There is also the possibility of ATP compartmentation in subcellular organelles. Our current techniques can only determine the total tissue content of a particular metabolite and may not provide a good estimate of its concentration in a specific subcellular compartment. If a large concentration gradient of any compound across the mitochondrial membrane is present, the interpretation will be difficult.[369] Furthermore, it is not the sole concentration of ATP that is critical to cell survival, but it is the rate of ATP turnover that is more important to the survival of the cells.[370] This will gradually change when our knowledge of biochemical and immunological mechanisms of organ damage improves.

Further improvement of preservation techniques is critical with the increased use of marginal and suboptimal donor organs and organs from nonheart-beating donors (NHBD). However, organ preservation is only one ring in the chain of organ damage, beginning before retrieval and ending at restoration of blood supply. Many of the organs from NHBD are severely damaged during the long period of hypotension before retrieval, making preservation more critical.

True long-term organ preservation for months or years would be a valuable future development. Such techniques would transform transplantation into elective surgery, probably eliminate recipient waiting lists, make the process of time-consuming matching of donors and recipients more possible and provide time for immunological pretreatment of the recipients or donor organs.[371] It is unlikely that the current hypothermic storage or mechanical perfusion techniques will achieve this goal. However, hypothermic storage is simple and the most cost-effective. A gradual extension of current hypothermic preservation time will be possible when there is more understanding of the specific metabolic demands and physiologic requirements of different cells and organs. This would include the nature of hypoxic damage to cells and organs and the methods that avoid or reduce various kinds of damage. It may require a combined treatment of several drug categories to offset injuries caused by different mechanisms,[60,372,373] and the treatment is probably organ-specific.[374]

Acknowledgements

This study was supported in part by NIH grants HL64186, DK74566 and AR52984. The author wishes to thank Drs. Dongping Hua, Benjamin Chiang and William Ehringer for their experimental contributions; Jiusheng Ye for his HPLC work and Margaret A. Abby for her editorial assistance.

References

1. Opie LH. Cardiac metabolism-emergence, decline and resurgence. Part I. Cardiovasc Res 1992; 26:721-733.
2. Erecinska M, Wilson DF. Regulation of cellular energy metabolism. J Membr Biol 1982; 70:1-14.
3. Lin HY, Fox IH, Spychala J. ATP degradation and depletion of adenine nucleotides. In: Zelenock GB, ed. Clinical Ischemic Syndromes. St. Louis: Mosby, 1990.
4. Whitman G, Kieval R, Wetstein L et al. The relationship between global myocardial ischemia, left ventricular function, myocardial redox state and high energy phosphate profile. A phosphorus-31 nuclear magnetic resonance study. J Surg Res 1983; 35:332-339.
5. Michiels C. Physiological and pathological responses to hypoxia. Am J Pathol 2004; 164:1875-1882.
6. Reimer KA, Jennings RB, Hill ML. Total ischemia in dog hearts, in vitro. 2. High energy phosphate depletion and associated defects in energy metabolism, cell volume regulation and sarcolemmal integrity. Circ Res 1981; 49:901-911.
7. Bretschneider HJ, Hubner G, Knoll D et al. Myocardial resistance and tolerance to ischemia: Physiological and biochemical basis. J Cardiovasc Surg 1975; 16:241-260.
8. Opie LH. The Heart. Physiology and Metabolism. New York: Raven Press, 1991:
9. Reimer KA, Lowe JE, Rasmussen MM et al. The wave front phenomenon of ischemic cell death. I myocardial infarct size vs duration of coronary occlusion in dogs. Circulation 1977; 56:786-794.
10. van Wyk J, Liem DS, Eiseman B. Function of cadaver liver. Surgery 1965; 58:120-130.
11. van Wyk J, Tait I, Eiseman B. Function of livers left for graded periods within the cadaver. Surgery 1965; 58:374-380.
12. Mela L. Mitochondrial function in shock, ischemia and hypoxia. In: Cowley RA, Trump BF, eds. Pathophysiology of Shock, Anoxia and Ischemia. Baltimore: Williams and Wilkins, 1982.
13. McLean RF, Wong BI. Normothermic versus hypothermic cardiopulmonary bypass: Central nervous system outcomes. J Cardiothorac Vasc Anesth 1996; 10:45-52.
14. Kammermeier H. High energy phosphate of the myocardium: Concentration versus free energy change. Bas Res Cardiol 1987; 82(Suppl 2):31-36.
15. The oxidation of glucose. Glycolysis, the TCA cycle and oxidative phosphorylation. In: Meisenberg G, Simmons WH, eds. Principles of Medical Biochemistry. St. Louis: Mosby, 1998.
16. Oxygen transporters: hemoglobin and myoglobin. In: Meisenberg G, Simmons WH, eds. Principles of Medical Biochemistry. St. Louis: Mosby, 1998.

17. Silver IA. Changes in pO$_2$ and ion fluxes in cerebral hypoxia-ischemia. In: Reivich M, Coburn R, Lahiri S, Chance B, eds. Tissue Hypoxia and Ischemia. New York: Plenum Press, 1977.

18. Levitsky S, Feinberg H. Biochemical changes of ischemia. Ann Thorac Surg 1975; 20:21-29.

19. Hems DA, Bronsnan JT. Effects of ischaemia on content of metabolites in rat liver and kidney in vivo. Biochem J 1970; 120:105-111.

20. Amino acid metabolism. In: Meisenberg G, Simmons WH, eds. Principles of Medical Biochemistry. St. Louis: Mosby, 1998.

21. Integration of metabolism. In: Meisenberg G, Simmons WH, eds. Principles of Medical Biochemistry. St. Louis: Mosby, 1998.

22. Wollenberger A. The energy metabolism of the failing heart and the metabolic action of the cardiac glycosides. Pharmacol Rev 1949; 1:311-352.

23. Cain DF, Davies RE. Breakdown of adenosine triphosphate during a single contraction of working muscle. Biochem Biophys Res Commun 1962; 8:361-366.

24. Braasch W, Gudbjarnason S, Puri PS et al. Early changes in energy metabolism in the myocardium following acute coronary artery occlusion in anesthetized dogs. Circ Res 1968; 23:429-438.

25. Olson RE, Dhalla NS, Sun CN. Changes in energy stores in the hypoxic heart. Cardiology 1971; 56:114-124.

26. Scholz W, Albus U. Na$^+$/H$^+$ exchange and its inhibition in cardiac ischemia and reperfusion. Bas Res Cardiol 1993; 88:443-455.

27. Kahng MW. Intermediary metabolism. In: Cowley RA, Trump BF, eds. Pathophysiology of shock, anoxia and ischemia. Baltimore: Williams and Wilkins, 1982:

28. Podrazik RM, D'Alecy LG, Zelenock GB. Organ-specific physiology: A comparative review. In: Zelenock GB, ed. Clinical Ischemic Syndromes. St. Louis: Mosby, 1990.

29. Opie LH. The heart. Physiology and Metabolism. New York: Raven Press, 1991.

30. Taegtmeyer H. Energy metabolism of the heart: From basic concepts to clinical applications. Curr Probl Cardiol 1994; 19:62-113.

31. Jennings RB, Reimer KA, Steenbergen C. Complete global myocardial ischemia in dogs. Crit Care Med 1988; 16:988-996.

32. Lopaschuk GD. Treating ischemic heart disease by pharmacologically improving cardiac energy metabolism. Am J Cardiol 1998; 82:14K-17K.

33. Paul RJ. Functional compartmentalization of oxidative and glycolytic metabolism in vascular smooth muscle. Am J Physiol 1983; 244:C399-C409.

34. Lipasti JA, Nevalainen TJ, Alanen KA et al. Anaerobic glycolysis and the development of ischemic contracture in isolated rat heart. Cardiovasc Res 1984; 18:145-148.

35. Geisbuhler T, Altschuld RA, Trewyn RW et al. Adenine nucleotide metabolism and compartmentalization in isolated adult rat heart cells. Circ Res 1984; 54:536-546.

36. Jennings RB, Reimer KA. Lethal myocardial ischemic injury. Am J Pathol 1981; 102:241-255.

37. The cell surface. In: Cooper JM, ed. The Cell, A Molecular Approach. Washington, DC: ASM Press, 1997.

38. Weiss J, Hiltbrand B. Functional compartmentation of glycolytic versus oxidative metabolism in isolated rabbit heart. J Clin Invest 1985; 75:436-447.

39. Weiss JN, Lamp ST. Glycolysis preferentially inhibits ATP-sensitive K$^+$ channels in isolated guinea pig cardiac myocytes. Science 1987; 238:67-69.

40. Kristensen SR. Mechanisms of cell damage and enzyme release. Dan Med Bull 1994; 41:423-433.

41. Osbakken M, Ivanics T, Zhang D et al. Isolated cardiomyocytes in conjunction with NMR spectroscopy techniques to study metabolism and ion flux. J Biol Chem 1992; 267:15340-15347.

42. Higgins TJC, Bailey PJ. The effects of cyanide and iodoacetate intoxication and ischemia on enzyme release from the perfused rat heart. Biomed Biochim Acta 1983; 762:67-75.

43. Kristensen SR. A clinical appraisal of the association between energy charge and cell damage. Biomed Biochim Acta 1989; 1012:272-278.

44. Dora E. NAD pools in the brain cortex effect of reversible anoxic-anoxia and irreversible anoxic-ischemia. Adv Exp Med Biol 1984; 180:131-139.

45. Dora E, Tanaka K, Greenberg JH et al. Kinetics of microcirculatory, NAD/NADH and electrocorticographic changes in cat brain cortex during ischemia and recirculation. Ann Neurol 1986; 19:536-544.

46. Ozawa K. Energy metabolism. In: Cowley RA, Trump BF, eds. Pathophysiology of Shock, Anoxia and Ischemia. Baltimore: Williams and Wilkins, 1982.

47. Kobayashi K, Neely JR. Effects of ischemia and reperfusion on pyruvate dehydrogenase activity in isolated rat hearts. J Mol Cell Cardiol 1983; 15:359-367.

48. Carbohydrate metabolism. In: Meisenberg G, Simmons WH, eds. Principles of Medical Biochemistry. St. Louis: Mosby, 1998.

49. Mayes PA. Glycolysis and oxidation of pyruvate. In: Murray RK, Granner DK, Mayes PA, Rodwell VW, eds. Harper's Biochemistry. Norwalk: Appleton and Lange, 1990.
50. Hung VC, Lee JY, Zitelli JA et al. Topical tretinoin and epithelial wound healing. Arch Dermatol 1989; 125:65-69.
51. Smith RAJ, Porteous CM, Coulter CV et al. Selective targeting of an antioxidant to mitochondria. Eur J Biochem 1999; 263:706-716.
52. Coulter CV, Kelso GF, Lin TK et al. Mitochondrially targeted antioxidants and thiol reagents. Free Radic Biol Med 2000; 28:1547-1554.
53. Tennant R, Wiggers CJ. The effect of coronary occlusion on myocardial contraction. Am J Physiol 1935; 112:351-361.
54. Neely JR, Rovetto MJ, Whitmer JT et al. Effects of ischemia on function and metabolism of the isolated working rat heart. Am J Physiol 1973; 225:651-658.
55. Jennings RB, Hawkins HK, Lowe JE et al. Relation between high energy phosphate and lethal injury in myocardial ischemia in the dog. Am J Pathol 1978; 92:187-207.
56. Armiger LC, Gavin JB, Herdson PB. Mitochondrial changes in dog myocardium induced by neutral lactate in vitro. Lab Invest 1974; 31:29-33.
57. Neely JR, Grotyohann LW. Role of glycolytic products in damage to ischemic myocardium. Dissociation of adenosine triphosphate levels and recovery of function of reperfused ischemic hearts. Circ Res 1984; 55:816-824.
58. Gladden LB. Lactate metabolism: A new paradigm for the third millennium. J Physiol 2004; 558:5-30.
59. Dennis SC, Opie LH. Protons in ischemia: Where do they come from; Where do they go to? J Mol Cell Cardiol 1991; 23:1077-1086.
60. Opie LH. Myocardial ischemia-metabolic pathways and implications of increased glycolysis. Cardiovasc Drugs Ther 1990; 4:777-790.
61. Opie LH. Effects of regional ischemia on metabolism of glucose and fatty acids. Relative rates of aerobic and anaerobic energy production during myocardial infarction and comparison with effects of anoxia. Circ Res 1976; 38(Suppl I):52-74.
62. Levine RL. Ischemia: from acidosis to oxidation. FASEB J 1993; 7:1242-1246.
63. Usui M, Sakata H, Ishii S. Effect of fluorocarbon perfusion upon the preservation of amputated limbs. An experimental study. J Bone Joint Surg Br 1985; 67:473-477.
64. Haussinger D. The mutual interaction between cell volume and cell function: A new principle of metabolic regulation. Biochem Cell Biol 1991; 69:1-4.
65. Guyton AC. Physiology of the Human Body. Philadelphia: Saunders, 1979.
66. Flores J, DiBona DR, Frega N et al. Cell volume regulation and ischemic tissue damage. J Membr Biol 1972; 10:331-343.
67. MacKnight ADC, Leaf A. Regulation of cellular volume. Physiol Rev 1977; 57:510-573.
68. Downes G, Hoffman R, Huang J et al. Mechanism of action of washout solutions for kidney preservation. Transplantation 1973; 16:46-53.
69. Pegg DE. Principles of tissue preservation. In: Morris PJ, Tilney NL, eds. Progress in Transplantation. London: Churchill Livingstone, 1985.
70. Tranum Jensen J, Janse MJ, Fiolet JWT et al. Tissue osmolality, cell swelling and reperfusion in acute regional myocardial ischemia in the isolated porcine heart. Circ Res 1981; 49:361-381.
71. Walker WF, Macdonald JS, Pickard C. Hepatic vein sphincter mechanism in the dog. Br J Surg 1960; 48:218-220.
72. Southard JH, Belzer FO. Organ preservation. In: Flye MW, ed. Principles of Organ Transplantation. W.B. Saunders Co., Philadelphia 1989.
73. Ames A, III, Wright RL, Kowada M et al. Cerebral ischemia. II. The no-reflow phenomenon. Am J Pathol 1968; 52:437-453.
74. Grundmann R. Fundamentals of preservation methods. In: Toledo-Pereyra LH, ed. Basic Concepts of Organ Procurement, Perfusion and Preservation for Transplantation. New York: Academic Press, 1982.
75. Enerson DM. Cellular swelling: I. hypothermia graded hypoxia and the osmotic effects of low molecular weight dextran on isolated tissues. Ann Surg 1966; 163:169-174.
76. Batty PR, Hicks LG, DeWeese JA et al. Optimal osmolality for cold storage of the cardiac explant. J Surg Res 1990; 48:601-605.
77. Grum CM. Cellular energetics. In: Zelenock GB, ed. Clinical Ischemic Syndromes. St. Louis: Mosby, 1990.
78. Belzer FO, Hoffman R, Huang J et al. Endothelial damage in perfused dog kidney and cold sensitivity of vascular Na-K-ATPase. Cryobiology 1972; 9:457-460.

79. Johnson RWG. Kidney preservation. In: Catto GRD, ed. Clinical Transplantation. Boston: MTP Press Ltd., 1987.
80. Pavlock GS, Southard JH, Starling JR et al. Lysosomal enzyme release in hypothermically perfused dog kidneys. Cryobiology 1984; 21:521-528.
81. Jones DP. Mitochondrial dysfunction during anoxia and acute cell injury. Biomed Biochim Acta 1995; 1271:29-33.
82. Rauen U, de Groot H. New insights into the cellular and molecular mechanisms of cold storage injury. J Investig Med 2004; 52:299-309.
83. Neely JR, Feuvray D. Metabolic products and myocardial ischemia. Am J Pathol 1981; 102:282-291.
84. Maathuis MH, Leuvenink HG, Ploeg RJ. Perspectives in organ preservation. Transplantation 2007; 83:1289-1298.
85. Levy MN. Oxygen consumption and blood flow in the hypothermic, perfused kidney. Am J Physiol 1959; 197:1111-1114.
86. Buhl MR, Jensen MH. Metabolic inhibition. In: Karow AM Jr, Pegg DE, eds. Organ Preservation for Transplantation. New York: Marcel Dekker, Inc., 1981.
87. Wong KC. Physiology and pharmacology of hypothermia. West J Med 1983; 138:227-232.
88. Fuhrman FA. Oxygen consumption of mammalian tissues at reduced temperatures. In: Dripps RD, ed. The Physiology of Induced Hypothermia. Washington, DC: National Academy of Sciences, 1956.
89. Bretschneider HJ. Myocardial protection. Thorac Cardiovasc Surg 1980; 28:295-302.
90. Hearse DJ, Braimbridge MV, Jynge P. Protection of the Ischemic Myocardium: Cardioplegia. New York: Raven Press, 1981.
91. Karow AM Jr. Biophysical and chemical considerations in cryopreservation. In: Karow AM Jr, Pegg DE, eds. Organ Preservation for Transplantation. New York: Marcel Dekker, 1981.
92. Robertson R, Jacob SW. The preservation of intact organs. Adv Surg 1968; 3:75-159.
93. Carrel A. Results of the transplantation of blood vessels, organs and limbs. JAMA 1908; 51:1662-1667.
94. Carrel A. On the permanent life of tissues outside of the organism. J Exp Med 1912; 15:516-528.
95. Carrel A, Lindbergh CA. The culture of organs. Paul B. Hoeber, New York 1938;
96. Oudot J. Transplantation rénale. Presse Med 1948; 56:319-320.
97. Avramovici A. Les Transplantations du rein. Lyon Chir 1924; 21:734-753.
98. Parkinson D, Woodworth HC. Observations on vessel and organ transplants. Exp Med Surg 1947; 5:49-61.
99. Owens JC, Prevedel AE, Swan H. Prolonged experimental occlusion of thoracic aorta during hypothermia. Arch Surg 1955; 70:95-97.
100. Bigelow WG, Lindsay WK, Greenwood WF. Hypothermia. Ann Surg 1950; 132:849-866.
101. Archibald J, Cawley AJ. Some observations on transplantation of the canine kidney. Am J Vet Res 1956; 17:376-379.
102. Schloerb PR, Waldorf RD, Welsh JS. The protective effect of kidney hypothermia on total renal ischemia. SGO 1959; 109:561-565.
103. Lapchinsky AG. Recent results of experimental transplantation of preserved limbs and kidneys and possible use of this technique in clinical practice. Ann NY Acad Sci 1960; 87:539-571.
104. Proctor E, Matthews G, Archibald J. Acute orthotopic transplantation of hearts stored for 72 hours. Thorax 1971; 26:99-102.
105. Karow AM Jr, Webb WR. Cardiac storage with glycerol at zero centigrade. Arch Surg 1961; 83:719-720.
106. Barsamian EM, Win MS, Cady B et al. Preservation of the heart in vitro. In: Brest AN, ed. Heart Substitutes: Mechanical and Transplant. Springfield, Ill: Charles C Thomas, 1966.
107. Blumenstock DA, Lempert N, Morgado F. Preservation of the canine lung in vitro for 24 hours with the use of hypothermia and hyperbaric oxygen. J Thorac Cardiovasc Surg 1965; 50:769-774.
108. Toledo Pereyra LH, Condie RM, Hau T et al. Three-day hypothermic storage of canine lungs. Surg Forum 1977; 28:194-195.
109. Marshall VC. Organ preservation. In: Calne RY, ed. Clinical Organ Transplantation. Oxford: Blackwell Scientific Pub., 1971.
110. Human PA, Holl MJ, Vosloo S et al. Extended cardiopulmonary preservation: University of Wisconsin solution versus Bretschneider's cardioplegic solution. Ann Thorac Surg 1993; 55:1123-1130.
111. Fremes SE, Li RK, Weisel RD et al. The limits of cardiac preservation with University of Wisconsin solution. Ann Thorac Surg 1991; 52:1021-1025.
112. Karow AM Jr. The organ bank concept. In: Karow AM, Jr., Pegg DE, eds. Organ Preservation for Transplantation. New York: Marcel Dekker, Inc., 1981.
113. Collins GM, Halasz NA. Forty-eight hour ice storage of kidneys: Importance of cation content. Surgery 1976; 79:432-435.

114. Burg MB, Orloff J. Active cation transport by kidney tubules at 0 degree centigrade. Am J Physiol 1964; 207:983-988.
115. Armitage WJ. Heart. In: Karow AM, Jr., Pegg DE, eds. Organ Preservation for Transplantation. New York: Marcel Dekker, Inc., 1981.
116. Pegg DE. The biology of cell survival in vitro. In: Karow AM, Jr., Pegg DE, eds. Organ Preservation for Transplantation. Marcel Dekker, Inc., New York 1981.
117. Calne RY. Organ Grafts. Baltimore: The Williams and Wilkins Co., 1975.
118. Collins GM. Flush preservation. In: Pegg DE, Jacobsen IA, Halasz NA, eds. Organ Preservation: Basic and Applied Aspects. Boston: MTP Press, 1982.
119. Jacobsen IA, Kemp E, Buhl MR. An adverse effect of rapid cooling in kidney preservation. Transplantation 1979; 27:135-136.
120. Yoshioka T, Sugimoto H, Uenishi M et al. Prolonged hemodynamic maintenance by the combined administration of vasopressin and epinephrine in brain death: A clinical study. Neurosurgery 1986; 18:565-567.
121. Pegg DE, Jacobsen IA, Halasz NA. Organ Preservation. Basic and Applied Aspects. Boston: MTP Press, 1982.
122. Pegg DE. The principles of organ storage procedures. In: Pegg DE, Jacobsen IA, Halasz NA, eds. Organ Preservation: Basic and Applied Aspects. Boston: MTP Press, 1982.
123. Belzer FO, Downes GL. Kidney. In: Karow AM Jr, Abouna GJM, Humphries AL Jr, eds. Organ Preservation for Transplantation. Boston: Little, Brown and Co., 1974.
124. Reckendorfer H, Burgmann H, Spieckermann PG. Hepatic energy metabolism during hypothermic storage and after reperfusion. Evaluation of the University of Wisconsin and the Bretschneider solutions. Transplant Proc 1991; 23:1974-1975.
125. Cavallari A, Nardo B, Recordare A et al. Cold preservation of a human liver in University of Wisconsin solution for 36 hours: a morphological and biochemical study. Transplant Proc 1991; 23:2422-2423.
126. Sorrentino D, van Ness K, Ribeiro I et al. Functional and morphological features of isolated hepatocytes preserved in University of Wisconsin solution. Hepatology 1991; 14:331-339.
127. Holloway CMB, Harvey PRC, Strasberg SM. Viability of sinusoidal lining cells in cold-preserved rat liver allografts. Transplantation 1990; 49:225-229.
128. García Valdecasas JC, González FG, Grande L et al. Study of liver preservation: The use of University of Wisconsin or Euro-Collins solutions alone or in a combined method. Transplant Proc 1991; 23:2453-2455.
129. Storey KB, Storey JM. Metabolic rate depression in animals: Transcriptional and translational controls. Biol Rev Camb Philos Soc 2004; 79:207-233.
130. Muleme HM, Walpole AC, Staples JF. Mitochondrial metabolism in hibernation: Metabolic suppression, temperature effects and substrate preferences. Physiol Biochem Zool 2006; 79:474-483.
131. Webster KA. Hypoxia: life on the edge. Antioxid Redox Signal 2007; 9:1303-1307.
132. Glass ML. The enigma of aestivation in the African lungfish protopterus dolloi—Commentary on the paper by Perry et al. Respir Physiol Neurobiol 2008; 160:18-20.
133. Wang LCH. Energetic and field aspects of mammalian torpor: the Richardson's ground squirrel. In: Wang LCH, Hudson JW, eds. Strategies in Cold: Natural Torpidity and Thermogenesis. New York: Academic Press, 1978.
134. Lyman CP, O'Brien RC, Greene GC et al. Hibernation and longevity in the Turkish hamster Mesocricetus brandti. Science 1981; 212:668-670.
135. Barnes BM. Freeze avoidance in a mammal: body temperature below 0°C in an arctic hibernator. Science 1989; 244:1593-1595.
136. Macmillen RE. Aestivation in the cactus mouse, peromyscus eremicus. Comp Biochem Physiol 1965; 16:227-248.
137. Moraes G, Altran AE, Avilez IM et al. Metabolic adjustments during semi-aestivation of the marble swamp eel (synbranchus marmoratus, Bloch 1795)—A facultative air breathing fish. Braz J Biol 2005; 65:305-312.
138. Hudson NJ, Lavidis NA, Choy PT et al. Effect of prolonged inactivity on skeletal motor nerve terminals during aestivation in the burrowing frog, cyclorana alboguttata. J Comp Physiol A Neuroethol Sens Neural Behav Physiol 2005; 191:373-379.
139. Oeltgen PR, Blouin RA, Spurrier WA et al. Hibernation "trigger" alters renal function in the primate. Physiol Behav 1985; 34:79-81.
140. Oeltgen PR, Walsh JW, Hamann SR et al. Hibernation "trigger": Opioid-like inhibitory action on brain function of the monkey. Pharmacol Biochem Behav 1982; 17:1271-1274.
141. Green CJ, Fuller BJ, Ross B et al. Storage of organs from ground-squirrels during and after hibernation. The 20th Annual Meeting for The Society for Cryobiology, 1983; 149. Ref Type: Abstract

142. Spurrier WA, Dawe AR. Several blood and circulatory changes in the hibernation of the 13-lined ground squirrel, citellus tridecemlineatus. Comp Biochem Physiol 1973; 44A:267-282.
143. Riedesel ML, Steffen JM. Protein metabolism and urea recycling in rodent hibernators. Fed Proc 1980; 39:2959-2963.
144. Wolfe RR, Nelson RA, Wolfe MH et al. Nitrogen cycling in hibernating bears. 30th Annual Conference on Mass Spectrometry and Allied Topics, 1982; 426.
145. Kruman II, Kolaeva SG, Iljasova EN et al. Seasonal variations of DNA synthesis in intestinal epithelial cells of hibernating animals: I. DNA synthesis in intestinal epithelial cells of ground squirrel (citellus undulatus) during hibernation. Comp Biochem Physiol 1986; 83B:173-177.
146. Chien S, Oeltgen PR, Diana JN et al. Two-day preservation of major organs with autoperfusion multiorgan preparation and hibernation induction trigger. A preliminary report. J Thorac Cardiovasc Surg 1991; 102:224-234.
147. Chien S, Zhang F, Zhang Z et al. Two-day preservation of the lungs by normothermic perfusion. Acta Biomed Ateneo Parmense 1994; 65:165-179.
148. Wu G, Zhang F, Salley RK et al. Delta opioid (DADLE) extends hypothermic preservation time of the lung. J Thorac Cardiovasc Surg 1996; 111:259-267.
149. Chien S, Maley R, Oeltgen PR et al. Canine lung transplantation after more than 24 hours of normothermic preservation. J Heart Lung Transplant 1997; 16:340-351.
150. Zancanaro C, Malatesta M, Mannello F et al. The kidney during hibernation and arousal from hibernation. A natural model of organ preservation during cold ischaemia and reperfusion. Nephrol Dial Transplant 1999; 14:1982-1990.
151. Green C. Mammalian hibernation: lessons for organ preparation? Cryo Letters 2000; 21:91-98.
152. Mubagwa K, Flameng W. Adenosine, adenosine receptors and myocardial protection: An updated overview. Cardiovasc Res 2001; 52:25-39.
153. Cejalvo D, Lloris Carsi JM, Toledo Pereyra LH et al. Effect of adenosine and allopurinol on liver ischemia-reperfusion. Transplant Proc 1993; 25:3023-3024.
154. Fujino Y, Kuroda Y, Morita A et al. The effect of fasting and exogenous adenosine on ATP tissue concentration and viability of canine pancreas grafts during preservation by the two-layer method. Transplantation 1993; 56:1083-1086.
155. Dalziel HH, Westfall DP. Receptors for adenine nucleotides and nucleosides: subclassification, distribution and molecular characterization. Pharmacol Rev 1994; 46:449-466.
156. Gerasimovskaya EV, Ahmad S, White CW et al. Extracellular ATP is an autocrine/paracrine regulator of hypoxia-induced adventitial fibroblast growth. Signaling through extracellular signal-regulated kinase-1/2 and the Egr-1 transcription factor. J Biol Chem 2002; 277:44638-44650.
157. Ely SW, Berne RM. Protective effects of adenosine in myocardial ischemia. Circulation 1992; 85:893-904.
158. Galinanes M, Chambers DJ, Hearse DJ. Effect of sodium aspartate on the recovery of the rat heart from long-term hypothermic storage. J Thorac Cardiovasc Surg 1992; 103:521-531.
159. Okonski P, Szram S, Banach M et al. Effect of L-arginine on overhydration and ultrastructure preservation of rat's heart exposed to cold cardioplegic ischaemia. Ann Transplant 2003; 8:57-62.
160. Zhang SJ, Shi JH, Tang Z et al. Protective effects of glycine pretreatment on brain-death donor liver. Hepatobiliary Pancreat Dis Int 2005; 4:37-40.
161. Chu Y, Wu YC, Chou YC et al. Endothelium-dependent relaxation of canine pulmonary artery after prolonged lung graft preservation in University of Wisconsin solution: Role of L-arginine supplementation. J Heart Lung Transplant 2004; 23:592-598.
162. Salehi P, Spratlin J, Chong TF et al. Beneficial effects of supplemental buffer and substrate on energy metabolism during small bowel storage. Cryobiology 2004; 48:245-253.
163. Lundstam S. Kidney metabolism during hypothermic perfusion. Scand J Urol Nephrol 1977; 43(Suppl):5-29.
164. Unruh HW. Lung preservation and lung injury. Chest Surg Clin North Am 1995; 5:91-106.
165. Segel LD, Follette DM. Cardiac function and glycogen content after twenty-four-hour preservation with various metabolic substrates. J Heart Lung Transplant 1998; 17:299-305.
166. Kirk AJB, Colquhoun IW, Dark JH. Lung preservation: A review of current practice and future directions. Ann Thorac Surg 1993; 56:990-1000.
167. Date H, Matsumura A, Manchester JK et al. Evaluation of lung metabolism during successful twenty-four-hour canine lung preservation. J Thorac Cardiovasc Surg 1993; 105:480-491.
168. Okada Y, Kondo T. Impact of lung preservation solutions, Euro-Collins vs low-potassium dextran, on early graft function: a review of five clinical studies. Ann Thorac Cardiovasc Surg 2006; 12:10-14.
169. Seifart HI, Delabar U, Siess M. The influence of various precursors on the concentration of energy-rich phosphates and pyridine nucleotides in cardiac tissue and its possible meaning for anoxic survival. Bas Res Cardiol 1980; 75:57-61.

170. Im MJ, Hoopes JE. Improved skin flap survival with nicotinic acid and nicotinamide in rats. J Surg Res 1989; 47:453-455.
171. McDonagh PF, Laks H, Chaudry IH et al. Improved myocardial recovery from ischemia. Treatment with low-dose adenosine triphosphate-magnesium chloride. Arch Surg 1984; 119:1379-1384.
172. Robinson LA, Braimbridge MV, Hearse DJ. Creatine phosphate: An additive myocardial protective and antiarrhythmic agent in cardioplegia. J Thorac Cardiovasc Surg 1984; 87:190-200.
173. Makan NR. Induction of permeability change and restoration of membrane permeability barrier in transformed cell cultures. Exp Cell Res 1978; 114:417-427.
174. Rozengurt E, Heppel LA, Friedberg I. Effect of exogenous ATP on the permeability properties of transformed cultures of mouse cell lines. J Biol Chem 1977; 252:4584-4590.
175. Hultman J, Ronquist G, Forsberg JO et al. Myocardial energy restoration of ischemic damage by administration of phosphoenolpyruvate during reperfusion. A study in a paracorporeal rat heart model. Eur Surg Res 1983; 15:200-207.
176. Thelin S, Hultman J, Ronquist G et al. Enhanced protection of rat heart during ischemia by phosphoenolpyruvate and ATP in cardioplegia. Thorac Cardiovasc Surg 1986; 34:104-109.
177. Thelin S, Hultman J, Ronquist G et al. Metabolic and functional effects of phosphoenolpyruvate and adenosine triphosphate on rat hearts subjected to global ischemia. Scand J Thorac Cardiovasc Surg 1985; 19:237-245.
178. Thelin S, Hultman J, Jakobson S et al. Functional effects of phosphoenolpyruvate and ATP on pig hearts in cardioplegia and during reperfusion. An in vivo study with cardiopulmonary bypass. Eur Surg Res 1987; 19:348-356.
179. Damen J, Hitchcock JF. Reactive pulmonary hypertension after a switch operation. Br Heart J 1985; 53:223-225.
180. So PW, Fuller BJ. Enhanced energy metabolism during cold hypoxic organ preservation: Studies on rat liver after pyruvate supplementation. Cryobiology 2003; 46:295-300.
181. Regitz V, Azumi T, Stephan H et al. Biochemical mechanism of infarct size reduction by pyruvate. Cardiovasc Res 1981; 15:652-658.
182. Fox AC, Reed GE, Meilman H et al. Release of nucleosides from canine and human hearts as an index of prior ischemia. Am J Cardiol 1979; 43:52-58.
183. Mauser M, Hoffmeister HM, Nienaber C et al. Influence of ribose, adenosine and "AICAR" on the rate of myocardial adenosine triphosphate synthesis during reperfusion after coronary artery occlusion in the dog. Circ Res 1985; 56:220-230.
184. Harmsen E, de Tombe PP, de Jong JW et al. Enhanced ATP and GTP synthesis from hypoxanthine or inosine after myocardial ischemia. Am J Physiol 1984; 246:H37-H43.
185. Southard JH. Temperature effects and cooling. In: Zelenock GB, ed. Clinical ischemic syndromes. St. Louis: Mosby, 1990:
186. Maguire MH, Lukas MC, Rettie JF. Adenine nucleotide salvage synthesis in the rat heart: Pathways of adenosine salvage. Biochim Biophys Acta 1972; 262:108-115.
187. Parker JC, Smith EE, Jones CE. The role of nucleoside and nucleobase metabolism in myocardial adenine nucleotide regeneration after cardiac arrest. Circ Shock 1976; 3:11-20.
188. Pegg DE. Organ preservation. Surg Clin North Am 1986; 66:617-632.
189. St.Cyr J, Ward H, Kriett J et al. Long term model for evaluation of myocardial metabolic recovery following global ischemia. Adv Exp Med Biol 1984; 194:401-414.
190. St.Cyr JA, Bianco RW, Schneider JR et al. Enhanced high energy phosphate recovery with ribose infusion after global myocardial ischemia in a canine model. J Surg Res 1989; 46:157-162.
191. Ward HB, St.Cyr JA, Cogordan JA et al. Recovery of adenine nucleotide levels after global myocardial ischemia in dogs. Surgery 1984; 96:248-255.
192. Zimmer HG. Ribose accelerates the repletion of the ATP pool during recovery from reversible ischemia of the rat myocardium. J Mol Cell Cardiol 1984; 16:863-866.
193. Baldwin DR, McFalls EO, Jaimes D et al. Myocardial glucose metabolism and ATP levels are decreased two days after global ischemia. J Surg Res 1996; 35-38.
194. Tan ZT, Bhayana JN, Bergsland J et al. Concanavalin A, ribose and adenine resuscitate preserved rat hearts. J Cardiovasc Pharmacol 1997; 30:26-32.
195. Liu B, Clanachan AS, Schulz R et al. Cardiac efficiency is improved after ischemia by altering both the source and fate of protons. Circ Res 1996; 79:940-948.
196. Halestrap AP, Wang X, Poole RC et al. Lactate transport in heart in relation to myocardial ischemia. Am J Cardiol 1997; 80(3A):17A-25A.
197. Markov AK, Oglethorpe NC, Blake TM et al. Hemodynamic, electrocardiographic and metabolic effects of fructose diphosphate on acute myocardial ischemia. Am Heart J 1980; 100:639-646.
198. Belzer FO, Southard JH. Principles of solid-organ preservation by cold storage. Transplantation 1988; 45:673-676.

199. Southard JH, Belzer FO. Organ preservation. Annu Rev Med 1995; 46:235-247.
200. Wieland OH, Urumow T, Drexler P. Insulin, phospholipase and the activation of the pyruvate dehydrogenase complex: an enigma. Ann NY Acad Sci 1989; 573:274-284.
201. Denton RM, Midgley PJW, Rutter GA et al. Studies into the mechanism whereby insulin activates pyruvate dehydrogenase complex in adipose tissue. Ann NY Acad Sci 1989; 573:285-296.
202. Rao V, Merante F, Weisel RD et al. Insulin stimulates pyruvate dehydrogenase and protects human ventricular cardiomyocytes form simulated ischemia. J Thorac Cardiovasc Surg 1998; 116:485-494.
203. Lopaschuk GD, Wambolt RB, Barr RL. An imbalance between glycolysis and glucose oxidation is a possible explanation for the detrimental effects of high levels of fatty acids during aerobic reperfusion of ischemic hearts. J Pharmacol Exp Ther 1993; 264:135-144.
204. Wahr JA, Childs KF, Bolling SF. Dichloroacetate enhances myocardial functional and metabolic recovery following global ischemia. J Cardiothorac Vasc Anesth 1994; 8:192-197.
205. Wang P, Lloyd SG, Chatham JC. Impact of high glucose/high insulin and dichloroacetate treatment on carbohydrate oxidation and functional recovery after low-flow ischemia and reperfusion in the isolated perfused rat heart. Circulation 2005; 111:2066-2072.
206. Schofield RS, Hill JA. Role of metabolically active drugs in the management of ischemic heart disease. Am J Cardiovasc Drugs 2001; 1:23-35.
207. Li XL, Man K, Ng KT et al. Insulin in UW solution exacerbates hepatic ischemia/reperfusion injury by energy depletion through the IRS-2/SREBP-1c pathway. Liver Transpl 2004; 10:1173-1182.
208. Webb WR. Metabolic effects of fructose diphosphate in hypoxic and ischemia states. J Thorac Cardiovasc Surg 1984; 88:863-866.
209. Moresco RN, Santos RC, Alves Filho JC et al. Protective effect of fructose-1,6-bisphosphate in the cold storage solution for liver preservation in rat hepatic transplantation. Transplant Proc 2004; 36:1261-364
210. Sola A, De Oca J, Alfaro V et al. Protective effects of exogenous fructose-1,6-biphosphate during small bowel transplantation in rats. Surgery 2004; 135:518-526.
211. Sud V, Lindley SG, McDaniel O et al. Evaluation of fructose 1,6 diphosphate for salvage of ischemic gracilis flaps in rats. J Reconstr Microsurg 2005; 21:191-196.
212. Pan SJ, Combs AB. Effects of pharmacological interventions on emetine cardiotoxicity in isolated perfused rat hearts. Toxicology 1995; 97:93-104.
213. Rigobello MP, Galzigna L, Bindoli A. Fructose 1,6-bisphosphate prevents oxidative stress in the isolated and perfused rat heart. Cell Biochem Funct 1994; 12:69-75.
214. Park JY, Kim EJ, Kwon KJ et al. Neuroprotection by fructose-1,6-bisphosphate involves ROS alterations via p38 MAPK/ERK. Brain Res 2004; 1026:295-301.
215. Romsi P, Kaakinen T, Kiviluoma K et al. Fructose-1,6-bisphosphate for improved outcome after hypothermic circulatory arrest in pigs. J Thorac Cardiovasc Surg 2003; 125:686-698.
216. Nunes FB, Simoes Pires MG, Farias Alves Filho JC et al. Physiopathological studies in septic rats and the use of fructose 1,6-bisphosphate as cellular protection. Crit Care Med 2002; 30:2069-2074.
217. Martin FL, McLean AEM. Comparison of protection by fructose against paracetamol injury with protection by glucose and fructose-1,6 diphosphate. Toxicology 1996; 108:175-184.
218. Tavazzi B, Starnes JW, Lazzarino G et al. Exogenous fructose-1,6 diphosphate is a metabolizable substrate for the isolated normoxic rat heart. Bas Res Cardiol 1992; 87:280-289.
219. Takeuchi K, Hung CD, Friehs I et al. Administration of fructose-1,6-diphosphate during early reperfusion significantly improves recovery of contractile function in the postischemic heart. J Thorac Cardiovasc Surg 1998; 116:335-343.
220. Niu W, Zhang F, Ehringer W et al. Enhancement of hypothermic heart preservation with fructose-1,6-diphosphate. J Surg Res 1999; 85:120-129.
221. Chien S, Zhang F, Niu W et al. Fructose-1,6-diphosphate and a glucose-free solution enhances functional recovery in hypothermic heart preservation. J Heart Lung Transplant 2000; 19:277-285.
222. Wheeler TJ, McCurdy JM, denDekker A et al. Permeability of fructose-1,6-bisphosphate in liposomes and cardiac myocytes. Mol Cell Biochem 2004; 259:105-114.
223. DeBoer LWV, Ingwall JS, Kloner RA et al. Prolonged derangements of canine myocardial purine metabolism after a brief coronary artery occlusion not associated with anatomic evidence of necrosis. Proc Natl Acad Sci USA 1980; 77:5471-5475.
224. Schaper J, Mulch J, Winkler B et al. Ultrastructural, functional and biochemical criteria for estimation of reversibility of ischemic injury: A study on the effects of global ischemia on the isolated dog heart. J Mol Cell Cardiol 1979; 11:521-541.
225. Danesi R, Bernardini N, Marchetti A et al. Protective effects of fructose-1,6 diphosphate on acute and chronic doxorubicin cardiotoxicity in rats. Cancer Chemother Pharmacol 1990; 25:326-332.
226. Didlake R, Kirchner KA, Lewin J et al. Protection from ischemic renal injury by fructose-1,6-diphosphate infusion in the rat. Circ Shock 1985; 16:205-212.

227. Akimitsu T, White JA, Carden DL et al. Fructose-1,6 diphosphate or adenosine attenuate leukocyte adherence in postischemic skeletal muscle. Am J Physiol 1995; 269:H1743-H1751.
228. Sun J, Farias LA, Markov AK. Fructose 1,6 diphosphate prevents intestinal ischemic reperfusion injury and death in rats. Gastroenterology 1990; 98:117-126.
229. Mihas AA, Maliakkal RJ, Mihas TA et al. Effects of fructose-1,6 diphosphate on the activity of rat liver nitric oxide synthase in vitro. Pharmacology 1997; 54:43-48.
230. Cacioli D, Clivati A, Pelosi P et al. Hemorheological effects of fructose-1,6 diphosphate in patients with lower extremity ischemia. Curr Med Res Opin 1988; 10:668-674.
231. Hutcheson AE, Rao MR, Olinde KD et al. Myocardial toxicity of cyclosporin A: inhibition of calcium ATPase and nitric oxide synthase activities and attenuation by fructose-1,6-diphosphate in vitro. Res Commun Mol Pathol Pharmacol 1995; 89:17-26.
232. Cavallini L, Deana R, Francesconi MA et al. Fructose-1,6 diphosphate inhibits platelet activation. Biochem Pharmacol 1992; 7:1539-1544.
233. Lazzarino G, Cattani L, Costrini R et al. Increase of intraerythrocytic fructose-1,6-diphosphate after incubation of whole human blood with fructose-1,6-diphosphate. Clin Biochem 1984; 17:42-45.
234. Kirtley ME, McKay M. Fructose-1-6-bisohosphate, a regulator of metabolism. Mol Cell Biochem 1977; 18:141-149.
235. Rigobello MP, Deana R, Galzigna L. Enzymatic events following the interaction of fructose -1,6-diphosphate with red cell membranes. In: Levy E, ed. Advances in Pathology. Oxford: Pergamon Press, 1982.
236. Ehringer W, Niu W, Chiang B et al. Membrane permeability of fructose-1,6-diphosphate in lipid vesicles and endothelial cells. Mol Cell Biochem 2000; 210:35-45.
237. Ehringer WD, Chiang B, Chien S. The uptake and metabolism of fructose-1,6-diphosphate in rat cardiomyocytes. Mol Cell Biochem 2001; 221:33-40.
238. Ehringer WD, Su S, Chiang B et al. Destabilizing effects of fructose-1,6-bisphosphate on membrane bilayers. Lipids 2002; 37:885-892.
239. Hardin CD, Lazzarino G, Tavazzi B et al. Myocardial metabolism of exogenous FDP is consistent with transport by a dicarboxylate transporter. Am J Physiol Heart Circ Physiol 2001; 281:H2654-H2660.
240. Buchthal F, Deutsch A, Knappeis GG. Release of contraction and changes in birefringence caused by adenosine triphosphate in isolated cross striated muscle fibers. Acta Physiol Scand 1944; 8:271-287.
241. Woo YT, Manery JF, Riordan JR et al. Uptake and metabolism of purine nucleosides and nucleotides in isolated from skeletal muscle. Life Sci 1977; 21:861-876.
242. Fedelesova M, Ziegelhoffer A, Krause EG et al. Effect of exogenous adenosine triphosphate on the metabolic state of the excised hypothermic dog heart. Circ Res 1969; 24:617-627.
243. Hirasawa H, Soeda K, Ohtake Y et al. Effects of ATP-MgCl₂ and ATP-Na₂ administration on renal function and cellular metabolism following renal ischemia. Circ Shock 1985; 16:337-346.
244. Chaudry IH, Stephan RN, Dean RE et al. Use of magnesium-ATP following liver ischemia. Magnesium 1988; 7:68-77.
245. Dornald's Directories. Medical and Healthcare Marketplace Guide. Philadelphia: Dornald Data Networks, 2002.
246. Chaudry IH, Clemens MG, Baue AE. The role of ATP-magnesium in ischemia and shock. Magnesium 1986; 5:211-220.
247. Chaudry IH. Use of ATP following shock and ischemia. Ann N Y Acad Sci 1990; 603:130-140.
248. Chaudry IH. Does ATP cross the cell plasma membrane? Yale J Bio Med 1982; 55:1-10.
249. Chaudry IH. Editorial commentary: "The effect of ATP on survival in intestinal ischemia shock, hemorrhagic shock and endotoxin shock in rats". Circ Shock 1982; 9:629-631.
250. Parratt JR, Marshall RJ. The response of isolated cardiac muscle to acute anoxia: Protective effect of adenosine triphosphate and creatine phosphate. J Pharm Pharmacol 1974; 26:427-433.
251. Puisieux F, Fattal E, Lahiani M et al. Liposomes, an interesting tool to deliver a bioenergetic substrate (ATP). in vitro and in vivo studies. J Drug Target 1994; 2:443-448.
252. Cell and tissue structure. In: Meisenberg G, Simmons WH, eds. Principles of medical biochemistry. St. Louis: Mosby, 1998:
253. Morcos PA. Achieving efficient delivery of morpholino oligos in cultured cells. Genesis 2001; 30:94-102.
254. Price RJ, Kaul S. Contrast ultrasound targeted drug and gene delivery: an update on a new therapeutic modality. J Cardiovasc Pharmacol Ther 2002; 7:171-180.
255. Ludtke JJ, Sebestyen MG, Wolff JA. The effect of cell division on the cellular dynamics of microinjected DNA and dextran. Mol Ther 2002; 5:579-588.
256. McAllister DV, Allen MG, Prausnitz MR. Microfabricated microneedles for gene and drug delivery. Annu Rev Biomed Eng 2000; 2:289-313.

257. Pitarresi G, Pierro P, Giammona G et al. Beads of acryloylated polyaminoacidic matrices containing 5-Fluorouracil for drug delivery. Drug Deliv 2002; 9:97-104.

258. Raghuvanshi RS, Singh O, Panda AK. Formulation and characterization of immunoreactive tetanus toxoid biodegradable polymer particles. Drug Deliv 2001; 8:99-106.

259. Jiao YY, Ubrich N, Marchand Arvier M et al. Preparation and in vitro evaluation of heparin-loaded polymeric nanoparticles. Drug Deliv 2001; 8:135-141.

260. Abazinge M, Jackson T, Yang Q et al. Comparison of in vitro and in vivo release characteristics of sustained release ofloxacin microspheres. Drug Deliv 2000; 7:77-81.

261. Lawrence MJ, Rees GD. Microemulsion-based media as novel drug delivery systems. Adv Drug Deliv Rev 2000; 45:89-121.

262. Ke WT, Lin SY, Ho HO et al. Physical characterizations of microemulsion systems using tocopheryl polyethylene glycol 1000 succinate (TPGS) as a surfactant for the oral delivery of protein drugs. J Control Release 2005; 102:489-507.

263. Ostro MJ. Liposomes. Sci Am 1987; 256:102-111.

264. Gregoriadis G, Florence AT. Liposomes in drug delivery. Clinical, diagnostic and ophthalmic potential. Drugs 1993; 45:15-28.

265. Ramos C, Bonato D, Winterhalter M et al. Spontaneous lipid vesicle fusion with electropermeabilized cells. FEBS Lett 2002; 518:135-138.

266. Stromberg A, Ryttsen F, Chiu DT et al. Manipulating the genetic identity and biochemical surface properties of individual cells with electric-field-induced fusion. Proc Natl Acad Sci USA 2000; 97:7-11.

267. Omori N, Maruyama K, Jin G et al. Targeting of post-ischemic cerebral endothelium in rat by liposomes bearing polyethylene glycol-coupled transferrin. Neurol Res 2003; 25:275-279.

268. Takeuchi H, Kojima H, Yamamoto H et al. Evaluation of circulation profiles of liposomes coated with hydrophilic polymers having different molecular weights in rats. J Control Release 2001; 75:83-91.

269. Gabizon AA. Stealth liposomes and tumor targeting: one step further in the quest for the magic bullet. Clin Cancer Res 2001; 7:223-225.

270. Kaneda Y, Saeki Y, Morishita R. Gene therapy using HVJ-liposomes: The best of both worlds? Mol Med Today 1999; 5:298-303.

271. Mouritsen OG, Jorgensen K. A new look at lipid-membrane structure in relation to drug research. Pharm Res 1998; 15:1507-1519.

272. Lasic DD, Papahadjopoulos D. General introduction. In: Lasic DD, Papahadjopoulos D, eds. Medical Applications of Liposomes. New York: Elsevier, 1998:

273. Gugliotti M, Chaimovich H, Politi MJ. Fusion of vesicles with the air-water interface: the influence of polar head group, salt concentration and vesicle size. Biochim Biophys Acta 2000; 1463:301-306.

274. Jones MN. The surface properties of phospholipid liposome systems and their characterization. Adv Colloid Interface Sci 1995; 54:93-128.

275. Novick RJ, Veldhuizen RAW, Possmayer F et al. Exogenous surfactant therapy in thirty-eight hour lung graft preservation for transplantation. J Thorac Cardiovasc Surg 1994; 108:259-268.

276. Wittwer T, Franke UF, Fehrenbach A et al. Donor pretreatment using the aerosolized prostacyclin analogue iloprost optimizes post-ischemic function of nonheart beating donor lungs. J Heart Lung Transplant 2005; 24:371-378.

277. Higgins RSD, Letsou GV, Sanchez JA et al. Improved ultrastructural lung preservation with prostaglandin E1 as donor pretreatment in a primate model of heart-lung transplantation. J Thorac Cardiovasc Surg 1993; 105:965-971.

278. Izuishi K, Fujiwara M, Hossain MA et al. Protective effect of intraportal prostaglandin E1 on prolonged cold preserved rat liver. Transplant Proc 2003; 35:130-131.

279. Kimblad PO, Steen S. Eliminating the strong pulmonary vasoconstriction caused by Euro-Collins solution. Ann Thorac Surg 1994; 58:728-733.

280. Cheung JY, Bonventre JV, Malis CD et al. Calcium and ischemic injury. N Engl J Med 1986; 314:1670-1676.

281. Trocha M, Szelag A. The role of calcium and calcium channel blocking drugs in damage to the liver preserved for transplantation. Ann Transplant 2004; 9:5-11.

282. Marzi I, Walcher F, Buhren V. Improvement of liver preservation by the calcium channel blocker nisoldipine. An experimental study applying intravital microscopy to transplanted rat livers. Transpl Int 1992; 5(Suppl 1):S395-S397.

283. Shirakura R, Matsuda H, Nakata S et al. Prolonged preservation of cadaver heart with Belzer UW solution: 24-hour storage system for asphyxiated canine hearts. Eur Surg Res 1990; 22:197-205.

284. Adachi T, Suzuki MK, Limura O. Effects of verapamil on myocardial stunning in xanthine-oxidase deficient hearts: pretreatment vs post-ischemic treatment. Bas Res Cardiol 1994; 89:16-28.

285. Lee BH, Seo HW, Yi KY et al. Effects of KR-32570, a new Na+/H+ exchanger inhibitor, on functional and metabolic impairments produced by global ischemia and reperfusion in the perfused rat heart. Eur J Pharmacol 2005; 511:175-182.

286. Wang QD, Pernow J, Sjoquist PO et al. Pharmacological possibilities for protection against myocardial reperfusion injury. Cardiovasc Res 2002; 55:25-37.

287. Masereel B, Pochet L, Laeckmann D. An overview of inhibitors of Na(+)/H(+) exchanger. Eur J Med Chem 2003; 38:547-554.

288. Gao L, Hicks M, Macdonald PS. Improved preservation of the rat heart with celsior solution supplemented with cariporide plus glyceryl trinitrate. Am J Transplant 2005; 5:1820-1826.

289. Rabkin DG, Curtis LJ, Weinberg AD et al. Na+/H+ exchange inhibition and antioxidants lack additive protective effects after reperfusion injury in the working heterotopic rat heart isograft. J Heart Lung Transplant 2005; 24:386-391.

290. Wahlers T, Hirt SW, Haverich A et al. Future horizons of lung preservation by application of a platelet-activating factor antagonist compared with current clinical standards. J Thorac Cardiovasc Surg 1992; 103:200-205.

291. Corcoran PC, Wang Y, Katz NM et al. Platelet activating factor antagonist enhances lung preservation in a canine model of single lung allotransplantation. J Thorac Cardiovasc Surg 1992; 104:66-72.

292. Novick RJ, Menkis AH, McKenzie FN. New trends in lung preservation: A collective review. J Heart Lung Transplant 1992; 11:377-392.

293. Castellvi J, Borobia FG, Figueras J et al. Effect of the platelet-activating factor antagonist BN-52021 on liver preservation (4 degrees): Experimental study in isolated reperfused rat liver model. Transplant Proc 2002; 34:50-52.

294. Takada Y, Fukunaga K, Taniguchi H et al. Energy metabolism of hepatic allografts subjected to prolonged warm ischemia and pharmacologic modulation with FK506 and platelet activating factor antagonist. Transplant Proc 1998; 30:3694-3695.

295. Kim JD, Baker CJ, Roberts RF et al. Platelet activating factor acetylhydrolase decreases lung reperfusion injury. Ann Thorac Surg 2000; 70:423-428.

296. Scherer M, Demertzis S, Langer F et al. C1-esterase inhibitor reduces reperfusion injury after lung transplantation. Ann Thorac Surg 2002; 73:233-238.

297. Bergamaschini L, Gobbo G, Gatti S et al. Endothelial targeting with C1-inhibitor reduces complement activation in vitro and during ex vivo reperfusion of pig liver. Clin Exp Immunol 2001; 126:412-420.

298. Schmidt A, Tomasdottir H, Bengtsson A. Influence of cold ischemia time on complement activation, neopterin and cytokine release in liver transplantation. Transplant Proc 2004; 36:2796-2798.

299. Saugstad OD. Does nitric oxide prevent oxidative mediated lung injury? Acta Paediatr 2000; 89:905-907.

300. Jheon S, Lee YM, Sung SW et al. Pulmonary preservation effect of nitroglycerine in isolated rat lung reperfusion model. Transplant Proc 2004; 36:1933-1935.

301. Loehe F, Preissler G, Annecke T et al. Continuous infusion of nitroglycerin improves pulmonary graft function of nonheart-beating donor lungs. Transplantation 2004; 77:1803-1808.

302. Quintana AB, Lenzi HL, Almada LL et al. Effect of S-nitrosoglutathione (GSNO) added to the University of Wisconsin solution (UW): II) Functional response to cold preservation/reperfusion of rat liver. Ann Hepatol 2002; 1:183-191.

303. Weinberger B, Laskin DL, Heck DE et al. The toxicology of inhaled nitric oxide. Toxicol Sci 2001; 59:5-16.

304. Colasanti M, Suzuki H. The dual personality of NO. Trends Pharmacol Sci 2000; 21:249-252.

305. Chen T, Zamora R, Zuckerbraun B et al. Role of nitric oxide in liver injury. Curr Mol Med 2003; 3:519-526.

306. Anggard E. Nitric oxide: mediator, murderer and medicine. Lancet 1994; 343(8907):1199-1206.

307. Belge C, Massion PB, Pelat M et al. Nitric oxide and the heart: Update on new paradigms. Ann N Y Acad Sci 2005; 1047:173-182.

308. Faure JP, Petit I, Zhang K et al. Protective roles of polyethylene glycol and trimetazidine against cold ischemia and reperfusion injuries of pig kidney graft. Am J Transplant 2004; 4:495-504.

309. Jones BU, Serna DL, Smulowitz P et al. Extended ex vivo myocardial preservation in the beating state using a novel polyethylene glycolated bovine hemoglobin perfusate based solution. ASAIO J 2003; 49:388-394.

310. Ahmed I, Ahmad N, Attia MS et al. Protective effects of polyethylene glycol (20 mol/L) in phosphate-buffered sucrose for rat liver preservation. Transplant Proc 2001; 33:3713-3715.

311. Hauet T, Mothes D, Goujon JM et al. Protective effect of polyethylene glycol against prolonged cold ischemia and reperfusion injury: study in the isolated perfused rat kidney. J Pharmacol Exp Ther 2001; 297:946-952.

312. Gibelin H, Hauet T, Eugene M et al. Beneficial effects of addition of polyethylene glycol to extracellular type solutions to minimize ischemia/reperfusion injuries in an isolated-perfused rat liver model. Transplant Proc 2002; 34:768.

313. Genesca M, Sola A, Azuara D et al. Apoptosis inhibition during preservation by fructose-1,6-diphosphate and theophylline in rat intestinal transplantation. Crit Care Med 2005; 33:827-834.

314. Aussedat J, Ray A, Kay L et al. Improvement of long-term preservation of isolated arrested rat heart: beneficial effect of the antiischemic agent trimetazidine. J Cardiovasc Pharmacol 1993; 21:128-135.

315. Baumert H, Goujon JM, Richer JP et al. Renoprotective effects of trimetazidine against ischemia-reperfusion injury and cold storage preservation: a preliminary study. Transplantation 1999; 68:300-303.

316. Hauet T, Bauza G, Richer JP et al. Trimetazidine added to University of Wisconsin during 48-hour cold preservation improves renal energetic status during reperfusion. Transplant Proc 2000; 32:496-497.

317. Fujioka H, Yoshihara S, Tanaka T et al. Enhancement of posthypoxic contractile and metabolic recovery of perfused rat hearts by dl-propranolol: possible involvement of nonbeta-receptor mediated activity. J Mol Cell Cardiol 1991; 23:949-962.

318. Shirakura R, Matsuda H, Nakano S et al. Cardiac function and myocardial performance of 24-hour-preserved asphyxiated canine hearts. Ann Thorac Surg 1992; 53:440-444.

319. Kaneko H, Schweizer RT. Venous flushing with vasodilators aids recovery of vasoconstricted and warm ischemic injured pig kidneys. Transplant Proc 1989; 21:1233-1235.

320. Satoh S, Stowe NT, Inman SR et al. Renal vascular response to vasodilators following warm ischemia and cold storage preservation in dog kidneys. J Urol 1993; 149:186-189.

321. Rubin Y, Skutelsky E, Anihai D et al. Improved hypothermic preservation of rat heart by furosemide. J Thorac Cardiovasc Surg 1995; 110:523-531.

322. Petracek MR, Lawson JD, Johnson HK. Three simple solutions for renal flush and preservation. South Med J 1981; 73:321-324.

323. Collins GM, Green RD, Boyer D et al. Protection of kidneys from warm ischemic injury: dosage and timing of mannitol administration. Transplantation 1980; 29:83-84.

324. Vlahakes GJ, Giamber SR, Rothaus KO et al. Hyperosmotic mannitol and collateral blood flow to ischemic myocardium. J Surg Res 1989; 47:438-466.

325. Cooper JD, Vreim CE. NHLBI workshop summary: biology of lung preservation for transplantation. Am Rev Respir Dis 1992; 146:803-807.

326. Omasa M, Fukuse T, Matsuoka K et al. Effect of green tea extracted polyphenol on ischemia/reperfusion injury after cold preservation of rat lung. Transplant Proc 2003; 35:138-139.

327. Chen F, Nakamura T, Wada H. Development of new organ preservation solutions in Kyoto University. Yonsei Med J 2004; 45:1107-1114.

328. El Wahsh M. Liver graft preservation: an overview. Hepatobiliary Pancreat Dis Int 2007; 6:12-16.

329. McCabe RE, Stevens LE, Fitzpatrick HF. Corticosteroids in organ preservation. Transplant Proc 1975; 7:113-116.

330. Liu X, Drognitz O, Neeff H et al. Apoptosis is caused by prolonged organ preservation and blocked by apoptosis inhibitor in experimental rat pancreatic grafts. Transplant Proc 2004; 36:1209-1210.

331. Wu GH, Tomei LD, Bathurst IC et al. An anti-apoptotic compound improves hypothermic rabbit heart preservation. 15th Annual Meeting of the American Society of Transplant Physicians 1996; 162. Ref Type: Abstract

332. Wu GH, Tomei LD, Bathurst IC et al. Enhancement of hypothermic rat liver preservation with an antiapoptotic compound. 22nd Annual Scientific Meeting of the American Society of Transplant Surgeons 1996; 258(Abstract).

333. Lokkegaard H, Bilde T, Gyrd Hansen N et al. Kidney preservation for 24 hours after one hour of warm ischemia. Preliminary report. Acta Med Scand 1971; 190:451-452.

334. Stringham JC, Paulsen KL, Southard JH et al. Improved myocardial ischemic tolerance by contractile inhibition with 2,3-butanedione monoxime. Ann Thorac Surg 1992; 54:852-860.

335. Stowe DF, Boban M, Graf BM et al. Contraction uncoupling with butanedione monoxime versus low calcium or high potassium solutions on flow and contractile function of isolated hearts after prolonged hypothermic perfusion. Circulation 1994; 89:2412-2420.

336. Lopukhin SY, Southard JH, Belzer FO. University of Wisconsin solution containing 2,3-butanedione-monoxime extends myocardium preservation time. Transplant Proc 1993; 25:3017-3018.

337. Dahl NA, Balfour WM. Prolonged anoxic survival due to anoxia pre-exposure: Brain ATP, Lactate and pyruvate. Am J Physiol 1964; 207:452-456.

338. Murry CE, Jennings RB, Reimer KA. Preconditioning with ischemia: A delay of lethal cell injury in ischemic myocardium. Circulation 1986; 74:1124-1136.

339. Vuorinen K, Ylitalo K, Peuhkurinen K et al. Mechanisms of ischemic preconditioning in rat myocardium. Roles of adenosine, cellular energy state and mitochondrial F_1F_0-ATPase. Circulation 1995; 91:2810-2818.
340. Schulz R, Post H, Sakka S et al. Intraischemic preconditioning. Increased tolerance to sustained low-flow ischemia by a brief episode of no-flow ischemia without intermittent reperfusion. Circ Res 1995; 76:942-950.
341. Wusteman MC, Jacobsen IA, Pegg DE. A new solution for initial perfusion of transplant kidneys. Scand J Urol Nephrol 1978; 12:281-286.
342. Morton WA, Parmentier EM, Petschek HE. Study of aggregate formation in region of separated blood flow. Thrombos Diathes Haemorrh 1975; 34:840-854.
343. Saniabadi AR, Lowe GDC, Madhok R et al. A critical investigation into the existence of circulating platelet aggregates. Thromb Haemost 1986; 56:45-49.
344. Billingham RE. The passenger cell concept in transplantation immunology. Cell Immunol 1971; 2:1-12.
345. Kallerhoff M, Blech M, Kehrer G et al. Short-term perfusion and "equilibration" of canine kidneys with protective solutions. Urol Res 1987; 15:5-12.
346. HÖlscher M, Groenewoud AF. Current status of the HTK solution of Bretschneider in organ preservation. Transplant Proc 1991; 23:2334-2337.
347. Shiraishi T, Igisu H, Shirakusa T. Effects of pH and temperature on lung preservation: A study with an isolated rat lung reperfusion model. Ann Thorac Surg 1994; 57:639-643.
348. Kallerhoff M, Holscher M, Kehrer G et al. Effects of preservation conditions and temperature on tissue acidification in canine kidneys. Transplantation 1985; 39:485-489.
349. Kunikata S, Ishii T, Nishioka T et al. Measurement of viability in preserved kidneys with ^{31}P NMR. Transplant Proc 1989; 21:1269-1271.
350. Xu H, Lee CY, Clemens MG et al. Pronlonged hypothermic machine perfusion preserves hepatocellular function but potentiates endothelial cell dysfunction in rat livers. Transplantation 2004; 77:1676-1682.
351. Robicsek F, Sanger PW, Taylor FH. Simple method of keeping the heart "alive" and functioning outside of the body for prolonged periods. Surgery 1963; 53:525-530.
352. Robicsek F, Pruitt JR, Sanger PW et al. The maintenance of function of the donor heart in the extracorporeal stage and during transplantation. Ann Thorac Surg 1968; 6:330-342.
353. Hardesty RL, Griffith BP. Autoperfusion of the heart and lungs for preservation during distant procurement. J Thorac Cardiovasc Surg 1987; 93:11-18.
354. Adachi H, Fraser CD, Kontos GJ et al. Autoperfused working heart-lung preparation versus hypothermic cardiopulmonary preservation for transplantation. J Heart Transplant 1987; 6:253-260.
355. Kontos GJ Jr, Borkon AM, Baumgartner WA et al. Improved myocardial and pulmonary preservation by metabolic substrate enhancement in the autoperfused working heart-lung preparation. J Heart Transplant 1988; 7:140-144.
356. Zhang Z, Proffitt G, Salley RK et al. Autoperfused heart-lung preparation: one reason for unsuccessful lung preservation. Acta Biomed Ateneo Parmense 1994; 65:115-131.
357. Chien S, Oeltgen PR, Diana JN et al. Extension of tissue survival time in multiorgan block preparation with a delta opioid DADLE ([D-Ala^2, D-Leu^5]-enkephalin). J Thorac Cardiovasc Surg 1994; 107:964-967.
358. Kaplan E, Diehl JT, Peterson MB et al. Extended ex vivo preservation of the heart and lungs. Effects of acellular oxygen-carrying perfusates and indomethacin on the autoperfused working heart-lung preparation. J Thorac Cardiovasc Surg 1990; 100:687-698.
359. Glick DB, Cronin CS, Jaquiss RDB et al. The extended use of the autoperfusing heart-lung preparation for electrophysiologic studies. Surg Forum 1993; 44:309-311.
360. Minten J, Segel LD, Van Belle H et al. Differences in high-energy phosphate catabolism between the rat and the dog in a heart preservation model. J Heart Lung Transplant 1991; 10:71-78.
361. de Perrot M, Keshavjee S. Lung preservation. Semin Thorac Cardiovasc Surg 2004; 16:300-308.
362. Chien S, Zhang F, Niu W et al. Comparison of university of wisconsin, euro-collins, low-potassium dextran and krebs-henseleit solutions for hypothermic lung preservation. J Thorac Cardiovasc Surg 2000; 119:921-930.
363. Taylor DO, Edwards LB, Boucek MM et al. Registry of the international society for heart and lung transplantation: Twenty-second official adult heart transplant report-2005. J Heart Lung Transplant 2005; 24:945-955.
364. Takahashi A, Braimbridge MV, Hearse DJ et al. Long-term preservation of the mammalian myocardium. Effect of storage medium and temperature on the vulnerability to tissue injury. J Thorac Cardiovasc Surg 1991; 102:235-245.

365. Havel M, Owen AN, Simon P. Basic principles of cardioplegic management in donor heart preservation. Clin Ther 1991; 13:289-303.
366. Jeevanandam V, Auteri JS, Sanchez JA et al. Improved heart preservation with University of Wisconsin solution: Experimental and preliminary human experience. Circulation 1991; 84(Suppl III):324-328.
367. Taegtmeyer H, Russell RR. Biochemistry of the heart. In: Garfein OB, ed. Current Concepts in Cardiovascular Physiology. New York: Academic Press, 1990.
368. Dunn RB, Griggs DM Jr. Transmural gradients in ventricular tissue metabolites produced by stopping coronary blood flow in the dog. Circ Res 1975; 37:438-445.
369. Williamson JR, Ford C, Illingworth J et al. Coordination of citric acid cycle activity with electron transport flux. Circ Res 1976; 38(Suppl I):39-51.
370. Hohorst HJ, Reim M, Bartels H. Studies on the creatine kinase equilibrium in muscle and the significance of ATP and ADP levels. Biochem Biophys Res Commun 1962; 7:142-146.
371. Jacobsen IA, Pegg DE. Kidney. In: Karow AM Jr, Pegg DE, eds. Organ Preservation for Transplantation. New York: Marcel Dekker, Inc., 1981.
372. Wechsler AS, Abd Elfattach AS, Murphy CE et al. Myocardial protection. J Card Surg 1986; 1:271-306.
373. Tomasek JJ, Gabbiani G, Hinz B et al. Myofibroblasts and mechano-regulation of connective tissue remodelling. Nat Rev Mol Cell Biol 2002; 3:349-363.
374. Wang G, Reader J, Hynd J et al. Improved heart and lung preservation in a rat model. Transpl Int 1990; 3:206-211.

CHAPTER 7

Kidney Preservation

Luis H. Toledo-Pereyra,* Juan M. Palma-Vargas and Alexander H. Toledo

Introduction

Optimal conditions for organ procurement and preservation are fundamentally based on the suppression of the metabolism by hypothermia (0-5°C).[1-4] Recent progress obtained in kidney preservation can be directly traced to the control of the damage caused by prolonged hypothermia such as cell swelling,[5,6] nucleotide decay with subsequent loss of energy sources[7] and reperfusion injury.[8] Various substances have been used to overcome these harmful effects, including impermeable anions, colloids and protective drugs including free radical scavengers, calcium channel blockers and allopurinol among others.[9-12] Hypothermia remains the most single effective factor in successful kidney preservation. As indicated in previous chapters, two methods of preservation, hypothermic storage and hypothermic perfusion, are frequently used in kidney transplantation. This chapter will study the clinical issues and experimental findings relevant to kidney preservation.

Hypothermic Storage

Kidney hypothermic storage has been used extensively and is an excellent method for clinical transplantation. Most transplant centers in the world have used simple cold storage for kidney preservation. After the organ is removed from the donor, the kidney is flushed with a cold preservation solution developing rapid core cooling, reaching temperatures between 0°C and 5°C during storage prior to transplantation. The obvious advantages of this preservation technique are its simplicity and low cost. However, its reliability is dependent on the length of preservation and the type of solution used. If kidneys are obtained from heart-beating cadaver donors with minimal warm ischemia, simple cold storage after flushing is usually reliable for 24 hours. With the use of the appropriate flushing solutions, the duration of preservation can be extended to 36 hours.

When any degree of warm ischemia is involved, the effectiveness of each preservation method is decreased. Simple cold storage is effective for optimal harvested kidneys, whereas organs damaged by warm ischemia or hypotension are not as well preserved by this method as they are by the perfusion technique.

A large number of solutions have been developed for use in clinical hypothermic storage. Collins et al[13] in 1969 described a crystalloid solution that they used successfully to preserve dog kidneys on ice for up to 30 hours. The glucose and the impermeant anion sulfate (SO_4^-) in this solution may have been a primary reason for preservation success since the renal tubules are relatively impermeable to glucose under hypothermic conditions and sulfate is a poorly reabsorbed inorganic anion (Table 1). Both compounds improve kidney preservation by decreasing cell swelling as observed with depletion of energy stores. In 1976, the Eurotransplant Organization agreed on a standardized preservation solution based on a modified Collins solution that omitted $MgSO_4$, but had an increased concentration of dextrose resulting in a higher osmolarity. This solution was named Euro-Collins and has been widely used for simple flush preservation, with successful results in the

*Corresponding Author: Luis H. Toledo-Pereyra—Departments of Research and Surgery
Michigan State University, Kalamazoo Center for Medical Studies, Kalamazoo, Michigan
49008, USA. Email: toledo@kcms.msu.edu

Organ Preservation for Transplantation, Third Edition, edited by Luis H. Toledo-Pereyra.
©2010 Landes Bioscience.

Table 1. Composition of some common cold storage solutions for kidney preservation

Substance mM/L	Collins	Euro-Collins	Sacks	Hypertonic Citrate
Na⁺	10	9.3	15	78
K⁺	115	107	143	84
Mg²⁺	30	—	16	40
Cl⁻	15	14	15	1
HCO₃⁻	10	10	38	—
SO₄⁻	30	—	—	40
PO₄⁻	47	60	120	—
Mannitol	—	—	—	150
Citrate	—	—	—	54
Osmolarity	320	340	430	400
pH	7.0	7.3	7.0	7.1

clinical setting for up to 50 hours[14] (Table 1). In an attempt to overcome some limitations of the Collins-type solutions, the hypertonic citrate solution was created in 1976.[15,16] This solution also contains high concentrations of potassium and magnesium, but citrate replaces phosphate and mannitol replaces glucose (Table 1). In 1981, Toledo-Pereyra and associates introduced a series of modifications to the silica gel fraction that allowed them to increase the length of preservation.[17] A few years later, in 1986, Belzer's group from the University of Wisconsin introduced a preservation solution (UW) originally developed for preservation of the pancreas[18] and later applied to the kidney.[19] Like the Collins and Euro-Collins, UW solution had a high K⁺ formulation (Table 2). It contained MgSO₄, but at lower concentration than Collins. Much of the PO₄⁻ and all of the Cl⁻ in Collins and Euro-Collins were replaced by the organic anion, lactobionate. Raffinose, at a moderate concentration, was included as an impermeant sugar. Hydroxyethyl starch was used as a colloid. Pharmacologic additions included glutathione, adenosine, allopurinol, insulin, dexamethasome and an antibiotic.[20]

In 1980, Bretschneider[21] introduced the initial concept of the protective effect of the histidine-tryptophan-ketoglutarate (HTK) solution in myocardial tissue. Several years later, in 1999, de Boer and his associates[22] completed a Eurotransplant randomized multicenter study demonstrating similar results in kidney preservation when comparing UW versus HTK solutions. Interestingly, enough, both solutions showed identical results with delayed graft function in 33% of the transplants. Around the same time, another European study by Roels and his group[23] showed conflicting data by advancing inferior kidney preservation results when preserving these organs for more than 24 hours with HTK solution. Eight years later, in 2006, Agarwal and his colleagues[24] demonstrated a considerable clinical benefit of cold storing kidneys in HTK solution as compared to UW solution. At present, it would be fair to say that clinicians in the United States and Europe, consider both solutions equally effective for cold kidney preservation, depending on their use and experience accumulated with multiple organ procurement.

In an excellent review on organ preservation, Maathuis and his associates[25] analyzed many of the important solutions used in organ preservation in general. In our case, we are focusing in this chapter, on those solutions applied to kidney transplantation. (Tables 1, 2 and 3). In addition, to what has already been discussed, other solutions utilized in kidney preservation are the phosphate buffer solution (PBS),[26] the Celsior solution, introduced in 1994 for heart preservation[27] and expanded into clinical kidney preservation in 2001[28] and the Institut George Lopez

Table 2. **Composition of the University of Wisconsin (UW) solution for kidney hypothermic storage**

Sodium (mmol/L)	30
Potassium (mmol/L)	125
Magnesium (mmol/L)	5
Sulphate (mmol/L)	5
Lactobionate (mmol/L)	100
Phosphate (mmol/L)	25
Raffinose (mmol/L)	30
Adenosine (mmol/L)	5
Glutathione (mmol/L)	3
Allopurinol (mmol/L)	1
Insulin (U/L)	100
Dexamethasome (mg/L)	8
Bactrim (ml)	0.5
HES (g/L)	50
Osmolality (mmol/Kg)	320
pH	7.4

solution,[29] developed by the Lyon group in 2002.[30] The IGL-1 solution utilizes the extracellular composition of Celsior and the colloid effect of the UW solution where polyethylene glycol is used instead of hydroxyethyl starch. The Celsior solution combines some of the principles associated with the UW and HTK solutions as well[25,31] (Tables 1, 2 and 3). A clear superiority of these new solutions over the previously accepted standard solutions, such as UW and HTK, has not been demonstrated yet.

Recently, in 2007, Lopez-Neblina and his group[32] introduced a new solution (HBS) tested on rat kidneys for hypothermic storage. The composition of the solution was unique in that it included macromolecules, high-energy cellular substrates and a mixture of antiproteolytic amino acids, antioxidants and anti-inflammatory compounds. It was interesting to observe that the HBS solution was demonstrated to be superior to the UW and HTK solutions when the three of them were compared in the same model of isotransplantation of preserved kidneys in inbred Brown Norway rats. Further studies, in larger animals are needed to prove the potential advantageous role of this new preparation for kidney hypothermic storage.

Presently, UW and HTK are the most commonly used solutions (Figs. 1 and 2).[25,31,33] In general, the efficacy of each solution is dependent upon its composition, the length of warm ischemia and preservation time. Both are different, however, in that HTK has histidine as a buffer and impermeant and in UW the impermeants are saccharides and lactobionate.[25,31] Other differences in buffers, electrolytes, antioxidants and additives are obvious.[25] Several investigators have determined that to be protective in preventing storage damage and avoiding hypothermic-induced cell swelling, all solutions, upon reperfusion, need to restore osmotic stability with maintenance of cell viability.[25,31,34-36]

Results from human trials have shown that kidneys preserved in UW or HTK solutions yielded better renal graft outcome than those preserved in Euro-Collins solution.[22] However, this difference has not been as significant when compared to other data, as the one seen in the results of the European multicenter trial[37] and from the UCLA Transplant Registry[38] (Fig. 3).

Table 3. *Composition of phosphate-buffered sucrose (PBS), histidine-tryptophan-ketoglutarate (HTK), Celsior and Institut George Lopez (IGL-1) solutions*

Substance	PBS	HTK	Celsior	IGL-1
PEG-35 (g/L)	—	—	—	1
Histidine (mM/L)	—	198	30	—
Lactobionate (mM/L)	—	—	80	100
Mannitol (mM/L)	—	38	60	—
Raffinose (mM/L)	—	—	—	30
Sucrose (mM/L)	140	—	—	—
KH$_2$PO$_4$ (mM/L)	—	—	—	25
NaH$_2$PO$_4$ (mM/L)	13	—	—	—
Na$_2$ HPO$_4$ (mM/L)	56	—	—	—
Calcium (mM/L)	—	0.0015	0.25	0.5
Chloride (mM/L)	—	32	42	—
Magnesium (mM/L)	—	4	13	—
Magnesium sulphate (mM/L)	—	—	—	5
Potassium (mM/L)	—	9	15	25
Sodium (mM/L)	125	15	100	120
Allopurinol (mM/L)	—	—	—	1
Glutathione (mM/L)	—	—	3	3
Tryptophan (mM/L)	—	2	—	—
Adenosine (mM/L)	—	—	—	5
Glutamic acid (mM/L)	—	—	20	—
Ketoglutarate (mM/L)	—	1	—	—

From Maathuis et al.[25]

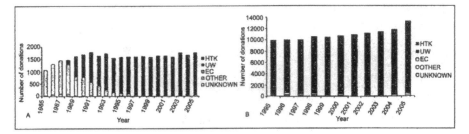

Figure 1. Use of cold storage solutions in Eurotransplant region in deceased donors from 1985-2008. Used with permission of Lippincott Williams and Wilkins, from Maathuis et al, Transplantation 2007; 83:1289-1298.

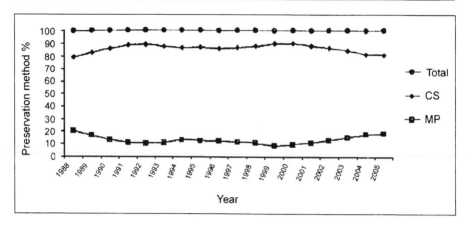

Figure 2. Use of cold storage solution in the United States in deceased donors from 1985-2005 (based on Organ Procurement and Transplantation Network data of October 2006). Used with permission of Lippincott Williams and Wilkins, from Maathuis et al, Transplantation 2007; 83:1289-1298.

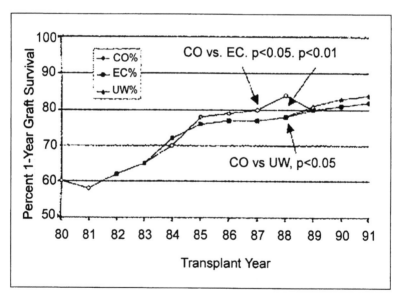

Figure 3. Early historical results on the effect of different cold storage solutions on transplant outcome. No significant differences in survival rates were noted among these three popular cold storage solutions, except in 1987 and 1988, during which time kidneys preserved in Collins solution enjoyed significantly higher graft survival. However, Collins solution was used in less than 20% of these cases in 1987 and 1988 and this result may be primarily a center-related effect. Obtained with permission from Zhou YC, Cecka JM. Preservation. Clinical Transplant 1992, Terasaki and Cecka, eds. UCLA Tissue Typing Laboratory, Los Angeles, California.

Hypothermic Perfusion

The second alternative method of renal preservation is continuous hypothermic perfusion. Two basic types of perfusion systems have been used in transplantation, hypothermic pulsatile perfusion and hypothermic nonpulsatile perfusion.

Continuous hypothermic perfusion preservation of canine kidneys was introduced in 1967 by Belzer and associates.[39] Since then, pulsatile preservation has been used clinically and it made cadaver renal transplantation a semi-elective procedure.

All perfusion machines work via the same mechanism, with the perfusate being delivered to the renal artery continuously by a pulsatile or roller pump. The venous effluent is collected and cooled in a heat exchanger, oxygenated by a membrane oxygenator, in some cases filtered and then passed back to the kidney (Fig. 4). A rising perfusion pressure is often a reliable indicator of kidney damage; however, this method is not foolproof and should be used in coordination with other standards, since nonviable kidneys do not always show evidence of rising perfusion pressure.

Depending on the perfusion machine used, in general, perfusion pressures of less than 60 mmHg are commonly used as the acceptable normal limit, since a higher perfusion pressure may damage the kidney.[40] With normal levels of perfusion pressure, flow rates are usually maintained at about 1 ml/min/g and a pulse rate at 60 beats per minute.[40] The concentration of oxygen dissolved in the perfusate does not appear to be too critical for effective preservation, thus many groups have eliminated the membrane oxygenator from the perfusion machine. While we have shown that monitoring levels of dissolved oxygen is non-essential to a perfusate's success, the exact composition of the ideal perfusate has yet to be determined. Many questions remain concerning the need for amino acids, free fatty acids and substrates, as well as the appropriate pH, osmolarity and the ratio of sodium to potassium in the perfusate.

With minor variations, most perfusate solutions used for continuous perfusion contain electrolyte concentrations resembling extracellular fluids, substrates for anaerobic metabolism and osmotically active but poorly permeable substances such as glucose and mannitol.[41,42] Hyperosmolarity appears to offer some protection because it forces fluid from the interstitial to the intravascular space. Maintenance of a physiological pH during perfusion has also been the goal of perfusion technologists. There is some early evidence that kidneys perfused at a pH of 7.4 function better than kidneys perfused at acid or alkaline pH.

Cryoprecipitated plasma (CPP) was for many years the standard solution for kidney preservation. However, because of complicated steps needed to facilitate its manufacturing, less enthusiasm was oriented to its production. At the University of Minnesota in 1973, Toledo-Pereyra et al developed a novel perfusate, silica gel fraction of plasma (SGF), which was prepared by treating plasma

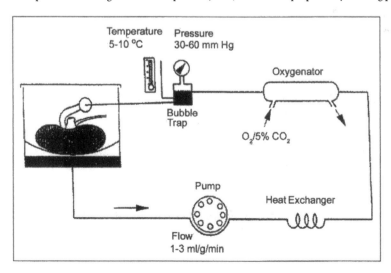

Figure 4. Diagrammatic scheme of a continuous hypothermic perfusion system for kidney preservation. Reprinted with permission from Marshall et al. Preservation. In Morris PJ, ed. Kidney Transplantation, 4th ed. Pennsylvania: WB Saunders Company, 1994:97.

with silica.[43] This perfusate, capable of removing unstable lipoproteins and other lipid materials, as shelf-stable for long periods of time and yielded good preservation of kidneys for 72 hours.[44]

New synthetic stable albumin-based perfusates were also clinically introduced in 1973 replacing other preservation solutions.[45] The use of human serum albumin (HAS) or a plasma protein fraction (PPF) allowed for preservation times of 48-72 hours, similar to SGF.[46] However, due to its sensitive chemical structure, albumin can undergo protein denaturation within the changing environment of perfusion preservation.[47] Thus, it is not surprising that this particular perfusate has been implicated in renal dysfunction during long-term perfusion.[48]

With the introduction of hydroxyethyl starch (HES), a new synthetic colloid as part of the University of Wisconsin preservation solution, better long-term preservation was obtained in comparison to the traditional albumin perfusate. HES offered several advantages over human serum albumin, including ease of preparation, long shelf-life and stability, no batch variation, increased osmotic activity and potentially lower cost. Furthermore, its completely synthetic composition eliminated the risk of immunogenicity.[49,50]

In order to further enhance the effectiveness of perfusates some modifications were made to the UW solution. The electrolyte content was modified from an intracellular (high K[+]- low Na[+] in the original UW solution) to an extracellular formulation (low K[++]- high Na[+]). Calcium (1.5 mM) was also added to the modified perfusate. Additional variations included replacement of lactobionate, the major anion from the standard UW perfusate, by gluconate which resulted in a superior impermeable anion composition (Table 4). These substitutions resulted in a significant improvement in the continuous preservation of kidneys (up to 5 days) using the high Na[+]- gluconate solution[51] as compared to machine perfusion with the high K[+]- lactobionate solution without calcium.[52]

Other perfusates have been utilized for HPP with similar results except for noted exceptions.[53-58] In any event, the Belzer perfusate has been modified to include gluconate (Belzer-I), albumin gluconate (Belzer-II), or to have additives of the type of prostaglandin E1, nitroglycerine and polyethylene glycol-superoxide dismutase.[53] With this last modification, Guarrera and his group improved the 6-month survival of cadaver perfused kidneys to 96% compared to 87%, using the Belzer albumin gluconate perfusate and 90%, using the regular Belzer machine perfusate.[53] In a previous study from the same laboratories, Polyak demonstrated that the combination of Belzer I perfusate and PGE1 improved the overall efficiency of cadaver kidney transplantation.[54] Others have used different perfusates, including the utilization of the Celsior solution and the addition of polyethylene glycol to it when successfully perfusing pig kidneys.[58] The aforementioned modifications and their overall impact on organ viability when using different preservation techniques reflected fundamental distinctions between the mechanism of injury during cold storage as compared to the one seen after continuous machine perfusion.

Table 4. Composition of the University of Wisconsin (UW) solution for continuous kidney machine perfusion (HES-gluconate)

Sodium (mmol/L)	140
Potassium (mmol/L)	25
Magnesium (mmol/L)	5
Gluconate (mmol/L)	85
Phosphate (mmol/L)	25
Glucose (mmol/L)	10
Gluthatione (mmol/L)	3
HES (g/L)	50
Osmolality (mmol/L)	350
pH (0°C)	7.1

Advantages and Disadvantages of Preservation Methods

Hypothermic Storage vs Hypothermic Perfusion

The characteristics and details of the various preservation methods for kidney transplantation have been reviewed before in this book. Briefly suffice it to day, that the previous controversy as to which method of preservation, hypothermic storage (HS) or hypothermic pulsatile perfusion (HPP) should be routinely utilized has diminished in recent times. Many parameters must be taken into consideration for an appropriate evaluation, specifically, the period of warm ischemia, donor pretreatment, the hemodynamic characteristics of the donor, the length of preservation and the preservation solution used.

In the United States, HPP was the standard protocol for kidney preservation through the 1970s, though few written protocols existed. However, a variety of studies showing little benefit of HPP over HS led to the nearly uniform adoption for HS by the early 1980s.[59] HS was logistically easier, much less costly than HPP and facilitated organ sharing. No highly trained personnel were needed. Furthermore, the switch from hospital-based organ procurement organizations to those serving multiple transplant centers also favored simpler preservation methods. However, on the other hand, simple HS may result in poorer preservation and it is associated with a significant rate of early posttransplant renal dysfunction, necessitating dialysis within the first week after transplantation.

In the cyclosporine era, delayed graft function (DGF) was associated with poorer graft and patient survivals.[60,61] One multi-center study in the United States indicated that the rate of acute tubular necrosis (ATN)—defined by the need of post-operative dialysis—for kidneys preserved with static HS method, regardless of the preservation solution, was approximately 27%.[33] We have seen similar results and even higher rates of ATN with the use of similar cold storage methods.[62]

In the last early 1990s, several publications showed improved early function up to 90%, with machine perfusion, in kidneys preserved for an average of 30 hours.[63-65] The higher immediate function rate achieved with machine perfusion—only used in less than 20% of the kidneys preserved in the United States—probably meant that fewer kidney recipients required postoperative dialysis and therefore reduced kidney transplant cost was a firm possibility.

At the turn of the century, in 2004, pulsatile machine perfusion with various new solutions showed improved graft function.[66] In 2007, Kwaitkowski and his group,[67] demonstrated that cadaver kidneys transplanted after machine perfusion had better graft survival and lower incidence of ATN. Matsuoka and his associates have shown similar results in expanded criteria donors.[68] So, there appears to be a trend toward improved results of machine preservation for nonmarginal and marginal kidney donors.[53-38,66-68] More studies ahead will allow us to define the definitive role of machine perfusion in kidney transplantation. A machine perfusion trial in collaboration with Eurotransplant, which began in November, 2005, should present some preliminary evidence in this regard.[69]

Another advantage using pulsatile perfusion systems is the ability to eliminate 75% of hepatitis C virus (HCV) from the organ with the original machine perfusion scheme. Additional viral depletion steps (dilution and or ultra-filtration) allow a further reduction of the final HCV innoculum in the procured organ, up to 99% of the virus being eliminated form the kidney.[70] This may allow for the practical reduction of the virus transmission and related liver disease in high risk recipients during renal transplantation.

Even though another chapter in this book analyzes the utilization of different perfusion machines for kidney preservation, we will refer to this topic. There are three kidney perfusion machines currently in clinical use. The old MOX-100 (Waters Medical Systems, Rochester, MN), the LifePort* Kidney Transporter (Organ Recovery Systems, Charleston, SC) (Fig. 5) and the RM3* Pulsatile Perfusion Kidney Preservation System (Waters Medical Systems, Rochester, MN) (Fig. 6). The latter two are by far the most frequently used in this country. Other individual units are still utilized in come pioneering centers in the United States and Europe. One of Us (LHT-P) has had an opportunity of working with the three previously described perfusion machines and finds

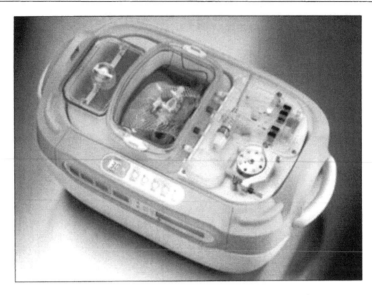

Figure 5. LifePort® kidney transporter from organ recovery systems. LifePort is a product of Organ Recovery Systems, Inc. (Photographer, Tom Petroff Photography).

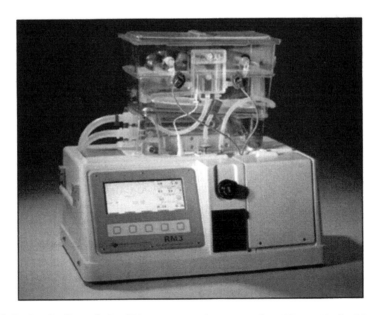

Figure 6. RM3 pulsatile perfusion kidney preservation system from Waters Medical Systems.

them to be readily accessible and easy to use. Preference is based on experience and dedication to learn the details of the operating system.

Many experimental systems of kidney machine perfusion have been utilized in the last 20 years. For instance, a new system called homeostatic perfusion apparatus (HPA) was developed by Rapaport's group.[1] The HPA consists of a monitoring and control circuit, a compliance chamber and a roller pump with adjustable flow rate. The monitoring and control circuit continuously measures the

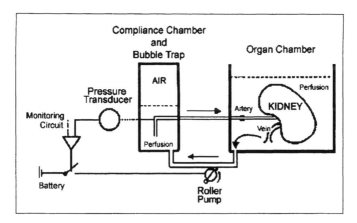

Figure 7. Schematic view of the homeostatic perfusion apparatus (HPA) developed by Rapaport and associates. Reprinted with permission from Yland MJ, et al. Transplant Proc 1993; 25:3087.

perfusion pressure and regulates pump action through a closed feedback loop (Fig. 7). The HPA provides perfusion at a controlled pressure instead of perfusion at a controlled flow as used in the MOX-100 console. In the MOX-100 machine, the pulse rate is usually constant at 60 beats per minute, whereas in the HPA the flow is reduced by decreasing the pulse rate. The HPA operates without an oxygenator and the entire perfusion circuit is placed in an ice chest. The low pulse rate provided by the HPA has been designed to prevent perfusion trauma to the endothelium and the microcirculation. Furthermore, the HPA perfusion machine is independent of organ size because perfusion is pressure controlled. Therefore, the HPA may by useful for machine preservation of other organs as well for improving the outcome of organs preserved by traditional machine perfusion systems. Another novel kidney machine device from the Groningen group appeared last year (2007).[72] This system provided improved viability especially when 30/20 mmHg pressures were used.[72]

In summary, kidney preservation has made important advances since the publication of the second edition of this book in 1997.[73] Basically, the role of kidney hypothermic storage and perfusion preservation is better understood and continuously more studies are adding enhanced evidence to the basis of improved preservation.

Pharmacological Strategies for Kidney Preservation

A number of other maneuvers targeted at different pathophysiologic processes have shown promise and are being evaluated in various systems. These procedures can be divided according to the time of implementation: before perfusion (donor), during perfusion (graft) and before transplantation (recipient).

Before Perfusion (Donor) Strategies

Preservation strategies differ according to the type of donor: living versus cadaveric. In living donors, the maneuvers are mainly focused to maintain extracellular fluid volume and to ensure adequate diuresis with the use of osmotic agents such as mannitol.

Brain death leads to hemodynamic deficiencies due to loss of brain stem function,[74,75] therefore potential kidney cadaveric donors require intravascular volume expansion and dopaminergic renal support, if necessary to maintain a systolic blood pressure above 100 mgHg and urinary output at 100 ml/hr. Furthermore, in order to prevent vasospasm during organ harvesting, vasoactive agents such as chlorpromazine, phenobanzamine and calcium-channel antagonists have been used experimentally with varying results.[76,77] Finally, correction of acid-base disturbances[78] and metabolic support (adenine nucleotides or precursors) in case of prolonged treatment can be considered for superior preservation.

During Perfusion (Graft) Strategies

Calcium has been shown to be a critical factor for cell function and it has also been implicated in renal ischemic injury following reperfusion.[79] In this regard, McAnulty et al demonstrated that the addition of 0.5 mM Ca^{+2} in a gluconate-based perfusate solution provided superior results as demonstrated by improved mitochondrial function and cell viability in comparison to the same solution with no added Ca^{+2}. Furthermore, calcium has been an important factor as an element for perfusion preservation of canine kidneys for up to five days.[80]

Phenothiazines, due to their inhibitory effects on calmodulin and phospholipases, have been of interest with regard to their potential for modifying the preservation injury.[81] Improvement of preservation solution was achieved by adding chlorpromazine or trifluoperazine to Collins solution used for canine kidneys[82] or by adding either chlorpromazine or trifluoperazine to a simplified lactobionate sucrose preservation solution used for rat livers.[83]

Furthermore, activation of proteases in lysosomes or cytosol could contribute to tissue damage during ischemia. According to one study, addition of a protease inhibitor cocktail, which included leupeptin, pepstatin A, phenylmethylsulfonyl fluoride and diisopropyl-fluorophosphate to the Euro-Collins solution, allowed for substantially improved survival of rat livers after 4 hours of cold storage.[84]

Moreover, polyethyleneglycol has been shown to offer protection of isolated hepatocytes and kidneys against cold injury.[25,29,30] Its protective effects could be attributable to the suppression of the inflammatory response, membrane stabilization and modification of immunogenicity.[25] Polyethyleneglycol in a simplified UW solution gave superior results to standard UW in models of cardiac and pancreas transplantation.[25,85]

Prior to Transplantation (Recipient) Strategies

Measures include attempts utilized to decrease damage mediated by reactive oxygen species (ROS) after reperfusion such as administration of indomethacin (to reduce cyclo-oxygenase mediated ROS) prior to harvesting,[86] treatment of rabbit donor with deferoxamine (to reduce iron-catalyzed free radical formation)[87] and the infusion of 20 mg of the free radical scavengers, superoxide dismutase and catalase, into the renal artery just before reperfusion in cold preserved porcine kidneys, substantially ameliorated the reperfusion injury improving renal function after transplantation.[88,89]

Treatment with calcium antagonists such as verapamil has been shown to decrease neutrophils, adherence to damaged endothelium preventing reperfusion damage after transplantation.[90,91] In a more recent study, Anaya-Prado and his group,[92] extensively reviewed the most important advances associated with ischemic injury, which is directly applicable to transplantation results.

Length of Preservation

There is a trend for cold-stored kidneys to exhibit higher acute tubular necrosis (ATN) rates with increased preservation time.[33] This trend is not as persistent in kidneys preserved by pulsatile perfusion.[93] Improvement in the quality of organ preservation has been noted with the use of new perfusates as previously indicated in this chapter.[53-58,66-68] The modification of the storage solution has provided improved results as well.[25-30]

Excellent immediate function of renal allografts has been reported when kidneys initially preserved by simple cold storage were later placed onto the perfusion machine using the Belzer's perfusate for long preservation (combined procedure), before transplantation.[94,95]

Cold ischemia time (CIT) has shown to have a significant impact on renal allograft outcome. Multivariate analysis of the UCLA[96] and UNOS Register reported an inferior 1 and 3 year survival rate of kidneys with prolonged CIT. There were significantly more cases of delayed graft function (ATN) among kidney with longer CIT. However, interestingly enough, longer CIT did not significantly affect survival. HLA matching effectively neutralized the effect of CIT, as reported previously.[97]

Furthermore, the South Eastern Organ Procurement Foundation,[98] in a retrospective study of 17,937 cadaveric renal transplants, using multivariate analyses, found that graft survival was adversely affected by recipient retransplantation, poor match and delayed graft function. Good outcome was more likely to occur in primary allograft recipients who received a well-matched organ with immediate graft function. It was interesting to note that kidneys preserved beyond 48 hours demonstrated one year allograft survival function not different from organs preserved for shorter periods of time, suggesting then that kidneys should no longer discarded for that reason alone.

This particular area has been debated for years in transplantation and still we do not have an answer as to the time limit of kidney preservation. We previously considered this issue and stated that kidneys preserved for periods longer than 30 hours should be carefully evaluated and only accepted when there are no other high risk factors present.[73]

Nonheartbeating Cadaveric Donors

Considering the overall shortage of organs for transplantation during the last few years, nonheartbeating donors (NHBD) have become a valuable source of kidney allografts.[99,100] In an attempt to standardize the criteria for procurement and transplantation of NHBD, clinicians from Maastricht University hospital have proposed a classification system consisting of four principal categories. This concept is based on different types of NHBD in which organ procurement could be feasible (Table 5).[101]

During the last few years several techniques have been developed to improve the viability of posttransplant function of organs obtained form NHBDs; total body cooling (TBC), in situ cooling and in situ perfusion. To shorten warm ischemia time and thus minimize ischemic damage to the kidneys, an emergency in situ perfusion protocol was implemented by Kootstra's group.[102] In short, the protocol in general, considers the following steps: the patient is suitable for kidney donation according to NHBD criteria shown in (Table 6), family consent is obtained for organ procurement. The in situ perfusion protocol is initiated by placement of a silicone, radiopaque double balloon triple lumen (DBTL) catheter (Porges AJ 6516) in the femoral artery. The size of the catheter is 16 charriere (CH), the total length is 90 cm and the distance between the thoracic and the aortic balloons is 23 cm. This special design allowed for selective perfusion of the renal arteries as depicted in (Fig. 8). After correct placement of the DBTL catheter determined by abdominal x-rays, the infusion system is connected to the catheter and perfusion with histidine-tryptophan-ketoglutarate (HTK) at 4°C is started. Perfusion flow starts at 250 ml/min with pressure of 80-100 mmHg. After 6 to 8 L of perfusion fluid, the flow can be reduced to 100 ml/min. The total amount of perfusion fluid usually needed to clear the venous effluent normally does not exceed 15 L. Finally, nephrectomy is performed in the operating room, preferably within 2 hours after starting the in situ perfusion. Subsequently, kidneys are placed in a machine perfusion system (for viability testing) until transplantation.

Table 5. The Maastricht categories for NHB donors

Category	Description
1	Dead on arrival
2	Unsuccessful resuscitation
3	Awaiting cardiac arrest
4	Cardiac arrest while brain dead

Reproduced with permission from Koostra G et al. Categories of non-heart beating donors. Transplant Proc 1995; 27:2893.

Table 6. Non-beating donor criteria

1. Total duration of circulatory arrest less than 30 minutes, excluding the time of effective resuscitation

2. Effective resuscitation no longer than 2 hours

3. Age <65 years

4. No signs of intravenous drug abuse

5. No systemic infection or evidence of sepsis

6. No history of kidney disease, uncontrolled hypertension or malignancies other than primary, nonmetastasizing, central nervous system tumors

Reproduced with permission from Heineman E et al. Non-heart beating donors: methods and techniques. Transplant Proc 1995; 27:2895.

Kowalski and her associates have also developed an alternative two step technique for organ procurement from NHBD.[103] First, the femoral arterial-venous system is cannulated for kidney flushing-perfusion at 70 mmHg with UW solution containing 4 mg of trifluoroperazine, 40 units of insulin, 16 mg dexamethasone and 20,000 units of heparin. The second step (in situ cooling) begins with the placement of two catheters which are inserted into the abdomen through specially designed laparoscopy ports for recirculation of 3-4 liters of sterile iced 0.9% saline through the abdominal cavity using a sterile, closed-circuit, refrigerated pump. Thereafter, the kidneys are harvested and placed on a pulsatile preservation machine—using the Belzer's solution—until transplantation.

The implementation of effective alternative organ procurement techniques for NHBD might allow for an increase in the number of grafts available for transplantation. According to some studies, the outcome of kidneys procured from NHBD showed a low primary nonfunction rate (<10%) with patient and graft survival rates that have not been significantly decreased,[104,105] However, because of the high incidence of delayed function (up to 60%), it is recommended to maintain cold ischemia times at a minimum and transplantation performed within 24 hours under this particular setting.[106]

A the number of NHBD is increasing a source of marginal kidneys for transplantation, more practical measures have been taken into consideration based on the down-time from cardiac arrest to the initiation of cold intra-aortic (HTK or UW) perfusate infusion. In my experience (LHT-P), times of 10 minutes of less to the beginning of cold flush administration are ideal for organ placement, time of 20-30 minutes or higher have become extremely difficult for coordinators to find a transplant program which would be willing to accept these kidneys even with acceptable machine perfusion parameters. As we understand better the effect of down-time on the NHBD pathophysiology, the transplant community would be wiling to accept kidneys or even other organs, which otherwise are not even considered today.

Predicting Factors of Renal Function

The use of pulsatile perfusion for kidney preservation provides an opportunity to quantitatively evaluate the suitability of an organ for transplantation, according to some authors, the best indicators of kidney viability are perfusate pressure and pulsatile flow.[70] There is a direct correlation between the integrity of the cortical microcirculation during preservation, reflected by a constant low perfusate pressure and subsequent life-sustaining function of the kidney after transplantation.[107] However, these variables usually remain satisfactory except for those kidneys with severe damage. Thus, these factors alone may not ultimately predict organ viability.

Another potential factor is the composition of the perfusate, which may reflect the kidney's metabolic tolerance. Fahy, et al, measured several parameters including metabolites such as lactate,

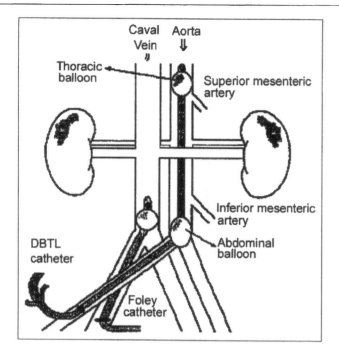

Figure 8. Schematic representation of the double-balloon triple-lumen (DBTL) catheter when inserted in the abdominal aorta for procurement of kidneys of nonheartbeating donors. Obtained from Heineman E et al. Transplant Proc 1995; 27:2895.

glucose, xanthine and hypoxanthine, as well as potassium, pH and enzymes and found no direct correlation between these parameters and overall kidney function.[108]

Preserved kidney biopsies, although controversial, represent another tool for assessment of organ viability. For example, Sells et al showed a positive correlation between tissue K^+/Na^+ ratios and kidney function after transplantation.[109] However, the extent of cellular damage is often more apparent after reperfusion, therefore limiting the predictive value of tissue biopsies. Furthermore, a similar study failed to show a correlation between nucleotide measurement in histological samples of preserved kidneys and ATP resynthesizing capacity upon revascularization, yet another limitation in assessing organ viability.

Kidneys from younger and older donors have inferior renal graft outcome. One of the explanations to consider was that younger and older kidneys might be more vulnerable to prolonged cold ischemia time. The survival of kidneys from donors under age 15 had significantly poorer outcome when cold ischemia time was greater than 36 hours.[110]

HLA matching should be the first step to achieve acceptable results and avoid poor outcomes with marginal kidneys. If a matched recipient is not available, the marginal donor kidney will benefit from a reduced cold ischemia time.

Other Possibilities for Improved Kidney Preservation

The use of oxygen (O_2) during kidney machine preservation or cold storage has shown beneficial effects.[25] From renal persufflation preservation utilized clinically in 10 kidneys in 1989[110] to the use of an oxygenated machine perfusion solution[112] and the possibility of utilizing perfluorocarbons (PFC) in renal preservation[113] are a few of the many potential alternatives for O_2 delivery. The administration of CO (carbon monoxide) and NO (nitric oxide) are worth considering in renal preservation[25,114,115] as well.

Normothermic recirculation of abdominal organs using a cardiopulmonary bypass protocol prior to cold storage resulted in reduction of primary graft dysfunction of kidneys obtained from NHBD.[116] The use of normothermic machine perfusion (NMP) is another feasible alternative when rescuing severely damaged kidneys for transplantation.[117,118,119]

As indicated by Jamieson and Friend in their recent comprehensive review of organ reperfusion and preservation,[31] donor preconditioning with pharmacological interventions such as hormonal replacement therapy, administration of adenosine deaminase inhibitor, N-acetyl cysteine, prostacyclin analogues and maneuvers to enhance heme-oxygenase-1 (HO-1) are all realistic and viable treatments. Gene transfer preconditioning, ischemic preconditioning and heat-shock preconditioning, ischemic are future considerations in their thoroughly presented analysis of these new possibilities.[31]

Conclusion

The progress made over the past several years in understanding acute ischemic injury has significantly improved renal preservation. Nevertheless, acute ischemic injury continues to be a problem in the transplanted organ as an important cause of delayed graft function. With improved understanding of the pathophysiological mechanism involved, strategies have been devised to minimize ischemic injury during preservation ex vivo. With further insight into the mechanisms of ischemic renal injury and new pharmacologic approaches that derive from this knowledge, better results can be obtained after preservation. Flushing and perfusing techniques, solutions and new additives continue to evolve and to improve. Optimal management of the heart-beating brain dead donor and the NHBD set the stage for good early function in the recipient. Attempts to rescue kidneys from less than optimal donors are required more frequently, as the need for organs continues to outstrip the supply. The effectiveness of a number of other components of preservation solutions, as well as the relative merits of continuous perfusion of the organ ex vivo, remains important. The future may see increasing use of therapy through the wide immunological window that more prolonged preservation can offer.

References

1. Levy MN. Oxygen consumption and blood flow in the hypothermic perfused kidney. Am J Physiol 1959; 197:1111-14.
2. Mitchell RM. Renal cooling and ischemia. Br J Surg 1959; 46:93-94.
3. Calne RY, Pegg DE, Pryse-Davies J et al. Renal preservation by ice cooling. An experimental study relating to kidney transplantation from cadavers. Br Med J 1963; 6:221.
4. Belzer FO, Southard JH. Principles of solid-organ preservation by cold storage. Transplantation 1988; 45:673.
5. McKnight ADC, Leaf A. Regulation of cellular volume. Physiol Rev 1977; 57:510.
6. Flores J, DiBona DR, Bech CH et al. The role of cell swelling in ischemic renal damage and the protective effect of hypertonic solute. J Clin Invest 1972; 51:118.
7. McCord JM. Oxygen-derived free radicals in post-ischemic tissue injury. N Eng J Med 1985; 312:158.
8. Cheung JY, Bonvengre JV. Calcium and ischemic injury. N Engl J Med 1986; 314:1670.
9. Toledo-Pereyra LH. Organ Preservation. I. Kidney and pancreas. J Surg Res 1981; 30:165.
10. D'Alessandro AM, Southard JH, Love RB et al. Organ preservation. Surg Clin North Am 1994; 74:5:1083.
11. Belzer FO, Southard JH. The University of Wisconsin organ preservation solution: components, comparisons and modifications. Transplant Rev 1993; 7:176.
12. Kumano K, Wang W, Masaki Y et al. Roles of solution impermeants during cold kidney storage. Transplant Proc 1994; 26:2410.
13. Collins BM, Bravo SM, Terasaki PI. Kidney preservation for transplantation. Initial perfusion and 30 hours ice storage. Lancet 1969; 2:1219-22.
14. Squifflet JP, Pirson Y et al. Safe preservation of human renal cadaver transplants by Euro-Collins solution up to 50 hours. Transplant Proc 1981; 13:693-96.
15. Ross H, Marshall VC, Escott ML. 72 hour canine kidney preservation without continuous perfusion. Transplantation 1976; 21:498.

16. Marshall VC, Ross H, Scott DF et al. Preservation of cadaver renal allografts: comparison of ice storage and machine perfusion. Med J Aust 1977; 2(11):353-6.
17. Toledo-Pereyra LH. A new generation of colloid solutions for preservation. Dial Transpl 1985; 14:143.
18. Wahlberg JA, Southard JH, Belzer FO et al. Development of a cold storage solution for pancreas preservation. Cryobiology 1986; 23:477.
19. Ploeg RJ, Goosens D, McAnulty et al. Successful 72-hour storage of dog kidneys in UW solution. Transplantation 1988; 45:935.
20. Southard JH, Belzer FO. The University of Wisconsin organ preservation solution: components, comparisons and modifications. Transplant Rev 1993; 4:176.
21. Bretschneider HJ. Myocardial protection. J Thorac Cardiovasc Surg 1980; 28:295.
22. de Boer J, De Meester J, Smits JM et al. Eurotransplant randomized multicenter kidney graft preservation study comparing HTK with UW and Euro-Collins. Transpl Int 1999; 12:447.
23. Roels L, Coosemans W, Dorzck J et al. Inferior outcome of cadaveric kidneys preserved for more than 24 h in histidine-tryptophan ketoglutarate solution. Leuven Collaborative Group for Transplantation. Transplantation 1998; 66:1660.
24. Agarwal A, Murdock P, Fridell JA. Comparison of histidine tryptophan ketoglutarate solution and University of Wisconsin solution in prolonged cold preservation of kidney allografts. Transplantation 2006; 81:480.
25. Maathius M, Leuvenink HGD, Ploeg RJ. Perspectives in organ preservation. Overview. Transplantation 2007; 83:1289.
26. Lam FT, Mavor AI, Potts DJ et al. Improved 72-hour renal preservation with phosphate-buffered sucrose. Transplantation 1989; 47:767.
27. Menasche P, Termignon JL, Pradier F et al. Experimental evaluation of Celsior, a new heart preservation solution. Eur J Cardiothorac Surg 1994; 8:207.
28. Faena A, Catena F, Nardo B et al. Kidney preservation with University of Wisconsin and Celsior solutions: a prospective multicenter randomized study. Transplantation 2001; 72:1274.
29. Ben Abdemmebi H, Steghens JP et al. A preservation solution with polyethylene glycol and calcium: A possible multi-organ liquid. Transpl Int 2002; 15:348.
30. Hauet T, Goujom JM, Baumert H et al. Polyethylene glycol reduces the inflammatory injury due to cold ischemia/reperfusion in autotransplanted pig kidneys. Kidney Int 2002; 62:654.
31. Jamieson RW, Friend PJ. Organ reperfusion and preservation. Frontiers in Bioscience 2008; 13:221.
32. Lopez-Neblina F, Toledo AH, Toledo-Pereyra LH. Evaluation of a noel cold storage solution (HBS) in a rat kidney transplant model. J Invest Surg 2007; 20:257.
33. Belzer FO. Evaluation of preservation of the intra-abdominal orgqans. Transplant Proc 1993; 25:2527.
34. Downes G, Hoffmann R, Huang J et al. Mechanism of action washout solutions for kidney preservation. Transplantation 1973; 16:45.
35. Green C, Pegg D. The effect of variations in electrolyte composition and Osmolarity of solutions for infusion and hypothermic storage of kidneys. Organ preservation II. Edinburgh: Churchill Livingstone, 1979; 86-101.
36. Andrews P, Coffey A. Factors which improve the preservation of nephron morphology during cold storage. Lab Invest 1982; 46:100.
37. Ploeg RJ, Van Bockel JH, Langendijk et al. Effect of preservation solution on results of cadaveric kidney transplantation. Lancet 1992; 340:129.
38. Koyama H, Cecka JM, Terasaki PI. A comparison of cadaver donor kidney storage methods: pump perfusion and cold storage solutions. Clin Transplant 1993; 7(2):199-205
39. Belzer F, Ashby B, Dunphy J. 24 hours and 72 hours preservation of canine kidneys. Lancet 1967; 2:536.
40. Gransmann R, Raab M, Mensel E et al. Analysis of the optimal perfusion pressure and flow rate of the renal vascular resistance and oxygen consumption in the hypothermic perfused kidney. Surgery 1975; 77:451.
41. Southard J, Belzer F. Organ Transplantation and Replacement. Philadelphia: Lippincott, 1990; 296.
42. Halasz N, Collins G. Simplification of perfusion preservation methods. Transplantation 1974; 17:534.
43. Toledo-Pereyra LH, Condie RM, Malmberg R et al. Long term kidney preservation with a new plasma perfusate. Forum Proc Clin Dial Transplant 1973; 3:88-90.
44. Toledo-Pereyra LH, Condie R, Simmons R et al. Better cadaver kidney preservation with the use of silica gel fraction of plasma. Transplant Proc 1977; 9:251.
45. Claes G, Blohme I. Experimental and clinical results of continuous albumin perfusion of kidneys. In: Pegg DE, ed. Organ preservation. Edinburgh: Churchill-Livingstone, 1972; 51.

46. Collste H. Bjorken C, Norr A. Renal preservation: Perfusion with albumin solutions. Arch Chir Scand 1971; 137:381.
47. Southard J, Senzing K, Hoffmann R et al. Denaturation of albumin: a critical factor in long term kidney preservation. J Surg Res 1981; 30:80.
48. Johnson R, Anderson M, Fear C et al. Evaluation of a new perfusate solution for kidney preservation. Transplantation 1972; 13:270.
49. Hoffmann R, Southard J, Lutz M et al. 72 hours preservation of dog kidneys using a purely synthetic perfusate containing hydroxy ethyl starch. Arch Surg 1983; 118:919.
50. Hoffmann R, Stratta RJ, D'Alessandro A et al. Combined cold storage perfusion preservation with a new synthetic perfusate. Transplantation 1989; 47:32.
51. McAnulty JF, Ploeg RJ, Southard JH et al. Successful five days perfusion preservation of the canine kidney. Transplantation 1989; 47:37.
52. McAnulty J, Vreugdenhil P, Southard J et al. Use of UW cold storage solution for machine perfusion of kidneys. Transplant Proc 1990; 22:458.
53. Guarrera JV, Polyak MM, Arrington B et al. Pushing the envelope in renal preservation; improved results with novel perfusate modifications for pulsatile machine perfusion of cadaver kidneys. Transpl Proc 2004; 37:2642.
54. Polyak MM, Arrington BO, Stubenbord WT et al. The influence of pulsatile preservation on renal transplantation in the 1990s. Transplantation 2000; 69:249.
55. Stratta RJ, Moore PS, Farney AC et al. Influence of pulsatile perfusion preservation on outcomes in kidney transplantation from expanded criteria donors. J Am Coll Surg 2007; 204:873.
56. Sellers MT, Gallichio MH, Hudson SL et al. Improved outcomes in cadaveric renal allografts with pulsatile preservation. Clin Transplant 2000; 14:543.
57. Wight J, Chilcott J, Holmes M et al. The clinical and cost-effectiveness of pulsatile machine perfusion versus cold storage of kidneys for transplantation retrieved from heart-beating and nonheart beating donors. Health Technol Access 2003; 70:1.
58. Maio R, Costa P, Figueiredo N et al. Evaluation of different preservation solutions utilized in machine perfusion of kidneys retrieved under cardiac arrest. An experimental study. Rev Port Cir Cardiotorac Vasc 2007; 14:149.
59. Spees EK, Vaughn WK, Mendez-Picon G et al. Preservation methods do not affect cadaver renal allograft outcome. The SEOPE prospective study 1977-1982. Transplant Proc 1984; 17:177.
60. Rosenthal JT, Danovitch GM, Wilkinson A et al. The high cost of delayed graft function in cadaveric renal transplantation. Transplantation 1991; 51:1115.
61. Howard RJ, Pfaff WW, Brunson ME et al. Increased incidence of rejection in patients with delayed graft function. Clin Transplant 1994; 8:527.
62. Toledo-Pereyra LH. Effect of multiple organ harvesting on subsequent renal function. Transplant Proc 1986; 28:434.
63. Barber WH, Laskow DA, Deierhoi MH et al. Comparison of simple hypothermic storage, pulsatile perfusion with Belzer's gluconate-albumin solution and pulsatile perfusion with UW solution for renal allograft preservation. Transplant Proc 1991; 23:2394.
64. Henry ML, Sommer BG, Tesi RJ et al. Improved immediate renal allograft function after initial simple cold storage. Transplant Proc 1990; 22:388.
65. Barber W, Hudson S, Deierhoi M et al. Pulsatile perfusion preservation. Early posttransplant dialysis requirement predicts rapid graft loss. Transplant Proc 1990; 2:446.
66. Guarrera JV, Polyak M, Arrington BO et al. Pulsatile machine perfusion with Vasosol solution improves early graft function after cadaveric renal transplantation. 2004; 77:1264.
67. Kwiatkowski A, Wszola M, Kosieradzki M et al. Machine perfusion preservation improves renal allograft survival. Am J Transplant 2007; 7:1942.
68. Matsuoka L, Shah T, Aswad S et al. Pulsatile perfusion reduces the incidence of delayed graft function in expanded criteria donor kidney transplantation. Am J Transplant 2006; 6:1473.
69. Ploeg RJ. Machine preservation tria. Available at: www.organpreservation.nl. Accessed 2008;
70. Zucker K Cirocco R, Roth D et al. Depletion of Hepatitis C virus from procured kidneys using pulsatile perfusion preservation. Transplantation 1994; 57:832.
71. Yland MJ, Nakayama Y, Abe Y et al. Organ preservation by a new pulsatile perfusion method and apparatus. Transplant Proc 1995; 27:1879.
72. Maathuis MJ, Manekeller S, van der Plaats A et al. Improved kidney function after preservation using a novel hypothermic machine perfusion device. Ann Surg 2007; 246:982.
73. Toledo-Pereyra LH. Organ Procurement and Preservation for Transplantation. Second Edition. Austin/ New York: Landes Bioscience/Chapman & Hall, 1997.
74. Muhlberg J, Wagner W, Rohling R et al. Hemodynamic and metabolic problems in the preparation for organ donation. Transplant Proc 1986; 18:11.

75. Wijnen RM, Van der Linden CJ. Donor treatment after pronouncement of brain death: a neglected intensive care problem. Transplant Int 1991; 4:186.
76. Jablonski P, Howden BO, Leslie E et al. Recovery of renal function after warm ischemia. 1. The effect of chlorpromazine and phenoxybenzamine. Transplantation 1983; 35:535.
77. Halasz NA. Pharmocological factors in organ preservation. In: Pegg DE, Jacobsen J, Halasz NA, eds. Organ preservation III. Basic and Applied Aspects. Boston: MIP Press, 1982; 151.
78. Chan I, Bore P, Ross B. Renal Preservation. In: Marberger M, Dreikorn K, eds. International perspectives in urology, Baltimore: Williams and Wilkins, 1981; 8:323.
79. Arnold PE, Van Putten VJ, Lumlertgul D et al. Adenosine nucleotide metabolism and mitochondrial Ca transport following renal ischemia. Am J Physiol 1986; 250:F357.
80. Mc Anulty JF, Ploeg RJ, Southard JH et al. Successful five-day perfusion preservation of the canine kidney. Transplantation 1989; 47:37.
81. Mittnacht S, Jr, Farber JL. Reversal effects of ischemic mitochondrial dysfunction. J Biol Chem 1981; 256:3199.
82. Ascari H, Sato K, Sonoda K et al. Protective effects of calmodulin inhibitor in canine kidney preservation. Transplant Proc 1987; 19:363.
83. Tokunaga Y, Wicomb WN, Concepcion W et al. Improved rat liver preservation using cholpromazine in a new sodium lactobionate sucrose solution. Transplant Proc 1991; 23:660.
84. Takei Y, Marzi I, Kauffman FC et al. Increase in survival time of liver transplants by protease inhibitors and calcium channel blocker, nisoldipine. Transplantation 1990; 50:14.
85. Zheng R, Lanza RP, Soon-Shiong P et al. Prolonged pancreas preservation using a simplified UW solution containing polyethylene glycol. Transplantation 1991; 51:63.
86. Gower JD, Healing G, Fuller BJ et al. Protection against oxidative damage in cold stored rabbit kidneys by deferoxamine and indomethacin. Cryobiology 1989; 26:309.
87. Gutteridge JMC, Halliwell B. Reoxygenation injury and antioxidant protection: a tale of two paradoxes. Arch Biochem Biophys 1990; 283:332.
88. Koyama I, Bulkley GB, Williams GM. The role of oxygen free radicals in mediating the reperfusion injury of cold preserved ischemic kidneys Transplantation 1985; 40:590.
89. Bosco PJ, Schweizer RT. Use of oxygen radical scavengers on autografted pig kidneys after warm ischemia and 48 hour perfusion preservation. Arch Surg 1988; 123:601.
90. Shapiro JI, Cheung C, Itabashi A et al. The effect of verapamil on renal function after warm and cold ischemia in the isolated perfused rat kidney. Transplantation 1985; 40:596.
91. Anaya-Prado R, Toledo-Pereyra LH, Lentsch AB et al. Ischemia/reperfusion injury. J Surg Res 2002; 105:248.
92. Nakamoto M, Shapiro JI, Mills SD et al. Improvement of renal preservation by verapamil with 24-hour cold perfusion in the isolated rat kidney. Transplantation 1988; 45:313.
93. Barber W, Deierhoi M, Philips M et al. Preservation by pulsatile perfusion improves early renal allograft function. Transplant Proc 1988; 5:865.
94. Mitchell H, Sommer B, Ferguson R. Improved immediate function of renal allotraft with Belzer perfusate. Transplantation 1988; 45:73.
95. Henry M, Sommer B, Tesi R et al. Improved immediate renal allograft function after initial simple cold storage. Transplant Proc 1990; 22:388.
96. Gjertson DW. Update: Center effects. In: Terasaki PI, ed. Clinical Transplants 1990. Los Angeles, UCLA Tissue Typing Laboratory 1990; 375.
97. Takemoto S, Terasaki PI, Cecka JM et al. Survival of nationally shared HLA-matched kidney transplant from cadaveric donors. N Engl J Med 1992; 327:834-38.
98. Peters TG, Shaver TR, Ames JE et al. Cold ischemia and outcome in 17,937 cadaveric kidney transplants. Transplantation 1995; 59:191.
99. Anaise D. The non-heartbeating cadaveric donor: a solution to the organ shortage crisis. UNOS Update 1992:33.
100. Mark O, Alan R, Erdal E et al. Nonheartbeating cadaveric organ donation. Annals of Surgery 1994; 4:578.
101. Kootstra G, Daemen JHC, Oomen APA. Categories of nonheart-beating donors. Transplant Proc 1995; 27:2893.
102. Heineman E, Daemen JHC, Kootstra G. Non-heart-beating donors: methods and techniques. Transplant Proc 1995; 27:2895.
103. Kowalski AE, Light JA, Ritchie WO et al. A new approach for increasing the organ supply. Clin Transplantation 1996; 10:653.
104. Schumpf R, Wever M, Weinreich T et al. Transplantation of kidneys from nonheart-beating donors: an update. Transplant Proc 1995; 27:2942.

105. Wijnen RH, Booster MH, Nieman FHM et al. Retrospective analysis of the outcome of transplantation of nonheart-beating donor kidneys. Transplant Proc 1995; 27:2945.
106. Koning OHJ, Van-Bockel JH, van-der-Woude FJ et al. Risk Factors for delayed graft function in University of Wisconsin solution preserved kidneys from multiorgan donors. Transplant Proc 1995; 27:752.
107. Tesi R, Elkhammas E, Daries E et al. Pulsatile kidney perfusion for preservation and evaluation use of high risk kidney donors to expand the donor pool. Transplantation 1989; 47:940.
108. Fahy G. The effect of cryoprotectant concentration on freezing damage in kidney slices. In: Toledo-Pereyra LH, ed. Basic Concepts in Organ Procurement, Perfusion and Preservation. New York: Academic Press, 1982:121.
109. Sells RA, Bore PJ, McLaughlin GA et al. A predictive test of renal viability. Transplant Proc 1977; 9:1557.
110. Ceka JM. Donor and preservation factors. In: Terasaki PI, ed. Clinical Transplantation. Los Angeles, UCLA Tissue Typing Laboratory 1988; 339.
111. Roles K, Foreman J, Pegg DE. A pilot clinical study of retrograde oxygen persufflation in renal preservation. Transplantation 1989; 48:339.
112. Manekeller S, Leuvenink H, Sitzia M et al. Oxygenated machine perfusion preservation of predamaged kidneys with HTK and Belzer machine perfusion solution. An experimental study in pigs. Transpl Proc 2005; 37:3274.
113. Matsumoto S, Kuroka Y. Perfluorocarbon for organ preservation before transplantation. Transplantation 2002; 74:1804.
114. Lopez-Neblina F, Paez AJ, Toledo AH et al. Role of nitric oxide in ischemia/reperfusion of the rat kidney. Circ Shock 1994; 44:91.
115. Triptara P, Patel NS, Webb A et al. Nitrite-derived nitric oxide protects the rat kidney against ischemia/reperfusion injury in vivo: Role for xanthine oxidoreductase. J Am Soc Nephrol 2007; 18:570.
116. Valero R, Cabrer C, Oppenheimer F et al. Normothermic recirculation reduces primary graft dysfunction of kidneys obtained from nonheart-beating donors. Transpl Int 2000; 13:303.
117. Brasile L, Stubenitsky BM, Booster MH et al. Overcoming severe renal ischemia: the role of ex vivo warm perfusion. Transplantation 2002; 73:897.
118. Maessen JG, van der Vusse GJ, Vork M et al. The beneficial effect of intermediate normothermic perfusion during cold storage of ischemically injured kidneys. A study of renal nucleotide homeostasis during hypothermia in the dog. Transplantation 2004; 77:1328.
119. Bagul A, Hosgood SA, Kaushick M et al. Experimental renal preservation by normothermic resuscitation perfusion with autologous blood. Brit J Surg 2008; 95:111.

CHAPTER 8

Pancreas Preservation

Shinichi Matsumoto,* Hirofumi Noguchi, Naoya Kobayashi,
Angelika Gruessner and David E.R. Sutherland

Pancreas preservation is an essential process prior to both pancreas transplantation and islet isolation. Traditionally, research on pancreas preservation focused on pancreas transplantation. Recently, we focused on islet isolation and transplantation and then created a new solution.

In this chapter, a history of pancreas preservation, clinical results related to pancreas preservation as documented in the International Pancreas Transplantation Registry (IPTR) data base and current advances in pancreas preservation for pancreas and islet transplantation are described.

History of Pancreas Preservation

Numerous experiments on pancreas preservation, with either machine perfusion or cold storage, have been published since the 1960s.[1-5] Hypothermic pulsatile machine perfusion has been widely used in clinical kidney transplantation. Initially developed by Carrel in 1938, this technique was used by Belzer et al to preserve kidneys for up to 3 days in 1967.[6] In experimental pancreas transplantation, several machines have been used for perfusion.[7-9] For canine segmental pancreas grafts, Florack et al compared simple cold storage in solution vs pulsatile machine flow perfusion.[10] Pancreas preservation failure rates with machine perfusion were 30% at 24 and 40% at 48 hours, as compared with 0% with simple cold storage in silica gel filtered (SGF) plasma for 24 to 48 hours. Therefore, machine perfusion has limited application in clinical pancreas transplantation.

Hypothermic storage in a fluid medium is the most widely used preservation method and several solutions have been used over the years. Basic concepts of hypothermic storage are (1) preventing hypothermia-induced cell swelling; 2) preventing expansion of the interstitial space; 3) maintaining the pH; 4) preventing free radical damage and ensuring adequate metabolic sustenance (including oxygen in some methods) and 5) supplying substrates for energy to drive the sodium pump.[11,12]

During the early 1970s, the Collins solution (Table 1) and its subsequent modifications were widely used.[13] The Collins solution consists of high concentrations of potassium, magnesium, phosphate, sulfate and glucose. Since magnesium led to precipitation, the Collins solution was modified by elimination of magnesium (Table 1). This modified Collins solution was named Euro-Collins solution, which became a standard preservation solution at Eurotransplant.[14]

Hyperosmolar plasma-based solutions were found to be superior to the simple intracellular electrolyte solutions and Collins solution that were widely used for kidney preservation.[15] Plasma based solutions were able to preserve canine pancreas grafts for between 48 and 72 hours, while Collins solution was effective for no more than 24 hours.[16,17] Clinically, through most of the 1980s, most centers that used Collins or Collins-like solutions limited human pancreas preservation to <12 hours.[18] At the University of Minnesota, silica-gel filtered (SGF) plasma solutions, made hyperosmolar with mannitol, were used to successfully preserve human pancreas grafts for up to

*Corresponding Author: Shinichi Matsumoto—54 Kawahara-cho Shogoin Sakyo-ku Kyoto
606-8507 Japan. Email: shinichi41@fujita-hu.ac.jp

Organ Preservation for Transplantation, Third Edition, edited by Luis H. Toledo-Pereyra.
©2010 Landes Bioscience.

Table 1. Collins and Euro-Collins solution

	Collins Solution	Euro-Collins Solution
Na⁺	10 mm/L	10 mm/L
K⁺	115 mm/L	116 mm/L
Cl⁻	15 mm/L	15 mm/L
Mg²⁺	30 mm/L	– mm/L
HCO₃⁻	10 mm/L	10 mm/L
SO₄⁻	30 mm/L	– mm/L
HPO₄²⁻	43 mm/L	43 mm/L
H₂PO₄⁻	15 mm/L	15 mm/L
Glucose	195 mm/L	200 mm/L
Osmolarity	320 mOsm/L	355 mOsm/L
pH	7.0	7.2

30 hours.[19-22] The drawback of plasma-based solutions was, of course, the risk of disease transmission with human viruses.

In 1987, the Wisconsin group introduced a new synthetic intracellular electrolyte solution in which lactobionate (a cellular impermeants) replaced chloride as the dominant anion and raffinose was used as the osmotic agent.[23] The unique composition of the UW solution (Table 2) is based on the following rationale: (1) impermeants (lactobionate and raffinose) to minimize hypothermia-induced cell swelling; (2) colloids hydroxyethyl starch (HES) to prevent expansion of the interstitial space; (3) adequate buffers (KH_2PO_4) to prevent intracellular acidosis; (4) free radical scavenger or inhibitor use (glutathione and allopurinol) to protect from the injury of these unstable radicals and (5) Adenosine triphosphate substrate (ATP) to provide the cells with energy and allow adequate function of a sodium pump. This solution was capable of preserving canine pancreases for up to 72 hours. University of Wisconsin (UW) solution was introduced clinically for all organs shortly therafter.[24] After UW solution became commercially available, it virtually replaced all other solutions for pancreas preservation; since 1990, UW solution has been used to preserve 95% to 100% of pancreas transplants annually worldwide.

Table 2. UW solution

Na⁺	30 mm/L
K⁺	125 mm/L
Mg²⁺	5 mm/L
SO₄⁻	5 mm/L
PO₄⁻	25 mm/L
Lactobionate	100 mm/L
Raffinose	30 mm/L
Adenosine	5 mm/L
Allopurinol	3 mm/L
Glutathione	3 mm/L
Insulin	100 U/L
Dexamethasone	8 mg/L
Penicillin	0.13 g/L
Hydroxyethyl stach	50 g/L
Osmolarity	330 mOsm/L
pH	7.4

Clinical Pancreas Transplant Results

As of December 31, 2004, more than 23,000 pancreas transplant had been reported to IPTR, more than 17,000 in the US.[25] The total number of pancreas transplants reported to UNOS from October 1987 to June 2004 was 15,333, including 11,898 simultaneous pancreas-kidney (SPK) transplants (78%), 2,427 pancreas after kidney (PAK) transplants (16%) and 1,008 pancreas transplants alone (PTA)(7%); in addition, a total of 264 pancreas transplants combined with at least one organ other than the kidney were reported. An analysis of US pancreas transplant performed between 1988 and 2003 showed a progressive improvement in outcome, with pancreas transplant graft survival rates going from 75% at 1 yr for 1988-1989 to 85% for 2002-2003 SPK cases, from 55% to 78% for PAK cases and from 45% to 77% for PTA cases.

Deceased donor pancreas preservation times have changed little in successive eras. In all categories (SPK, PAK, PTA), preservation times have ranged up to 36 h. Over the whole time period, 75% of the grafts were preserved for <18 h and only 5% were preserved >24 h. Median preservation times in 2002-2003 were 14.3 h for solitary pancreas (PAK and PTA) and 12 h for SPK transplants. Preservation times are significantly longer in the solitary than in the SPK category because a higher proportion of solitary pancreases are transported.

University of Minnesota and other centers reported that preservation >20 h in UW solution was associated with complications (leaks, thrombosis, wound infection and pancreatitis).[26,27]

Current Advances in Pancreas Preservation

Since oxygen supply is one of the key elements of organ preservation, Idezuki et al in the 1960s used a hyperbaric hypothermic oxygen chamber, but this method was too complicated to become popular.[28] Perfluorocarbons (PFCs), which store and release high levels of oxygen, were introduced by Kuroda in 1988 as a component of the two-layer method (TLM) of pancreas preservation.[29] The TLM consists of two kinds of solutions: one is PFC and the other is a standard organ preservation solution (typically UW). The PFC solution has a higher specific gravity (1.95) and a hydrophobic property, so it settles at the bottom of the organ preservation solution with a clear demarcation between the two. The pancreas is suspended at the interface of the two solutions. Pancreases preserved in the TLM are oxygenated through the PFC and substrates are supplied by the UW.[30,31] The TLM has the capability to store canine pancreases for 96 h and UW for 72 h.[32] During TLM preservation, pancreas grafts are oxygenated through the PFC and continuously generate ATP.[33,34] ATP is used to drive the sodium/potassium pump and to generate proteins, including heat shock protein 32 (hemeoxygenase-1) and heat shock protein 70.[35,36] The TLM also improves the blood circulation in the pancreas after transplantation.[37,38]

The TLM was used clinically for whole-pancreas transplantation with favorable results.[39] In this first clinical trial of 10 cases, there were no adverse events related to the TLM. Furthermore, the morphologic quality of the human pancreas grafts after reperfusion was better than pancreases stored in UW alone. In addition, there were no acute rejection episodes of pancreases preserved by the TLM. However, it was revealed that there was no significant difference in expression of MHC class I and II on isolated islets from fresh pancreases and from pancreases preserved with the TLM in the rodent model.[40] In addition, a long-term follow-up study revealed that the TLM had no significant impact on pancreas preservation.[41] Therefore, the TLM is not widely used for pancreas preservation before pancreas transplantation. Nonetheless, this first clinical trial made it possible to use the TLM for pancreas preservation before clinical islet isolation.

Currently, new solutions—Celsior solution and histidine-tryptophan-ketoglutarate (HTK) solution—were introduced for pancreas preservation (Table 3). After a mean cold ischemia time of 624 minutes (range 360 to 945 minutes) for UW vs 672 minutes (range 415 to 1005 minutes) for Celsior solution (P = NS), the two solutions showed similar safety and efficacy profiles for pancreas transplantation.[42] And early graft function and complications were comparable with HTK solution and UW solution for pancreas allograft preservation with a mean cold ischemia time of 15.1 +/- 2.1 h.[43] Since both Celsior solution and HTK solution have low potassium concentration,

Table 3. Celsior solution and HTK solution

	Celsior Solution	HTK Solution
Na⁺	100 mm/L	15 mm/L
K⁺	15 mm/L	10 mm/L
Mg^{2+}	13 mm/L	4 mm/L
Ca^{2+}	0.25 mm/L	-
Cl⁻	41 mm/L	50 mm/L
Histidine	30 mm/L	198 mm/L
Lactobionate	80 mm/L	-
Glutamic acid	20 mm/L	-
Mannitol	60 mm/L	-
Tryptophan	-	2 mm/L
α-ketoglutarate	-	1 mm/L
Osmolality	320 mOsm/kg	310 mOsm/L
pH	7.3	7.3

prolonged preservation time might cause hypothermia-induced cell swelling. However, with their limited preservation time, both solutions could replace the expensive UW solution.

Pancreas Preservation before Islet Isolation and Transplantation

Current Status of Islet Transplantation

In 2000, the University of Alberta demonstrated that allogeneic islet transplantation could cure type 1 diabetes with islets isolated from two or more human donor pancreases.[44] Their results stimulated a clinical trial of islet transplantation in the US and Europe. Currently, there has been an exponential increase in clinical islet transplantation activity, with 471 patients transplanted at 43 international institutions.[45] The University of Alberta, University of Minnesota and University of Miami demonstrated that 82% of 118 recipients of completion transplants were insulin-independent within the first year after transplantation.[45] Yet the University of Alberta demonstrated progressive loss of insulin independence over time, leaving 50% of patients still insulin-free at 3 years. Still, more than 80% patients continue to demonstrate persistent islet function at 5 years, with effective prevention of recurrent hypoglycemia or severe lability, combined with correction in HbA_{1c}.[46]

Introduction of TLM Prior to Islet Isolation

Since the Edmonton protocol needs two or more donor pancreases, it is important to improve the efficacy of the islet isolation process to achieve successful single-donor islet transplantation.

There was a clear advantage of pancreas preservation with the TLM before islet isolation in the canine model.[47] However, several groups have been skeptical of this potential benefit for human pancreases to the difference in thickness between the canine and human pancreas.

In 2000, we presented the advantage of TLM preservation prior to human islet isolation using pancreases from brain-dead donors for the first time.[48] In this experiment, human pancreases were preserved for a long period in order to show the difference between the TLM and UW. This experiment clearly demonstrated the benefit of the TLM for human pancreases before islet isolation. Next, we conducted a randomized comparison between the TLM and UW for short-term and long-term preservation before islet isolation with the Edmonton islet isolation protocol.[49] This experiment demonstrated that the TLM improved the islet isolation even for short-term preservation with the Edmonton islet isolation protocol.

The University of Alberta conducted experiments on the resuscitation after cold ischemic injury by the TLM. They isolated islets from human pancreases as soon as possible or preserved the human pancreas for an additional 3 hours with the TLM before isolation. They demonstrated that the additional TLM preservation improved the islet yields (349,000 IE TLM vs 214,000 IE, UW) and the success rate of transplantation (71% TLM vs 36%, UW).[50]

The University of Miami addressed the question of the benefit of the TLM for the marginal donor. They isolated islets from the pancreases of older donors (>50 years) in 15 cases with the TLM and 18 cases with UW. They demonstrated that the TLM improved the islet yields (306,000 IE vs 149,000 IE, UW) and the success rate of transplantation (53%, TLM vs 11%, UW) with marginal donors.[51]

The favorable results from the University of Alberta and University of Miami stimulated the desire to use the TLM before islet isolation at other islet isolation centers. However, the necessity of continuous oxygenation hindered the prevalence of the TLM. We postulated that since PFC is able to contain extremely high levels of oxygen, continuous oxygenation is not necessary once the PFC is fully oxygenated. Therefore, we developed oxygen static-charge TLM (static TLM) and examined the effect of this system (Fig. 1).[52] We demonstrated that static TLM before islet isolation was as efficient as the regular TLM and significantly better than UW. Currently, many islet isolation centers use static TLM before islet isolation.

The University of Minnesota demonstrated successful single-donor islet transplantation with the TLM. They were able to cure 12 out of 14 type 1 diabetic patients with the islets from single-donor pancreases preserved by the TLM for less than 5 hours.[53] Since they didn't compare with the UW, the benefit of the TLM was unclear; however, they were able to show the best islet transplantation results with the TLM.

The results from the University of Alberta, University of Miami and University of Minnesota and our results pushed the TLM to become the standard pancreas preservation method before islet isolation.

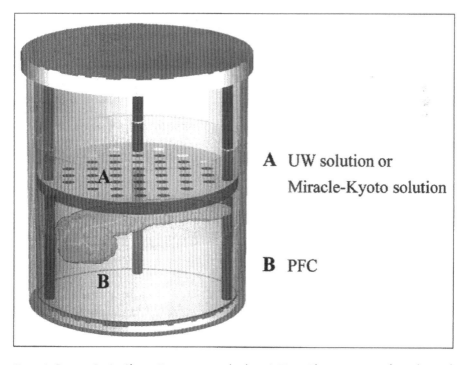

A UW solution or Miracle-Kyoto solution

B PFC

Figure 1. Oxygen Static-Charge Two-Layer Method (static TLM): The pancreas graft was located between two solutions, A) UW solution or Miracle-Kyoto solution and B) oxygenated PFC.

The TLM clearly improved human islet yields; however, the mechanisms are not clear. It was demonstrated that the TLM maintained the zymogen granules and, in sharp contrast, UW destroyed them.[54] Destroyed zymogen granules could lead to the activation of trypsin. Activated trypsin might cause the overdigestion of isolated islets or break the pancreatic duct. Since both the inhibition of trypsin activity[55] and conservation of the pancreatic duct contributed to an increase in islet yields,[56] maintenance of zymogen granules is one of the mechanisms of the TLM.

Current Advances in Pancreas Preservation before Islet Isolation

In Japan, it is illegal to use heart-beating brain-dead donors for islet transplantation. Therefore, islet donors must be either nonheart-beating-donors (NHBD) or living donors.[57] Both situations are challenging, since NHBDs must overcome warm ischemic injury and living donors need a perfect isolation procedure without failure.

In order to examine whether the TLM could overcome warm ischemic injury, we conducted canine experiments. After 90 minutes of warm ischemic injury, none of the canine pancreas grafts were viable; however, when we preserved the damaged grafts with the TLM for 24 to 48 hours, all grafts became viable.[58] In addition, the TLM could overcome warm ischemic injury after 5 h at $20°C$ and after 24 h at $4°C$ but not at $37°C$.[59] Our primate experiment also supported that the TLM could overcome warm ischemic injury.[60]

Before initiating clinical islet transplantation with NHBDs, we optimized pancreas preservation focused on islet isolation in a porcine model. At first, we introduced ductal injection for pancreatic duct preservation.[56] We avoided UW solution for ductal injection since UW solution inhibited collagenase activity and had a negative effect on human islet isolation results.[61] We made a new solution using a trypsin inhibitor, ulinastatin[62] (Mochida Pharmaceutical Company, Tokyo, Japan), along with ET-Kyoto solution[63] (Kyoto Biomedical, Kyoto, Japan) for this purpose. This solution was named "Miracle-Kyoto solution" since the brand name of ulinastatin is Miraclid (Table 4). We also use the Miracle-Kyoto solution for a component of the TLM.

After about 20 minutes of warm ischemia, the pancreas grafts were preserved for two hours with PFC plus either UW solution (UW group) or Miracle-Kyoto solution (Miracle-Kyoto group). Islet yields were double with the Miracle-Kyoto group, as compared with the UW group.[64] In addition, those islets from single NHBD pancreases in the Miracle-Kyoto group successfully reversed diabetes in surgically diabetic pigs.[65]

After our porcine experiments, we used the Miracle-Kyoto solution for clinical islet isolation. We have isolated islets from 18 human pancreases from NHBDs. These pancreases were preserved using the TLM consisting of the Miracle-Kyoto solution and PFC with ductal protection. Since the Miracle-Kyoto solution has an extracellular component, we limited cold preservation time to less than 5 hours in order to avoid hypothermia-induced cell swelling. Double balloon catheters were inserted before cardiac arrest combined with kidney retrieval in 15 cases.[66] In those 15 cases, average islet yields were $451,333 \pm 38,765$ IE: 14 islet preparations were transplanted into seven

---

Table 4. Miracle-Kyoto solution

Ulinastatin (Miraclid)	50,000 U/L
Na^+	100 mm/L
K^+	44 mm/L
Gluconate	100 mm/L
Trehalose	120 mm/L
PO_4^-	25 mm/L
Hydroxyethyl stach	30 g/L
Osmolarity	370 mOsm/L
pH	7.4

type 1 diabetic patients, for a transplantation rate of 93% (14/15). The first human islet transplant in Japan occurred on April 7, 2004 and this patient became insulin-independent after the second islet transplant, which was another first in Japan.[64,67] All seven transplanted patients have better glycemic control after islet transplantation, with positive C-peptide levels and reduced insulin requirements. Thus, the Miracle-Kyoto solution made it possible to perform islet transplantation with NHBDs.

We performed the first successful living donor islet transplant for the treatment of brittle diabetes.[68] A reliable islet isolation method is essential, since it is not acceptable for islet isolation to fail with living donor pancreases. Our current islet isolation method for NHBDs with the Miracle-Kyoto solution made us comfortable to perform islet isolation with living donors.

Distal pancreatectomy with splenectomy was performed using open laparotomy. The resected pancreas was perfused via the splenic artery with the Miracle-Kyoto solution and ductal injection was also performed. The pancreas graft was brought to our cell processing facility and the cold ischemic time was 44 minutes. More than 400,000 IE islets were isolated and these islets were transplanted immediately. The recipient has been insulin-independent posttransplant for more than 5 months at this time point. HbA$_{1c}$ levels before transplantation were 9.9% and have now been reduced to 6.2%. The recipient has not suffered hypoglycemic unawareness posttransplant. The donor's clinical course was uneventful; she was discharged 18 days after the operation without complications. She returned to her job within one month. Her HbA$_{1c}$ levels have been between 5.1% and 5.4%. Both the recipient's and donor's oral glucose tolerance test (OGTT) result at 37 post operative days was normal.[66]

The Miracle-Kyoto solution thus enabled us to perform reliable islet isolation with a living donor.

Conclusion

Pancreas grafts can be preserved prior to whole-pancreas transplantation for up to 30 h with UW solution, but more complications occurred when preserved more than 20 h. The TLM can be applied safely; however, it has no proven significant benefit as compared with UW storage, before whole-pancreas preservation.

Optimized pancreas preservation before islet isolation is different from that before whole-pancreas preservation. The TLM significantly improved islet isolation results, as compared with UW storage. However, the preservation period should be minimized: ideally, less than 5 h. Pancreatic ductal protection seems important for islet isolation and the Miracle-Kyoto solution is suitable for this purpose. The Miracle-Kyoto solution enabled us to isolate islets from NHBDs and perform living donor islet transplantation.

Pancreas preservation prior to islet isolation is still an important research focus for improving islet yields and expanding the preservation period.

References

1. Belzer FO, Southerd JH. Principles of solid-organ preservation by cold storage. Transplantation 1988; 45:673-676.
2. Idezuki Y, Feemster JA, Dietzman RH et al. Experimental pancreatico-duodenal preservation and transplantation. Surg Gynecol Onsted 1968; 126:1002-14.
3. Brynger H. Twenty-four-hour preservation of the duct-ligated canine pancreatic allograft. Eur Surg Res 1975; 7:341-354.
4. Baumgartner D, Sutherland DER, Najarian JS. Studies on segmental pancreas autotransplantation in dogs: Technique and preservation. Transplant Proc 1980; 12:163-171.
5. Florack G, Sutherland DER, Heil JE et al. Long-term preservation of segmental pancreas auto grafts, Surgery 1982; 92:260-269.
6. Belzer FO, Ashby BS, Dunphy JE. 24-hour and 72-hour preservation of canine kidneys. Lancet 1967; 2:536-538.
7. Baumgartner D, Sutherland DER, Najarian JS. Studies on segmental pancreas autotransplants in dogs: technique and preservation. Transplant Proc 1980; 12:163-171.
8. Brynger H. Twenty-four-hour preservation of the duct-ligated canine pancreatic allograft. Eur Surg Res 1975; 7:341-354.

9. Toledo-Pereyra LH, Valjee KD, Chee M et al. Preservation of the pancreas for transplantation. Surg Gynecol Obstet 1979; 148:57-61.
10. Florack G, Sutherland DER, Heil JE et al. Preservation of canine segmental pancreatic autografts: cold storage versus pulsatile machine perfusion. J Surg Res 1983; 34:493-504.
11. Toledo-Pereyra LH, Paez-Rollys AJ, Palma-Vargas JM. Science of organ preservation. In: Toledo-Pereyra LH, ed. Organ Procurement and Preservation for Transplantation. Second Edition. Austin/New York: Landes Bioscience/Chapman & Hall, 1997:1-16.
12. Kandaswamy R. Pancreas preservation. In: Gruessner RWG, Sutherland DER, ed. Transplantation of the Pancreas. New York: Springer-Verlag 2004: 119-125.
13. Collins GM, Bravo-Shugarman M, Terasaki PI. Kidney preservation for transportation. Initial perfusion and 30 hours' ice storage. Lancet 1969; 2:1219-1222.
14. Dreikorn K, Horsh R, Rohl L. 48- to 96-hour preservation of canine kidneys by initial perfusion and hypothermic storage using the Euro-Collins solution. Euro Urol 1980; 6:221-224.
15. Sutherland DER, Morrow CE, Florack G et al. Cold storage preservation of islet and pancreas grafts as assessed by in vivo function after transplantation to diabetic hosts. Cryobiology 1983; 20:138-150.
16. Abouna GM, Heil JE, Sutherland DER et al. Factors necessary for successful 48-hour preservation of pancreas grafts. Transplantation 1988; 45:270-274.
17. Heise JW, Sutherland DER, Heil JE et al. 72-hours preservation of pancreatic autotransplants in dogs using a urinary drainage technique. Transplant Proc 1988; 20:1029-1030.
18. Sutherland DER, Moudry KC, Dunn DL et al. Pancreas-transplant outcome in relation to presence or absence of end-stage renal disease, timing of transplant, surgical technique and donor sources. Diabetes 1989; 38(Suppl 1):10-12.
19. Abouna GM, Sutherland DER, Florack G et al. Function of transplanted human pancreatic allografts after preservation in cold storage for 6 to 26 hours. Transplantation 1987; 43:630-636.
20. Florack G, Sutherland DER, Heise JW et al. Successful preservation of human pancreas grafts for 28 hours. Transplant Proc 1987; 19:3882-3885.
21. Florack G, Sutherland DER, Morel P et al. Effective preservation of human pancreas grafts. Transplant Proc 1989; 21:1369-1371.
22. Morel P, Moudry-Munns KC, Najarian J et al. Influence of preservation time on outcome and metabolic function of bladder-drained pancreas transplants. Transplantation 1990; 49:294-303.
23. Wahlberg J, Southard JH, Belzer FO. Development of a cold storage solution for pancreas preservation. Cryobiology 1986; 23:477-482.
24. Belzer FO. Clinical organ preservation with UW solution. Transplantation 1989; 47:1097-9.
25. Gruessner AC, Sutherland DER. Pancreas transplant outcomes for United States (US) and non-US cases as reported to the United Network for Organ Sharing (UNOS) and the International Pancreas Transplant Registry (IPTR) as of 2004. Clin Transplant 2005; 19:433-455.
26. Humar A, Kandaswamy R, Drangstveit MB et al. Prolonged preservation increases surgical complications after pancreas transplants. Surgery 2000; 127:545-551.
27. Stratta RJ. Donor age, organ import and cold ischemia: Effect on early outcomes after simultaneous kidney-pancreas transplantation. Transplant Proc 1997; 29:3291-3292.
28. Idezuki Y, Dietzman RH, Feemster JA et al. Successful twenty-four hour preservation of pancreatico-duodenal allograft with hypothermia and hyperbaric oxygen. Surg Forum 1968; 19:221-222.
29. Kuroda Y, Kawamura T, Suzuki Y et al. A new, simple method for cold storage of the pancreas using perfluorochemical. Transplantation 1988; 46:457-460.
30. Matsumoto S. Clinical application of perfluorocarbons for organ preservation. Artificial Cells, Blood Substitutes and Biotechnology 2005; 33:75-78
31. Matsumoto S, Kuroda Y. Perfluorocarbon for organ preservation before transplantation. Transplantation 2000; 74:1804-1809.
32. Fujino Y, Kuroda Y, Suzuki Y et al. Preservation of canine pancreas for 96 hours by a modified two-layer (UW solution/perfluorochemical) cold storage method. Transplantation 1991; 51:1133-1135.
33. Matsumoto S, Kuroda Y, Hamano M et al. Direct evidence of pancreatic tissue oxygenation during preservation by the two-layer method. Transplantation 1996; 62:1667-1670.
34. Kuroda Y, Fujino Y, Kawamura T et al. Mechanism of oxygenation of pancreas during preservation by a two-layer (Euro-Collins' solution/perfluorochemical) cold-storage method. Transplantation 1990; 49:694-696.
35. Tanioka Y, Kuroda Y, Kim Y et al. The effect of ouabain (inhibitor of an ATP-dependent Na+/K+ pump) on the pancreas graft during preservation by the two-layer method. Transplantation 1996; 62:1730-1734.
36. Matsumoto S, Fujino Y, Suzuki Y et al. Evidence of protein synthesis during resuscitation of ischemically damaged canine pancreas by the two-layer method. Pancreas 2000; 20:411-414.

37. Kuroda Y, Fujita H, Matsumoto S et al. Protection of canine pancreatic microvascular endothelium against cold ischemic injury during preservation by the two-layer method. Transplantation 1997; 64:948-953.
38. Matsumoto S, Kuroda Y, Suzuki Y et al. Thromboxane A2 synthesis inhibitor OKY046 ameliorates vascular endothelial injury of pancreas graft during preservation by the two-layer UW solution/perfluorochemical method at 20 degrees C. Transplant Proc 1997; 29:1359-1362.
39. Matsumoto S, Kandaswamy R, Sutherland DER et al. Clinical application of the two-layer (University of Wisconsin solution/perfluorochemical plus O_2) method of pancreas preservation before transplantation. Transplantation 2000; 70:771-774.
40. Toyama H, Takada M, Tanaka T et al. Characterization of islet-infiltrating immunocytes after pancreas preservation by two-layer (UW/perfluorochemical) cold storage method. Transplant Proc 2003; 35: 1503-1505.
41. Hiraoka K, Kuroda Y, Suzuki Y et al. Outcomes in clinical pancreas transplantation with the two-layer cold storage method versus simple storage in University of Wisconsin solution. Transplant Proc 2002; 34:2688-2689.
42. Boggi U, Signori S, Vistoli F et al. University of Wisconsin solution versus Celsior solution in clinical pancreas transplantation. Transplant Proc 2005; 37:1262-1264.
43. Potdar S, Malek S, Eghtesad B et al. Initial experience using histidine-tryptophan-ketoglutarate solution in clinical pancreas transplantation. Clin Transplant 2004; 18:661-665.
44. Shapiro AM, Lakey JR, Ryan EA et al. Islet transplantation in seven patients with type 1 diabetes mellitus using a glucocorticoid-free immunosuppressive regimen. N Engl J Med 2000; 343:230-238.
45. Shapiro AM, Lakey JRT, Paty BW et al. Strategic opportunities in clinical islet transplantation. Transplantation 2005; 79:1304-1307.
46. Ryan EA, Paty BW, Senior PA et al. Five-year follow-up after clinical islet transplantation. Diabetes 2005; 54:2060-2069.
47. Tanioka Y, Sutherland DE, Kuroda Y et al. Excellence of the two-layer method (University of Wisconsin solution/perfluorochemical) in pancreas preservation before islet isolation. Surgery 1997; 122:435-441.
48. Matsumoto S, Qualley S, Rigley T et al. Prolonged preservation of the human pancreas prior to islet isolation using the two-layer (University of Wisconsin solution [UW]/perfluorocarbon) method. Transplantation 2000; 69(Supp):213.
49. Matsumoto S, Qualley S, Goel S et al. Effect of the two-layer (University of Wisconsin solution-perfluorochemical plus O_2) method of pancreas preservation on human islet isolation, as assessed by the edmonton isolation protocol. Transplantation 2002; 74:1414-1419.
50. Tsujimura T, Kuroda Y, Kin T et al. Human islet transplantation from pancreases with prolonged cold ischemia using additional preservation by the two-layer (UW solution/perfluorochemical) cold-storage method. Transplantation 2002; 74:1687-1691.
51. Ricordi C, Fraker C, Szust J et al. Improved human islet isolation outcome from marginal donors following addition of oxygenated perfluorocarbon to the cold-storage solution. Transplantation 2003; 75:1524-1527.
52. Matsumoto S, Rigley T, Qualley S et al. Efficacy of the oxygen-charged static two-layer method for short-term pancreas preservation and islet isolation from nonhuman primate and human pancreata. Cell Transplant 2002; 11:769-777.
53. Hering BJ, Matsumoto I, Sawada T et al. Impact of two-layer pancreas preservation on islet isolation and transplantation. Transplantation 2002; 74:1813-1816.
54. Iwanaga Y, Suzuki Y, Okada Y et al. Ultrastructural analyses of pancreatic grafts preserved by the two-layer cold-storage method and by simple cold storage in University of Wisconsin solution. Transpl Int 2002; 8:425-430.
55. Matsumoto S, Rigley T, Reems J et al. Improved islet yields from Macaca nemestrina and marginal human pancreata after two-layer method preservation and endogenous trypsin inhibition. Am J Transplant 2003; 3:53-63.
56. Sawada T, Matsumoto I, Nakano M et al. Improved islet yield and function with ductal injection of University of Wisconsin solution before pancreas preservation. Transplantation 2003; 75:1965-1969.
57. Matsumoto S, Tanaka K, Strong DM et al. Efficacy of human islet isolation from the tail section of pancreas for the possibility of living donor islet transplantation. Transplantation 2004; 78:839-843.
58. Kuroda Y, Morita A, Fujino Y et al. Restoration of pancreas graft function preserved by a two-layer (University of Wisconsin solution/perfluorochemical) cold storage method after significant warm ischemia. Transplantation 1993; 55:227-228.
59. Kuroda Y, Matsumoto S, Fujita H et al. Resuscitation of ischemically damaged pancreas during short-term preservation at 20 degrees C by the two-layer (University of Wisconsin solution/perfluorochemical) method. Transplantation 1996; 61:28-30.

60. Stevens RB, Matsumoto S, Lawrence O et al. Ischemically damaged human pancreases (PX) can be resuscitated by the two-layer method before islet isolation: implications for clinical islet transplantation. Am J of Transplant 2001; 1(Suppl):321.
61. Contractor HH, Johnson PR, Chadwick DR et al. Robertson GS, London NJ. The effect of UW solution and its components on the collagenase digestion of human and porcine pancreas. Cell Transplant 1995; 4:615-619.
62. Tsujino T, Komatsu Y, Isayama H et al. Ulinastatin for pancreatitis after endoscopic retrograde cholangiopancreatography: A randomized, controlled trial. Clin Gastroenterol Hepatol 2005; 3:376-383.
63. Chen F, Nakamura T, Wada H. Development of new organ preservation solutions in Kyoto University. Yonsei Med J 2004; 45:1107-1114.
64. Matsumoto S, Okitsu T, Iwanaga Y et al. Successful islet transplantation from nonheart-beating donor pancreata using modified Ricordi islet isolation method. Transplantation. (in press).
65. Ikeda H, Kobayashi N, Tanaka Y et al. A newly developed bioartificial pancreas successfully controls blood glucose in totally pancreatectomized diabetic pigs. Tissue Eng (in press).
66. Nagata H, Matsumoto S, Noguchi H et al. Procurement of the human pancreas for pancreatic islet transplantation from marginal cadaver donors. Transplantation (in press).
67. Matsumoto S, Tanaka K. Pancreatic islet cell transplantation using nonheart-beating-donors (NHBDs). J of Hepato-Biliary-Pancreatic Surgery 2005; 12:227-230.
68. Matsumoto S, Okitsu T, Iwanaga Y et al. Insulin independence after living-donor distal pancreatectomy and islet allotransplantation. Lancet 2005; 365:1642-1644.

CHAPTER 9

Liver Preservation

James V. Guarrera*

Introduction

L iver transplantation (LTx) is the standard treatment for end stage liver disease. Current practices in liver preservation are described spanning donor management, liver procurement techniques, preservation techniques and use of extended criteria donor (ECD) livers. Safe expansion of the donor pool to include elderly, steatotic and donor after cardiac death livers is one of the most pressing challenges facing the field today. Preclinical and clinical developments in liver preservation solutions and agents show promising results to target specific mechanisms and reduce ischemia-reperfusion injury. Combined with exciting developments in machine perfusion, the liver preservation field is poised for significant advances in the coming years, targeting expanded access to ECD livers, greater preservation times, enhanced graft integrity and ultimately, improved patient outcomes.

Brief History of Liver Transplantation and Organ Preservation

Liver transplantation (LTx) is the standard of care for patients with end stage liver disease (ESLD). Thomas Starzl performed the first orthotopic liver transplantation in a human in 1963, with extended survival reported in 1967.[1] Despite the development of complex surgical techniques, liver transplantation remained experimental through the 1970s, with one year patient survival around 25%.[2]

In 1969 Collins described a simple cold storage organ preservation method. Collins developed a preservation solution with a high level of potassium to mimic the intracellular composition of the kidney. It also contained a high level of glucose which acted as an effective osmotic agent. This suppressed cell swelling. Magnesium was added to act as a membrane stabilizer, but in the presence of phosphate, the magnesium phosphate formed a precipitate. To eliminate the problem, a modified Collins solution was developed in Europe (Euro-Collins) that omitted the magnesium and used mannitol in place of glucose. These solutions are effective for preservation of kidneys past 24-36 hours with the simple cold storage preservation method. Further experimental work in liver preservation showed that livers preserved with Collins solution would not routinely tolerate cold ischemia time (CIT) in excess of 8 hours (3).

The introduction of Cyclosporine A by Sir Roy Calne in 1983 and the advent of University of Wisconsin solution (UW) resulted in markedly improved patient outcomes in organ transplantation. The 1980's saw liver transplantation become a standard treatment for both adult and pediatric patients.

Initially, cold storage liver preservation was achieved with Euro-Collins solution with a maximal cold ischemia time of 6-8 hours.[3] UW solution was developed by Drs. James Southard and Folkert Belzer (Fig. 1) and was originally based on food preservatives. Early preclinical studies in perfusion proved cumbersome and inconsistent results, Belzer and his colleagues turned to simple

*James V. Guarrera—Division of Abdominal Organ Transplantation, Columbia University College of Physicians and Surgeons, 622 West 168th Street, PH 14 Center, Room 202, New York, NY 10032, USA. Email: jjg46@columbia.edu

Organ Preservation for Transplantation, Third Edition, edited by Luis H. Toledo-Pereyra.
©2010 Landes Bioscience.

Figure 1. Folkert O. Belzer MD. Used with permission from Media Solutions and University of Wisconsin School of Medicine and Public Health.

cold storage. The first solution tested was the UW solution which provided consistent 72 hour preservation of the canine pancreas.

In 1987 UW was used for the first time clinically for liver preservation. Initially CIT was kept less than six hours with operations were performed on an emergency basis. Later that year a patient with ESLD was prepared for liver transplantation. The liver was procured in Texas and because the recipient required several procedures, the transplant was scheduled for 8:00 am the next day. The liver was successfully transplanted with 20 hours of cold ischemia time. The recipient was extubated the next day and transferred to the transplant unit on the second postoperative day with excellent liver function. Thus scheduled liver transplantation became a reality. UW revolutionized LTx by lengthening of safe cold ischemia time and making it into a "semi-elective" procedure.[4-7] Advances brought about by UW and Cyclosporine A were further bolstered by steadily improving anesthesia, critical care and surgical techniques.

Current State of Liver Transplantation

Currently, one year patient and graft survival for liver transplantation approaches 90% and 85%,[8] respectively and outcomes continue to improve (Fig. 2). Excellent results have led to exponentially increasing demand for liver transplants within the last decade.[9] Unfortunately the demand for LTx has not been paralleled by supply with annual donor growth linear of <10%. Application of LTx is fundamentally limited by the existing donor-pool. This organ scarcity spurred the development and boom of living donor liver transplantation, which has plateaued due to concerns about donor safety.

Liver preservation has seen little progress since the advent of UW solution. HTK solution has become a viable choice for cold storage preservation but outcomes have only been equivalent. Thus the time is ripe for improvements in organ preservation which might allow safe utilization of an extended range of organs predisposed to poor early function when combined with cold ischemia.

Management of the Brain Dead Donor

Organ preservation begins when the potential donor is admitted to the hospital. Management of the injuries associated with brain death have a significant impact on the number of organs recovered and transplanted, as well as patient outcomes. Unfortunately, the management of potential organ donors is underrepresented in the literature and transplantation research.

While liver- specific changes associated with brain death are not well described, brain death leads to loss of autonomic tone and results in disturbed vascular autoregulation, with diminished oxygen and nutrient delivery to organs and tissues.[10] As a result, organ donors routinely require cardiovascular support with inotropes and/or vasopressors. Central venous access, continuous

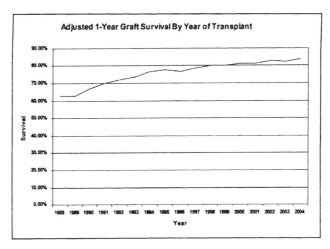

Figure 2. Improving outcomes after liver transplantation in the last two decades. Used with permission from 2006 Annual Report of the U.S. Organ Procurement and Transplantation Network and the Scientific Registry of Transplant Recipients: Transplant Data 1996-2005. Health Resources and Services Administration, Healthcare Systems Bureau, Division of Transplantation, Rockville, MD.

arterial pressure measurement should be utilized to guide the management. If cardiovascular and volume status is unclear, a Swan-Ganz catheter should be utilized without hesitation. Both initial and late circulatory changes can lead to ischemic liver damage prior to recovery, which may impair graft quality. Periods of warm ischemia followed by the necessary cold ischemia during organ preservation will result in increased severity of reperfusion injury. Overwhelming reperfusion injury is associated with primary nonfunction or severe early allograft dysfunction.

Hormonal changes associated with the autonomic storm increase circulating catecholamines. In addition, hypothalamic-pituitary axis dysfunction leads to neurogenic diabetes insipidus (DI) and a marked decrease in levels of thyroid hormones and cortisol.

DI may lead to profound hypernatremia which has been associated with primary nonfunction and poor early function of liver grafts. Hypernatremia is a frequent finding within the donor population that has a negative impact upon function of hepatic as well as extra-hepatic allografts.[11] Hypernatremia may result from aggressive treatment of cerebral edema, decreased antidiuretic hormone secretion secondary to cerebral ischemia, or inadequate donor fluid management. It is postulated that allograft function is impaired due to shifting of hepatocellular osmolality associated with the hypertonic state. During normalization of hypernatremia, intracellular water may rapidly accumulate resulting in cell swelling and injury.[11]

Donor hypernatremia has been shown to be a predictor postLTX graft dysfunction. Avolio et al. were the first to report a direct correlation between donor serum sodium concentration and peak serum aminotransferase following LTX.[12] Gonzalez et al derived a similar conclusion from a study of 168 consecutive LTX over a 3-year period, in which multivariate analysis revealed donor hypernatremia to be the strongest predictor of early graft dysfunction.[11] In an analysis of 649 LTX procedures performed among 11 Spanish transplant centers, donor plasma sodium exceeding 155 mmol/L at procurement (n = 103) was independently associated with an increased rate of retransplantation and decreased actuarial graft survival in a Kaplan-Meier analysis.[13] Markmann et al analyzed 1393 consecutive LTX performed at the University of California Los Angeles from 06/92 through 01/98 and determined donor serum sodium ≥ 170 meq/dL to be an independent prognostic value in predicting graft survival after primary LTX.[14] One-year graft survival of donor serum sodium <170 meq/dL was 75% versus 61% for donor serum sodium ≥ 170 meq/dL (p = 0.008) with a relative graft failure risk of 1.96 (p = 0.003).[14] One recent report refutes the other literature showing no negative impact on early graft function with sodium >160.[15] Whenever possible, appropriate resuscitation to correct donor hypernatremia in the setting of the intensive care unit should be optimized prior to recovery.

Depletion of high-energy stores has been reversed by a combination of T3, cortisol and insulin administration, suggesting that hormonal changes are the major cause of mitochondrial dysfunction with diminished cellular energy production.[16] These mitochondrial disturbances are further compounded by cold ischemic injury. Hormonal replacement in the form of a "cocktail" is now routinely practiced in most organ procurement organizations.

In summary, brain death may cause a variety of insults which result in end-organ damage. Pharmacologic support is critical to optimize perfusion, maintain homeostasis and prevent ongoing parenchymal challenge to the potential liver graft. The goal should be to deliver advanced critical care to all organ donors and all donor organs. Aggressive predonation resuscitation and management of potential organ donors is important to ensuring the best possible transplant outcomes and should be considered as an important part of organ preservation.

Procurement Techniques

Procurement followed by hypothermic preservation of the liver has been studied since the early 1960s beginning with early experiments by Starzl.[17] Liver cooling, followed by in situ perfusion of the liver with cold Ringer's lactate solution was the earliest description of liver preservation. The standard recommended temperature for cold storage is 4°C. Below this point as you approach 0°C the liver can easily freeze which will result in coagulative necrosis on reperfusion. Temperatures significantly above 4°C in static conditions will be associated with increased metabolic activity, ATP depletion, lactic acid buildup and mitochondrial disturbances resulting in severe parenchymal and endothelial injury.

Many liver procurement techniques have been described in the literature. The author prefers a modified rapid flush which allows prompt control of the aorta and assures the ability to begin cold preservation rapidly should the donor become hemodynamically unstable. Adequate dissection should then be performed to clearly understand the vascular anatomy. More than 75% of the cooling during liver harvesting occurs from the vascular flush rather than topical cooling with ice. In situ aortic and portal flush is recommended, with some surgeons reserving portal flush for the backtable.

In a randomized prospective trial with 20 donors in each arm, Chui reported equivalent graft outcomes for those procured between aortic flush alone and a combined in situ aortic and portal flush.[18] As the portal vein is responsible for over 75% of blood flow, common sense would suggest augmenting the aortic flush with an in-situ portal flush would be beneficial and improve cooling. In a prospective study, D'Amico et al [19] showed superior results with dual in-situ flush compared to aortic in-situ flush alone for "suboptimal" grafts. In addition we have encountered situations where the celiac axis is stenotic or the aortic clamp is only partially occluding the aorta. In these situations, in-situ portal flush would seem to be crucial for optimal preservation the graft. Different preservation solutions will be discussed later in the chapter.

Pre-reperfusion Flush of the Liver Allograft

There is no good evidence of any benefit to a particular method of flush of the liver graft prior to implantation. Colloid or crystalloid flush with or without a blood venting is a safe and practical technique which will minimize risk of post reperfusion hyperkalemia and hemodynamic disturbances. This flush may occur on the backtable or prior to completing the portal vein anastomosis.

Unlike the kidney, the liver's substantial volume requires a pre-reperfusion flush out is required if UW is used due to the potential for lethal ventricular arrhythmia from a rapid influx of K^+ from the residual solution in the liver.[22] Flush out can occur with blood, colloid, albumin, crystalloid solution or plasmalyte. The author recommends a similar flush even with HTK or Celsior solution which does not contain a high concentration of K^+. The significant buildup of waste products within the liver and straight reperfusion will likely liberate "evil humors" into the systemic circulation which may be deleterious. This is likely proportional to the preservation time.

Back-table flush of liver grafts preserved with HTK with 1-2 liters of fresh HTK prior to anastomosis is also a safe option and has been used at our center.

Liver Reperfusion and Preservation Injury

Overview of I/R Injury Mechanisms

Hepatic I/R injury is a complex phenomenon. An excellent in depth review of the topic was published by Lichtman and Lemasters.[20] Briefly, cold ischemia followed by reperfusion liberates vasoactive and injurious compounds from the donor liver and from the gut. These compounds ultimately lead to parenchymal and endothelial cell injury. The oxidative stress associated with reperfusion results in direct free radical damage to the liver. TNF-α and IL-1 are the two proinflammatory cytokines most commonly implicated with hepatic reperfusion. TNF-α production especially from activated Kupffer cells induces IL-1 production and activates apoptotic pathways that result in cell death. Elevated Plasma TNF-α levels after reperfusion have been reported by several groups. Both mediators induce production of IL-8, a chemoattractant for neutrophils. IL-6 has a dual function participating in both inflammatory responses and hepatic regeneration. TNF-α, IL-6 and IL-8 activate neutrophils which then upregulate cell surface adhesion molecules CD11/CD18 which cause strong neutrophil adherence to the endothelium where it causes toxic free radical injury. Similarly these cytokines activate ICAM-1 on the vascular endothelium which further facilitates neutrophil "sticking". In addition to liver graft injury, the reperfusion effects on endothelial cells may be found in other organs especially in the lungs which may result in pulmonary complications after transplant. IL-10 in contrast counteracts the inflammatory response and is protective in hepatic I/R.

Reperfusion-dependent events aggravate further ischemia-induced parenchymal injury either through prolongation of focal ischemia (no-reflow) or from the action of proinflammatory mediators (reflow paradox).[21] These initial interactions cause the so-called "rolling effect," which leads to progressive slowing of leukocyte traffic along the vascular wall, adherence of these circulating cells to the endothelium and their ultimate infiltration into graft tissue. Upregulation of early adhesion molecules (selectins) trigger subsequent events leading to leukocyte sticking and deploying toxic reactive oxygen species which further endothelial and parenchymal damage. This microvascular injury has been associated with primary graft nonfunction.

Numerous strategies to counteract ischemia-reperfusion injury in the liver include better preservation agents, pharmacological intervention to the donor, machine perfusion and techniques such as ischemic conditioning. Each of these approaches will be further discussed within the chapter.

Clinical Implications of Reperfusion/Preservation Injury

Early Manifestations of Poor Preservation

While surgical techniques vary from center to center, reperfusion disturbances are universal and a combination of both the recipient, the quality of the liver and a constellation of intraoperative variables. The so called "reperfusion syndrome" is a phenomenon characterized by acidosis and hypotension in part from the buildup of acid during portal and vena cava clamping. Veno-venous bypass appears to attenuate this phenomenon although it is associated with other complications. Liberation of "evil humors" result in hemodynamic disturbances, hypotension requiring increased vasopressor utilization. Ventricular arrythmias, predominantly from hyperkalemia and bradyarrthmias are also common. Poor preservation and impending primary nonfunction is usually characterized by refractory coagulopathy, acidosis and "stiff" liver in spite of normal central venous pressure. Congestion associated with reperfusion may be combated with intravenous nitroglycerine which is a potent venodilator not to mention a nitric oxide precursor. Unloading a stiff congested liver is important for good initial liver function. In our center we utilize N-Acetylcysteine routinely for elderly, steatotic donors and those with prolonged CIT. Early data analysis suggests lower peak hepatocellular injury markers and shorter ICU stay (unpublished data, Columbia University).

Late Manifestations of Preservation Injury

Biliary cast syndrome (BCS), also known as ischemic cholangiopathy, is associated with poor preservation, DCD donors or arterial stenoses (Figs. 3-5).[22] It was first described by Starzl in 1977[23]

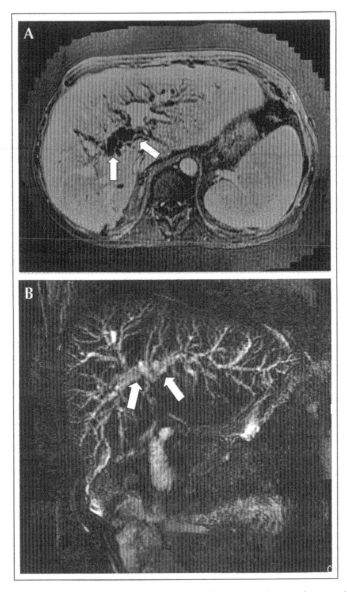

Figure 3. MRI (A) and MRCP (B): Intrahepatic duct dilatation with irregular peripheral ducts and filling defects caused by biliary cast syndrome. The liver allograft was from a 32-year old DCD. The patient ultimately underwent retransplantation after a temporizing Roux-en-Y lepaticojejunostomy.

and is characterized by intrahepatic biliary strictures, ductal dilatation, intrahepatic abscesses and biliary anastomotic leakage. While its incidence improved with better preservation and recovery techniques, increasing use of DCD and older livers has seen a recent resurgence of BCS.

Imaging with MRI/MRCP (Fig. 3A,B), ERCP or percutaneous transhepatic cholangiography (PTC) (Fig. 5) will confirm the diagnosis. There have been some successful reports of endoscopic management in focal cases. Unfortunately, most cases are diffuse and involve the

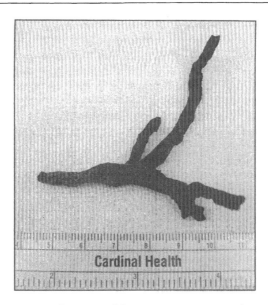

Figure 4. Biliary cast surgically removed from case in Figure 2. The cast recapitulates the intrahepatic and extrahepatic duct systems including the cystic duct stump.

intrahepatic and extrahepatic biliary tree of both lobes and are not compatible with long term survival. Sloughing of the biliary mucosa produced a cast which recapitulates the bile ducts affected (Fig. 4). Even if surgical therapy improves the situation, recurrent cholangitis due to small duct plugging with development of resistant organisms is typical. Supportive antibiotics, aggressive nutritional support, maximal biliary drainage and early retransplantation is the most prudent intervention.

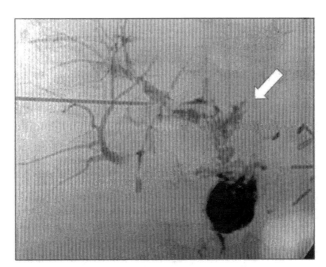

Figure 5. Percutaneous transhepatic cholangiogram of an infant with characteristic appearance of BCS (arrow) of the left lobe of the whole liver allograft associated with left hepatic artery thrombosis. The patient subsequent left lobectomy with a high Roux-en-Y anastomosis to the right sided ducts.

Table 1. Donor characteristics that can affect severity of preservation injury in liver transplantation

Elderly donors (>70 years)	• More susceptible to ischemic endothelial injury
	• Decreased ATP availablilty on reperfusion
	• Less tolerant of prolonged cold ischemia
	• May have decreased synthetic function and regenerative capacity
Underlying liver histopathology	• Macrosteatosis→ predisposes to poor initial function and primary nonfunction
	• Ischemic changes/necrosis
	• Significant alcohol abuse→ steatohepatitis
	• Hepatitis B and C activity/portal inflammation
	• Fibrosis→ may be associated with hepatits C or alcohol abuse and may affect long term outcomes
Ischemia associated with donor injury	• Donation after cardiac death→ frequently profound ischemia injury
	• High dose vasopressors
	• Prolonged or uncorrected hypoxemia or acidosis
Biochemical changes	• Hypernatremia
	• Rising transaminases or bilirubin

Extended Criteria Donors: Characteristics That Are Synergistic with Preservation Injury

Expansion of donor criteria to include allografts with an additional risk of delayed function has been advocated for over a decade.[24-27] During this period, numerous single-center reports have identified predictors of potentially poor allograft function. Ploeg et al first described factors which are independently associated with a higher incidence of primary nonfunction (PNF) and initial poor function (IPF).[28] These factors include reduced-size liver, donor macrosteatosis, older donor age and prolonged cold ischemia times. PNF requires emergent retransplanatation or the patient will die. The morbidity associated with IPF includes longer ICU stay, infectious complications, prolonged cholestasis and iatrogenic injuries from prolonged ICU stay. Table 1 summarizes donor characteristics that act synergistically with preservation injury to cause early graft dysfunction in liver transplantation.

Donor characteristics demonstrated to yield a higher risk of PNF and IPF: age over 60 years,[29-40] hypernatremia exceeding 155 meq/l,[35,41] macrovesicular steatosis exceeding 40%,[31,35,37,38,42-47] cold ischemia time exceeding 12 hours,[24,31,35,37,41] partial-liver allografts (split-liver transplantation, reduced-liver transplantation, adult-to-adult living donor liver transplantation)[34,37] and donation after cardiac death.[31,34,48,49] These represent liver transplant scenarios where reducing preservation injury is of paramount importance.

Elderly Donors

Acceptable donor age has expanded in response to population demographics, donor demographics and increasing organ scarcity. Donor age ≥65 years represents the largest expanding component of the current donor-pool. Recognition of unique physiologic and anatomic characteristics associated with older liver allografts is prerequisite to their successful utilization. Older liver allografts have a lower tolerance for cold preservation.[50] Endothelial injury from cold ischemia occurs earlier in older allografts; increasing the risk of inflammation, thrombosis and T-cell mediated rejection.[31,51] Adenosine tri-phosphate (ATP) synthesis is decreased postreperfusion.[51] This may impede synthetic function and regenerative capacity.[31,51,52] Characteristic anatomic features of older liver allografts include capsular fibrosis, smaller size, darker texture, increased steatosis and arterial atherosclerosis.[50]

Underlying Liver Histopathology: Steatosis

Steatosis is categorized as microvesicular and macrovesicular based upon histologic appearance. Microvesicular steatosis is diffuse, intracellular lipid vacuolization associated with altered hepatic physiology from sepsis, prolonged hospitalization, enteral starvation, or total parental nutrition.[50,53] Microvesicular steatosis is reversible and is frequently absent in functional allografts within one-week postLTX. Consistent data examining the effect of microvesicular steatosis on hepatic allograft function have not emerged.[54,55] While microvesicular steatosis in of itself does not impair the graft it should be regarded as a potential marker of stress of the donor.

Macrovesicular steatosis is a combination of intracellular lipid vacuolization and extracellular lipid depots associated with underlying obesity, inflammation and cellular injury. Lipoperoxidation and cellular dropout are replaced by adipocytes that disrupt the histologic architecture of hepatic sinusoids.[56] Macrovesicular steatosis is sub-categorized by histologic volume percent as mild (10-30%), moderate (30-60%) and severe (>60%). Precise grading of steatosis is difficult with significant inter-observer variability reported.[44,57,58] Pathologic estimation should be attempted only by an experienced pathologist in cooperation with a seasoned recovery physician and include only macrovesicular steatosis, as the presence of microvesicular steatosis does not correlate with allograft function.

Macrovesicular fat has been associated with increased incidence of IPF and PNF in many published series. Transplantation of livers with mild steatosis (<30%) and no other negative prognostic indicators should yield acceptable results when cold ischemia is minimized.[44,47] We believe that transplantation of severe steatosis allografts should be avoided as posttransplant function is unpredictable with reported incidences of PNF as high as 60% with current preservation techniques.[28,42,46,59]

The transplantation of allografts with moderate steatosis is the current clinical challenge.[44] The reported incidence of delayed early graft function approaches 35% as evidenced by peak transaminases, coagulopathy and increased transfusion requirements.[28,55,60] In this setting, procurement by an experienced donor surgeon is invaluable as this individual can provide histologic verification of the biopsy and accurate assessment of the allograft with respect to texture and appearance. Verran et al reported the largest experience to date of LTX with steatotic hepatic allografts.[47] In their series of 120 steatotic allografts transplanted between 01/86 and 12/00, initial poor function, early retransplantation and allograft loss correlated to increasing grades of hepatic allograft steatosis. The result was significantly lower patient survival among recipients of moderate and severe steatotoic allografts (78%) versus minimal or no hepatic allograft steatosis (87%). The most common cause of death among recipients of moderate and severe steatotic allografts was sepsis. Recipient age and status at LTX, in addition to HCV as the indication for LTx, correlated with donor steatosis to negatively impact allograft survival.[47] Therefore, the ultimate outcome of the allograft with moderate steatosis will be the result of biopsy findings, visual inspection by an experienced procurement physician, minimum cold ischemia time, the presence of additional donor risk factors and appropriate recipient matching, preferably in younger candidates without HCV.[44,47]

Length of Cold Ischemia Time

Cold preservation increases anaerobic metabolism and cellular acidosis. Metabolic activity is reduced with mitochondrial energy uncoupling. ATP stores are depleted with an accumulation of hypoxanthine, a substrate for the generation of toxic, reactive oxygen species during reperfusion.[50] Reperfusion following prolonged cold ischemia in human and animal models is associated with inflammatory changes within the allograft that include sinusoidal cell damage, complement activation, small vessel hypercoagulability and increased circulating levels of IL-6 and IL-8.[31,61,62] Prolonged cold ischemia time is an independent risk factor for the development of preservation injury and delayed graft function. Prolonged cold ischemia not only increases the incidence of short-term complications from allograft function, but also increases the incidence of long-term biliary complications.[13,63] The precise threshold for significant cold preservation injury varies with the individual allograft; however, general guidelines have emerged from the literature. In allografts from otherwise healthy donors who are not over age 60 years, the threshold for reduced allograft function secondary to prolonged cold ischemia lies between 14 and 16 hours.

In a report of 315 LTx procedures from a European multicenter study group, Porte et al identified CIT > 16 hours associated with increased PNF and reduced long-term graft survival.[64] Additional reports concur that cold ischemia times exceeding 14-16 hours have a roughly two-fold increase in complications related to allograft function.[28,65-67] Hepatic allografts from older donors (age > 60 years) are much more sensitive to preservation injury and demonstrate optimal function when cold ischemia is under 8 hours.[68] In addition there seems to be a synergistic effect of prolonged cold *and* warm ischemia time on post-operative graft outcome.[67,69]

Ischemia Associated with Liver Injury: Donation after Cardiac Death

Donation after cardiac death (DCD) is a rapidly expanding component of the donor-pool that has recently received increased emphasis as part of the Health Resources Service Administration sponsored Organ Transplant Breakthrough Collaborative.[70] DCD is a fundamentally different recovery technique based upon cardiopulmonary criteria for death rather than neurologic criteria for death. Potential donors must undergo circulatory arrest, declaration of death and a period of observation to rule out spontaneous resuscitation *before* the procurement can begin. Procurements based upon neurologic brain death criteria that proceed with cardiopulmonary circulation until the surgeon induces arrest by aortic crossclamp, venting and cold preservation. In contrast, DCD allografts experience a significantly more warm-ischemic insult that initiates with terminal extubation resulting in hypoxia, cardiopulmonary collapse, circulatory arrest and concludes with absent blood flow at body temperature for an indeterminate period until the procurement can be performed and cold perfusion initiated. The potential negative effect of neuro-humoral events associated with cardiopulmonary collapse upon the liver is also unknown.

Cohesive single-center outcomes data in addition to UNOS Transplant Registry data[71,72] on DCD liver allograft function are just beginning to emerge to the point where broad recommendations on utilization are possible (Fig. 5, Table 1). Standardization of terminology, technique, incorporation of vasodilatory drugs, antioxidants, preservation solutions and necessity for anticoagulation in DCD need to be implemented.[71] Available data indicate the time from terminal extubation to initiation of cold preservation is optimal under 30 minutes[71] and the incidence of biliary complications increases significantly with extended warm ischemic times.[73] Cold ischemia time should be kept to less than 8 hours.[71] The upper age limit for optimum outcome utilizing current preservation technology is likely in the sixth decade. Machine preservation holds promise in resuscitating and improving utilization of DCD livers. Detailed interpretation of the donor hospitalization and expert assessment of allograft quality by the procurement physician are essential.

Physiologic, functional and procurement insults typically of trivial significance during procurement of neurologic death cadavers assume much greater significance in attempting to predict the outcome of hepatic allografts procured from DCD donors. In the case of DCD donors, we believe heparinization is of crucial importance prior to procurement.

Clinically Available Preservation Solutions

The composition of clinically available preservation solutions are shown in Table 2. UW has been the gold standard for liver preservation since 1987. UW allowed safe extension of liver preservation times and "elective" liver transplantation. UW has a high potassium and low sodium concentration with a phosphate buffering system. The components raffinose and lactobionate are impermeates which reduce cellular swelling and organ edema. Hydroxyethyl starch, a colloid, allows a more effective flush-out of blood. Allopurinol is a xanthine oxidase inhibitor that reduces

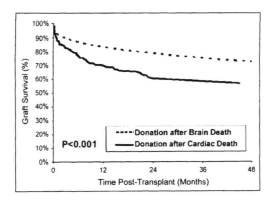

Figure 6. Recent outcomes of liver transplantation from DCD donors. Used with permission from reference 72.

Table 2. Components of UW, HTK and Celsior solutions

	UW	HTK	Celsior
Year of Introduction	1987	1975	1994
Source	Belzer (USA)	Bretschneider (Germany)	Pasteur-Mérieux (France)
Components			
Na^+ (mmol/L)	25-30	15	100
K^+ (mmol/L)	125-130	10	15
Mg^{2+} (mmol/L)	5	13	4
Ca^{2+} (mmol/L)	–	0.25	0.015
Mannitol (g/L)	–	30	60
Lactobionic acid (mmol/L)	100	–	80
HES (g/L)	50	–	–
Raffinose (mmol/L)	30	–	–
Histidine (mmol/L)	–	180	30
H_2PO_4/HPO_4 (mmol/L)	25	–	–
OH (mmol/L)	–	–	100
Glutathione (mmol/L)	3	–	3
Allopurinol (mmol/L)	–	1	–
Adenosine (mmol/L)	5	–	–
Acetone dicarboxylic acid (mmol/L)	–	1	–
Tryptophan (mmol/L)	–	2	–
Aminoglutaminic acid (mmol/L)	–	–	20
mOsm/L	320	310	320
pH	7.4	7.2	7.3

free radical burden and adenosine is a substrate for ATP generation. The original formulation calls for the addition of insulin, dexamethasone and penicillin. Some users add fresh glutathione for antioxidant effect. During adult procurement 2-4 liters of UW are used for an in situ aortic flush and usually 1 liter in the portal vein. For children, recommended flush volume is 50 ml/kg. Additional backtable flush of the portal vein and common bile duct are recommended.

The first clinical results of histidine-tryptophan-ketoglutarate (HTK) for liver preservation were reported in 1990.[74] HTK is as an extracellular solution introduced in 1987. HTK is a low viscosity solution, originally designed as a cardiopelgia solution with low sodium and potassium concentrations. The three major components of HTK are histidine which acts as a buffer, tryptophan which is a membrane stabilizer and ketoglutarate which is an energy substrate. In the past, recommended volumes were 8-10 liters for an in situ flush. However, recently some centers are using volumes only slightly higher than for UW flush.

A number of experimental comparisons between UW and HTK have supported UW as the superior preservation choice for longer cold ischemia times. Van Gulik et al compared UW and HTK for preservation of canine liver grafts. HTK preserved livers had progressive parenchymal swelling and enzyme markers of ischemic injury as cold ischemia time increased; all HTK grafts stored for 48 hours did not function, while all UW grafts stored for 48 hours functioned.[75] Another study focused on ultrastructural analysis of human endothelial cells, within a cell culture model, after hypothermic storage in UW, HTK or Euro-Collins. HTK and Euro-Collins stored endothelial cells had severe mitochondrial and cytoskeletal damage while the UW group's ultrastructure was well preserved. Integrity of endothelial cells have the clinical significance of being associated with PNF and early graft dysfunction.[76]

Celsior is the most recent of the clinically available solutions, is a hybrid, balancing the buffering capacity of HTK and the impermeants of UW. It is an extracellular solution with high Na^+/ low K^+. Impermeants include, lactobionate, similar to UW and mannitol, which prevents cellular edema. Histidine acts as a buffer and a high magnesium concentration prevenst calcium overload. Recommended volumes for flush include 60 ml/kg for aortic flush and 30 ml/kg for portal flush. A preclinical study comparing UW, HTK and Celsior in an animal model of cold storage reported highest bile production in UW livers and significantly lower injury markers and histological signs of injury in UW vs HTK at 8 hours CIT. UW showed significantly lower injury markers than Celsior at 16 hours of CIT.[77]

Comparison of UW, HTK and Celsior in Deceased Donor Liver Transplantation

UW, HTK and Celsior appear to be equivalent for most situations in deceased donor liver transplants, although UW remains the gold standard for longer preservation times.[77] Erhard et al conducted the first randomized comparison between UW and HTK over a decade ago and since then HTK has shown clinical equivalence when cold ischemia times are under 15 hours[78,79] however increased hepatocellular injury markers were noted in the HTK group.[78]

Mangus et al[80] recently reported a large retrospective study (n = 378) in the U.S. showing no significant differences in 1-, 6- and 12-month graft and patient survival when livers are preserved with UW or HTK. While there were higher post-operative AST, ALT and total bilirubin on Day 1 for the HTK group, these differences were resolved by Day 7 and did not appear to have clinical significance. In this study only 0.6 liters of additional HTK was used. A prospective study of UW and HTK in deceased donor liver transplantation (n = 102) showed comparable patient and graft survival, serum hepatic enzymes, prothrombin time and factor V variation, with patient follow-up ranging from 4 to 20 months. Interestingly, the study found more biliary complications in livers preserved with HTK solution.[81]

Another randomized, prospective trial (n = 80) compared preservation of livers with Celsior or UW solution.[82] Findings indicate that Celsior has comparable efficacy to UW for liver graft preservation within a median ischemia time of 7 hours. At Year 1 and Year 2 posttransplant, no significant differences were observed in liver function parameters, biliary complications and graft survival. These results verify an earlier pilot, randomized prospective trial.[83] Similarly, a small randomized prospective trial comparing HTK to Celsior is underway and preliminary results show clinical equivalence.[84]

Comparison of UW and HTK in Living Donor Transplantation

Living donor liver transplantation (LDLT) has shown comparable clinical performance of UW and HTK. One retrospective review of LDLT, with 42 UW-preserved and 44 HTK-preserved grafts showed no significant differences in posttransplant survival, liver function, biliary complication and acute rejection.[85] Jain et al prospectively studied 33 LDLTs, revealing comparable hepatic function. The HTK group had higher peak AST/ALT levels initially, consistent with some earlier deceased donor liver transplant findings.[86] However, the higher AST/ALT levels in HTK preserved LDLTs equilibrated by post-operative day 8 and did not have clinical relevance. Testa et al[87] prospectively verified the comparable clinical performance of UW and HTK for LDLT with a mean follow-up time of 13 ± 7 months posttransplantation.

Short preservation times, as well as rapid flush and cooling seem to make HTK a logical choice in LDLT. A higher biliary complication rate in LDLT is likely a strong impetus for using HTK based on the theoretical benefits of its improved biliary preservation. Still, there remains no prospective and randomized clinical trial for preservation solutions in LDLT.

Pharmacologic Treatment of the Donor

Some centers have suggested that treating the donor prior to organ recovery may improve organ viability. One agent used was N-acetyl-cysteine (NAC). The hypothesis was that NAC would improve hepatic glutathione levels which would protect against the oxidative stress of reperfusion. Unfortunately, a clinical benefit could not be demonstrated.[88] Phentolamine has been used as a vasodilator to improve delivery of perfusate after cross-clamp. There is one report of improved early kidney function but no literature in liver transplantation.[89] T-cell depleting antibodies and pharmacological interventions, such as Tacrolimus (FK-506), have been explored for pretreatment of the intestinal donor.[90-93] Apoptosis inhibitors may become important for donor pretreatment. Administration of a Caspase-3 and Caspase-7 inhibitor to rat donors improved graft survival and significantly decreased number of apoptotic endothelial cells, thereby improving microvascular perfusion. A cell protective effect was also suggested by an upregulation of BCL-2 at 7-days posttransplant.[94] In preclinical lung transplant models, prostaglandin E1 (PGE-1) administration ameliorates reperfusion injury[95] and improves ultrastructural lung preservation.[96] However, there are conflicting results and Vainikka et al conclude that donor pretreatment with PGE-1 does not have a beneficial effect on lung preservation.[97] A review of early outcomes after liver transplantation from donors that received PGE-1 was unable to tease out a benefit that was not likely related to the intrinsic quality of the donor.

Novel Liver Preservation Strategies

Solution modifications may improve preservation of extended criteria donors. The severe bottleneck in developing new preservation solution and other technologies includes development challenges between academia and industry, as well as issues of research and informed consent in organ procurement.

Apoptosis Inhibitors

A recent Phase II multi-centered, randomized clinical trial reported on IDN-6556 (Idun Pharmaceuticals, San Diego, CA, USA) a broad-spectrum caspase inhibitor to reduce I/R injury during liver transplantation.[98] The subject group receiving IDN-6556 in the preservation solution had significantly lowered AST/ALT levels from Days 1 through 7, as well as significantly lower levels of apoptosis on Day 7. Intravenous injections of IDN-6556 seemed to reverse these positive effects. A larger clinical trial is will be necessary to detect a clinical benefit.

Nitric Oxide Donors

Nitric oxide (NO) donors confer antioxidant protection during periods of warm ischemia[99,100] and protect hepatocyte mitochondria in culture.[101,102] Balance of the nitric oxide pathway upon reperfusion is of paramount importance in microvascular perfusion. Disordered NO balance at reperfusion is a strong predictor of the severity of end organ insults of reperfusion injury. Thus, NO donors have been an area of interest in organ preservation for some time. Ben Mosbah et al[103] examined the effect of UW solution supplemented with trimetazidine (TMZ) and aminoimidazole-4-carboxamide ribonucleoside (AICAR). TMZ was initially studied as an anti-ischemic drug in the heart and AICAR, as a well-known activator of AMPK. They found that TMZ and AICAR led to an increase in constitutive nitric oxide synthase (cNOS) in the liver via increasing AMPK. TMZ had beneficial effects on peak AST levels, bile output, LDH levels and a number of other hepatic function parameters in both normal and steatotic livers. AICAR added to UW solution showed a similar protective effect in normal and steatotic livers. NO donors may be beneficial in novel solution development.

Pentoxifylline

Improving preservation of DCD liver donors is a critical strategy for expanding the donor pool. Pentoxifylline suppresses inflammatory cytokines and decreases microcirculatory damaged by I/R events.[104] In an ex-vivo perfusion model with prolonged cold ischemia time (18-24 hours), pentoxifylline led to significant decreases in vascular resistance and hepatic enzyme release.[105] Pentoxifylline added to UW solution in porcine livers, exposed to 20 minutes of warm ischemia, conferred a protective effect and allowed prolongation of cold ischemia time to 16 hours versus 12 hours in the UW control group. The livers supplemented with pentoxifylline, had a 100% one-week survival and had improved hepatic function, including increased ATP, lower transaminases and improved microcirculation.[106] Further studies of pentoxifylline in liver preservation are warranted to characterize the potential benefits of this additive.

Antioxidants and Heat-Shock Proteins

Free radical scavengers may mitigate transplant I/R injury, as studied in various preclinical liver models.[107,108] Superoxide dismutase (SOD), which has been popularized in I/R literature, has the critical challenge of a 6 minute half-life.[109] In an I/R injury model, the portal vein and hepatic artery were occluded for 30 minutes, followed by analysis of injury 7 hours post reperfusion. Portal vein injection of polylipid nanoparticle genes led to liver SOD and catalase increases by 10-fold over controls. This was combined with significantly decreased ALT levels, restored glutathione stores and improved liver histology. These findings add to the growing body of studies targeting gene delivery as a protectant against transplant-relevant reperfusion injury.[110] While protective gene therapy is an intriguing preservation strategy, its clinical application seems futuristic at this point.

Curcumin is another antioxidant of interest with relevance to liver preservation via its activation of heat shock protein, HO-1. Heme oxygenase is the rate-limiting enzyme in the degradation of heme into carbon monoxide, iron and bilirubin. CO induces HO-1 and recent evidence suggests that the induction of HO-1 is associated with protection from ischemic-reperfusion injury. HO-1 limits leukocyte mediated microvascular injury through ICAM-1, decreased free radical injury and increases BCL-2 and iNOS expression. In hepatocyte cultures, curcumin induced HO-1 in hepatocytes. Elevated HO-1 levels are also associated with reduced oxidative stress and hepatic injury, as well as cryoprotection.[111] One study demonstrated that supplementation of Euro-Collins solution with curcumin led to significant benefits in an isolated rat perfusion model, suggesting comparable performance to standard UW solution. When curcumin was combined with UW solution, significant benefits, in terms of portal flow rates and increased bile production, were found for livers exposed to 24 hours of cold ischemia. Thus, curcumin may have a future role in extended liver preservation based on its ability to induce cryoprotective features of HO-1.[112]

IGL-1: Na$^+$/K$^+$ Inversion and PEG Substitution

IGL-1 is a novel perfusate based on the benefits of UW with modifications. High K$^+$ levels in UW have been criticized for inducing vasoconstriction and endothelial damage. In IGL-1, the Na$^+$/ K$^+$ concentration is inverted thus reducing the well known deleterious and vasoconstrictive effects of high K$^+$ on endothelial cells and the microvasculature. The colloid, hydroxyethyl starch (HES) is associated with microcirculatory red blood cell aggregation and plugging.[113] IGL-1 substitutes polyethylene glycol (PEG) for HES. PEG has been shown to have vasodilatory and antioxidant effects in addition to its colloid properties.

Compared to UW and in a rat OLT model, IGL-1 significantly reduced transaminase levels and caspase-3 activity at 24 hours. PEG substitution led to improved hepatic blood flow following reperfusion with fewer and smaller areas of hepatic necrosis, neutrophil infiltration and hemorrhage.[114] Beneficial effects of IGL-1 also extended to rat steatotic livers in an ex-vivo perfusion model, including lower transaminases and increased bile. Blockade of L-NAME implicated NO activity in the beneficial effects of IGL-1 preservation.[115] Early preclinical studies of IGL-1 show promise in the preservation of marginal livers.

Polysol

Originally developed for hypothermic machine perfusion (HMP), Polysol includes colloid, impermeants, buffers, free radical scavengers, amino acids and vitamins for improved energy and fat metabolism in the liver allograft. In an isolated perfusion model using steatotic rat livers, cold storage in Polysol was superior to HTK as measured by parenchymal AST and the integrity of the mitochondrial structures.[116] Using an ex vivo perfusion model, HMP Polysol, was superior to 24-hour cold storage with UW and HMP with UW-Gluconate.[117] Further studies in a porince OLT model demonstrated that HMP with Polysol was superior to cold storage with Celsior.[118]

Vasosol

Vasosol is a novel solution with enhanced vasodilatory and antioxidant capability. Vasosol CS consists of base UW solution supplemented with alpha-ketoglutarate as an energy substrate and mitochondrial stabilizer, Nitroglycerin and L-arginine as a nitric oxide precursor, N-acetylcysteine as a glutathione precursor and both nitroclycerin and prostaglandin E1 as a vasodilator and membrane stabilizer. Vasosol MPS is based on Belzer MPS with the modifications as in the cold storage formulation. In a small cohort of liver transplants, Vasosol CS showed significant improvement in early post transplant transaminases and early synthetic function.[119]

A preclinical pilot study in swine demonstrated safety and efficacy when Vasosol was used for liver HMP.[120] Vasosol MPS is currently being used in the first human clinical of liver HMP. This novel solution may in the future be useful for improved preservation of ECD livers.

Selective Matrix Metalloproteinase Inhibitors

Matrix metalloproteinases (MMP) play an important role in cold preservation of the liver.[121] Recently, an important pathway of I/R injury in steatotic livers was isolated to fibonectin- 4β1 integrin interactions, which selectively activate MMP-9.[122] Earlier studies used peptides with an RGD (Arg-Gly-Asp) motif to target FN-α4β1 interactions in steatotic livers, leading to significantly improved survival after OLT.[123] In subsequent study by this group, a cyclic RGD peptide targeting FN- α5β1 interactions, was administered via the portal vein of steatotic Zucker rat livers prior to and after cold storage. Along with inhibiting various pro-inflammatory responses, the RGD peptide improved functional parameters and histological preservation of steatotic liver grafts. 14-day survival of steatotic livers provided to lean recipients improved from 50% in the untreated group to 100% in treated peptide-treated group[124] Peptide therapy targeting leukocyte migration into the graft, via fibronectin-integrin intervention strategies, may provide a promising avenue for countering I/R injury, especially in the steatotic livers.

Hibernation: Intrinsic Protection from Ischemia-Reperfusion Injury

Hibernation is a downregulated state of metabolism found in some animals which helps conserve energy during periods of diminished food and oxygen consumption. This state is characterized by reduced core body temperature, O_2 consumption and greater homeostatic control over major ion gradients with decreasing temperature and hypoxia. An intriguing strategy to combat cold ischemia injury is the induction of genes controlling hibernation responses. Inuo et al examined whether D-Ala2-Leu5-enkephalin (DADLE), a hibernation inducer, could confer protective effects on hepatocytes during in vitro hypothermic storage injury. Significantly increased survival of hepatocytes as well as lowered release of ALT and LDH were observed in the study.[125]

Ischemic Preconditioning of Liver Grafts

In preclinical models, promising evidence has shown that brief exposure to ischemia prior to significant IR is protective. While the mechanisms are somewhat unclear, ischemic preconditioning induces genes and signaling pathways that are protective in ischemic-reperfusion injury, especially in the endothelium and through "inducible stress proteins" such as the heat shock proteins.[126] Ischemic preconiditioning is able to reduce leukocyte adhesion and upregulate NOS and other antioxidants. Recent studies on ischemic preconditioning (IPC) in liver grafts yielded conflicting results. Koneru et al[127] conducted a single-center randomized clinical trial involving the transplant of control (n = 51) and 10-minute IPC (n = 50) livers from deceased donors. Results showed IPC significantly increased early levels of AST and ALT, although this did not lead to adverse clinical consequences. Beneficially, the IPC livers were associated with higher systemic levels of IL-10 and a trend towards fewer moderate/severe rejections within 30 days. In another prospective trial, Azoulay et al, in contrast to the Koneru study, found that 10 minute IPC of the cadaveric liver (n = 46) led to significantly reduced post-operative peak AST and ALT levels compared to controls (n = 45). However, univariate analysis paradoxically indicated that IPC was significantly associated with initial poor function.[128] In a randomized trial by Amador et al[129] 10-minutes of IPC (n = 30; Control n = 30) led to lower post-operative AST levels, significantly improved biochemical markers of liver function including uric acid and HIF-1α levels, as well as a reduced need for re-operation. Other smaller trials by Jassem et al[130] and Cescon et al[131] demonstrated significant benefits in terms of reduced aminotransferases without other negative clinical parameters. Conflicting data indicate that larger clinical trials will be necessary to further interrogate IPC and assess clinical value, independently and in conjunction with other liver preservation strategies.

Machine Perfusion as a Liver Preservation Strategy

Currently hypothermic machine perfusion (HMP) is only clinically available for kidney preservation, where its use has become an increasingly standard practice. Growing evidence in the liver transplant area indicates that HMP may improve preservation of the liver compared to static cold storage. HMP provides continuous circulation of metabolic substrates for ATP generation, "wash-out" of waste products from direct endothelial-parenchymal contact. Improved delivery of metabolic substrates for ATP generation improves maintenance of ion channels and cytoskeletal elements and results in improved tissue ATP concentrations on reperfusion. In addition there is an opportunity to assess pretransplant organ viability and there is the potential for local pharmacologic intervention. An excellent comprehensive review of liver HMP was recently published by Schreinemachers et al.[132]

Hypothermic Machine Perfusion and Marginal Livers

The potential for better preservation and even resuscitation of marginal livers will be the impetus for applying HMP to liver transplantation. Recently, Bessems et al[133] showed that oxygenated, hypothermic machine perfusion significantly improves preservation of steatotic rat livers compared to static cold storage. Superior preservation of HMP with Polysol was also demonstrated for a rat model mimicking DCD donors.[134]

An interesting study was performed by Dutkowski et al,[135] whereby brief, hypothermic machine preservation at the end of cold storage was studied. HMP following cold storage may be a less cumbersome early application of liver HMP. Short-term 1-hour application of HMP after 10 hours of static cold storage reversed lipid peroxidation, LDH release and caspase activation. A model such as this would allow brief HMP during the recipient hepatectomy. HMP would be at the transplant center rather than the donor hospital which would be logistically easier. Further studies demonstrated similar results in rat livers exposed to warm ischemic time, thereby mimicking a DCD donor.[136] These studies represent some preliminary animal data for HMP in marginal grafts.

Development of a portable liver perfusion system will be an important step in encouraging widespread adoption of HMP of in liver transplantation. Two Dutch companies, Organ Assist (Groningen, The Netherlands) and Doorzand (Amsterdam, The Netherlands) are leading the march towards developing portable, user-friendly perfusion systems. To date these systems have not been used clinically. The Groningen HMP system (Organ Assist, Groningen, The Netherlands)[137] is compact, has disposable and reusable electronic elements and has been used extensively with excellent results in porcine liver transplants. Airdrive (Doorzand, Amsterdam, The Netherlands) is a portable and disposable device powered by compressed air. Preclinical trials have demonstrated efficacy.

The Columbia Hypothermic Machine Perfusion Clinical Trial

The first human clinical study of our ex-vivo liver HMP protocol is now underway at Columbia University. To date, 17 patients have been transplanted with livers that have been preserved with HMP. We have seen excellent patient and graft survival with no episodes of primary nonfunction, initial poor function, or vascular complications to date. Peak AST and prothrombin time are lower in the HMP group compared to cold stored control cases (Guarrera, JV, unpublished data). The study will also include mechanistic evaluation of HMP including, RT-PCR, immunohistochemistry and ultrastructural analysis to further characterize the mechanisms of protection in human HMP.

Conclusions—Ongoing Challenges and Future Directions

With a pipeline of new liver preservation agents and solutions, combined with landmark developments in HMP, the field of liver preservation is poised for explosive growth over the coming decade. Safe expansion of the current donor pool may be achieved by advances in liver preservation technology. At the same time, an ongoing challenge will be to continue to push promising preclinical developments into clinical trials and clinical application. If the field is able to meet this challenge, it is expected that the integration of machine perfusion with novel preservation solutions and targeted pharmacological intervention will enable more livers to be safely preserved for longer periods of time, with greater functional integrity and ultimately, improved organ availability, more lives saved and better outcomes for liver transplant recipients.

References

1. Starzl TE, Groth CG, Brettschneider L et al. Extended survival in 3 cases of orthotopic homotransplantation of the human liver. Surgery 1968; 63(4):549-563.
2. Jamieson NV, Friend PJ, Johnston PS et al. Orthotopic liver transplantation at Addenbrooke's Hospital Cambridge 1968 to 1991. Clin Transpl 1991; 119-125.
3. Todo S, Nery J, Yanaga K et al. Extended preservation of human liver grafts with UW solution. JAMA 1989; 261(5):711-714.
4. Cooper J, Rettke SR, Ludwig J et al. UW solution improves duration and quality of clinical liver preservation. Transplant Proc 1990; 22(2):477-479.
5. Stratta RJ, Wood RP, Langnas AN et al. The impact of extended preservation on clinical liver transplantation. Transplantation 1990; 50(3):438-443.
6. Gordon RD, Starzl TE. Changing perspectives on liver transplantation in 1988. Clin Transpl 1988; 5-27.
7. Kalayoglu M, Sollinger HW, Stratta RJ et al. Extended preservation of the liver for clinical transplantation. Lancet 1988; 1(8586):617-619.
8. Dousset B, Houssin D. Liver transplantation: the challenges of the 1990s. Biomed Pharmacother 1992; 46(2-3):79-83.
9. UNOS. United Network for Organ Sharing. http://www.UNOS.org.

10. Mertes P. Physiology of Brain Death. In: Tilney NL, Storm TB, Paul LC, eds. Transplantation Biology: Cellular and Molecular Aspects. Philadelphia: Lippincott, 1996; 275.
11. Gonzalez FX, Rimola A, Grande L et al. Predictive factors of early postoperative graft function in human liver transplantation. Hepatology 1994; 20(3):565-573.
12. Avolio AW, Agnes S, Magalini SC et al. Importance of donor blood chemistry data (AST, serum sodium) in predicting liver transplant outcome. Transplant Proc 1991; 23(5):2451-2452.
13. Figueras J, Busquets J, Grande L et al. The deleterious effect of donor high plasma sodium and extended preservation in liver transplantation. A multivariate analysis. Transplantation 1996; 61(3):410-413.
14. Markmann J, Markmann J, Markmann D et al. Preoperative factors associated with outcome and their impact on resource use in 1148 consecutive liver transplants. Transplantation 2001; 72(6):1113-1122.
15. Tector AJ, Mangus RS, Chestovich P et al. Use of extended criteria livers decreases wait time for liver transplantation without adversely impacting posttransplant survival. Ann Surg 2006; 244(3):439-450.
16. Novitzky D, Cooper DK, Reichart B. Hemodynamic and metabolic responses to hormonal therapy in brain-dead potential organ donors. Transplantation 1987; 43(6):852-854.
17. Kukral JC, Littlejohn MH, Williams RK et al. Hepatic function after canine liver transplantation. Arch Surg 1962; 85:157-165.
18. Chui AK, Thompson JF, Lam D et al. Cadaveric liver procurement using aortic perfusion only. Aust N Z J Surg 1998; 68(4):275-277.
19. D'Amico F, Vitale A, Gringeri E et al. Liver transplantation using suboptimal grafts: impact of donor harvesting technique. Liver Transpl 2007; 13(10):1444-1450.
20. Lichtman SN, Lemasters JJ. Role of cytokines and cytokine-producing cells in reperfusion injury to the liver. Semin Liver Dis 1999; 19(2):171-187.
21. Menger MD. Microcirculatory disturbances secondary to ischemia-reperfusion. Transplant Proc 1995; 27(5):2863-2865.
22. Cameron AM, Busuttil RW. Ischemic cholangiopathy after liver transplantation. Hepatobiliary Pancreat Dis Int 2005; 4(4):495-501.
23. Starzl TE, Putnam CW, Hansbrough JF et al. Biliary complications after liver transplantation: with special reference to the biliary cast syndrome and techniques of secondary duct repair. Surgery 1977; 81(2):212-221.
24. Briceno J, Lopez-Cillero P, Rufian S et al. Impact of marginal quality donors on the outcome of liver transplantation. Transplant Proc 1997; 29(1-2):477-480.
25. Hertl M, Malago M, Rogiers X et al. Surgical approaches for expanded organ usage in liver transplantation. Transplant Proc 1997; 29(8):3683-3686.
26. Mirza DF, Gunson BK, Da Silva RF et al. Policies in Europe on "marginal quality" donor livers. Lancet 1994; 344(8935):1480-1483.
27. Mor E, Klintmalm GB, Gonwa TA et al. The use of marginal donors for liver transplantation. A retrospective study of 365 liver donors. Transplantation 1992; 53(2):383-386.
28. Ploeg RJ, D'Alessandro AM, Knechtle SJ et al. Risk factors for primary dysfunction after liver transplantation—a multivariate analysis. Transplantation 1993; 55(4):807-813.
29. Alexander JW, Vaughn WK. The use of "marginal" donors for organ transplantation. The influence of donor age on outcome. Transplantation 1991; 51(1):135-141.
30. Busquets J, Xiol X, Figueras J et al. The impact of donor age on liver transplantation: influence of donor age on early liver function and on subsequent patient and graft survival. Transplantation 2001; 71(12):1765-1771.
31. Busuttil RW, Tanaka K. The utility of marginal donors in liver transplantation. Liver Transpl 2003; 9(7):651-663.
32. De Carlis L, Colella G, Sansalone CV et al. Marginal donors in liver transplantation: the role of donor age. Transplant Proc 1999; 31(1-2):397-400.
33. Detre KM, Lombardero M, Belle S et al. Influence of donor age on graft survival after liver transplantation—United Network for Organ Sharing Registry. Liver Transpl Surg 1995; 1(5):311-319.
34. Feng S, Goodrich NP, Bragg-Gresham JL et al. Characteristics associated with liver graft failure: the concept of a donor risk index. Am J Transplant 2006; 6(4):783-790.
35. Gruttadauria S, Cintorino D, Mandala L et al. Acceptance of marginal liver donors increases the volume of liver transplant: early results of a single-center experience. Transplant Proc 2005; 37(6):2567-2568.
36. Marino IR, Doyle HR, Aldrighetti L et al. Effect of donor age and sex on the outcome of liver transplantation. Hepatology 1995; 22(6):1754-1762.
37. Ploeg RJ, D'Alessandro AM, Hoffmann RM et al. Impact of donor factors and preservation on function and survival after liver transplantation. Transplant Proc 1993; 25(6):3031-3033.
38. Strasberg SM, Howard TK, Molmenti EP et al. Selecting the donor liver: risk factors for poor function after orthotopic liver transplantation. Hepatology 1994; 20(4 Pt 1):829-838.

39. Tisone G, Manzia TM, Zazza S et al. Marginal donors in liver transplantation. Transplant Proc 2004; 36(3):525-526.
40. Adam R, Astarcioglu I, Azoulay D et al. Liver transplantation from elderly donors. Transplant Proc 1993; 25(1 Pt 2):1556-1557.
41. Totsuka E, Fung U, Hakamada K et al. Analysis of clinical variables of donors and recipients with respect to short-term graft outcome in human liver transplantation. Transplant Proc 2004; 36(8):2215-2218.
42. Adam R, Reynes M, Johann M et al. The outcome of steatotic grafts in liver transplantation. Transplant Proc 1991; 23(1 Pt 2):1538-1540.
43. Briceno J, Padillo J, Rufian S et al. Assignment of steatotic livers by the Mayo model for end-stage liver disease. Transpl Int 2005; 18(5):577-583.
44. Selzner M, Clavien PA. Fatty liver in liver transplantation and surgery. Semin Liver Dis 2001; 21(1):105-113.
45. Soejima Y, Shimada M, Suehiro T et al. Use of steatotic graft in living-donor liver transplantation. Transplantation 2003; 76(2):344-348.
46. Todo S, Demetris AJ, Makowka L et al. Primary nonfunction of hepatic allografts with preexisting fatty infiltration. Transplantation 1989; 47(5):903-905.
47. Verran D, Kusyk T, Painter D et al. Clinical experience gained from the use of 120 steatotic donor livers for orthotopic liver transplantation. Liver Transpl 2003; 9(5):500-505.
48. Abt PL, Desai NM, Crawford MD et al. Survival following liver transplantation from nonheart-beating donors. Ann Surg 2004; 239(1):87-92.
49. Foley DP, Fernandez LA, Leverson G et al. Donation after cardiac death: the University of Wisconsin experience with liver transplantation. Ann Surg 2005; 242(5):724-731.
50. Busuttil R, Tanaka K. The utility of marginal donors in liver transplantation. Liver Transpl 2003; 9(7):651-663.
51. Tsukamoto I, Nakata R, Kojo S. Effect of ageing on rat liver regeneration after partial hepatectomy. Biochem Mol Biol Int 1993; 30(4):773-778.
52. Kimura F, Miyazaki M, Suwa T et al. Reduction of hepatic acute phase response after partial hepatectomy in elderly patients. Res Exp Med (Berl) 1996; 196(5):281-290.
53. D'Alessandro A, Kalayoglu M, Sollinger H et al. The predictive value of donor liver biopsies on the development of primary nonfunction after orthotopic liver transplantation. Trans Proc 1991; 23:1536-1537.
54. Fishbein TM, Fiel MI, Emre S et al. Use of livers with microvesicular fat safely expands the donor pool. Transplantation 1997; 64(2):248-251.
55. Urena MA, Ruiz-Delgado FC, Gonzalez EM et al. Assessing risk of the use of livers with macro and microsteatosis in a liver transplant program. Transplant Proc 1998; 30(7):3288-3291.
56. Selzner M, Clavien P. Fatty liver in liver transplantation and surgery. Semin Liver Dis 2001; 21(1):105-113.
57. Garcia Urena MA, Colina Ruiz-Delgado F, Moreno Gonzalez E et al. Hepatic steatosis in liver transplant donors: common feature of donor population? World J Surg 1998; 22(8):837-844.
58. Urena MA, Moreno Gonzalez E, Romero CJ et al. An approach to the rational use of steatotic donor livers in liver transplantation. Hepatogastroenterology 1999; 46(26):1164-1173.
59. D'Alessandro AM, Kalayoglu M, Sollinger HW et al. The predictive value of donor liver biopsies on the development of primary nonfunction after orthotopic liver transplantation. Transplant Proc 1991; 23(1 Pt 2):1536-1537.
60. Canelo R, Braun F, Sattler B et al. Is a fatty liver dangerous for transplantation? Transplant- Proc 1999; 31(1-2):414-415.
61. Schmidt A, Tomasdottir H, Bengtsson A. Influence of cold ischemia time on complement activation, neopterin and cytokine release in liver transplantation. Transplant Proc 2004; 36(9):2796-2798.
62. Shen XD, Gao F, Ke B et al. Inflammatory responses in a new mouse model of prolonged hepatic cold ischemia followed by arterialized orthotopic liver transplantation. Liver Transpl 2005; 11(10):1273-1281.
63. Scotte M, Dousset B, Calmus Y et al. The influence of cold ischemia time on biliary complications following liver transplantation. J Hepatol 1994; 21(3):340-346.
64. Porte RJ, Ploeg RJ, Hansen B et al. Long-term graft survival after liver transplantation in the UW era: late effects of cold ischemia and primary dysfunction. European Multicentre Study Group. Transpl Int 1998; 11 Suppl 1:S164-167.
65. Briceno J, Marchal T, Padillo J et al. Influence of marginal donors on liver preservation injury. Transplantation 2002; 74(4):522-526.
66. Hoofnagle JH, Lombardero M, Zetterman RK et al. Donor age and outcome of liver transplantation. Hepatology 1996; 24(1):89-96.
67. Piratvisuth T, Tredger JM, Hayllar KA et al. Contribution of true cold and rewarming ischemia times to factors determining outcome after orthotopic liver transplantation. Liver Transpl Surg 1995; 1(5):296-301.

68. Yersiz H, Shaked A, Olthoff K et al. Correlation between donor age and the pattern of liver graft recovery after transplantation. Transplantation 1995; 60(8):790-794.
69. Totsukali E, Fung JJ, Ishizawa Y et al. Synergistic effect of cold and warm ischemia time on postoperative graft outcome in human liver transplantation. Hepatogastroenterology 2004; 51(59):1413-1416.
70. Health Resources and Services Administration. www.hrsa.gov.
71. Bernat J. Report of a national conference on donation after cardiac death. Am J Transplant 2006; 6(2):281-291.
72. Merion RM, Pelletier SJ, Goodrich N et al. Donation after cardiac death as a strategy to increase deceased donor liver availability. Ann Surg 2006; 244(4):555-562.
73. D'Alessandro AM, Fernandez LA, Chin LT et al. Donation after cardiac death: the University of Wisconsin experience. Ann Transplant 2004; 9(1):68-71.
74. Gubernatis G, Pichlmayr R, Lamesch P et al. HTK-solution (Bretschneider) for human liver transplantation. First clinical experiences. Langenbecks Arch Chir 1990; 375(2):66-70.
75. van Gulik TM, Reinders ME, Nio R et al. Preservation of canine liver grafts using HTK solution. Transplantation 1994; 57(2):167-171.
76. Eberl T, Salvenmoser W, Rieger G et al. Ultrastructural analysis of human endothelial cells after hypothermic storage in organ preservation solutions. J Surg Res 1999; 82(2):253-260.
77. Straatsburg IH, Abrahamse SL, Song SW et al. Evaluation of rat liver apoptotic and necrotic cell death after cold storage using UW, HTK and Celsior. Transplantation 2002; 74(4):458-464.
78. Erhard J, Lange R, Scherer R et al. Comparison of histidine-tryptophan-ketoglutarate (HTK) solution versus University of Wisconsin (UW) solution for organ preservation in human liver transplantation. A prospective, randomized study. Transpl Int 1994; 7(3):177-181.
79. Pokorny H, Rasoul-Rockenschaub S, Langer F et al. Histidine-tryptophan-ketoglutarate solution for organ preservation in human liver transplantation-a prospective multi-centre observation study. Transpl Int 2004; 17(5):256-260.
80. Mangus RS, Tector AJ, Agarwal A et al. Comparison of histidine-tryptophan-ketoglutarate solution (HTK) and University of Wisconsin solution (UW) in adult liver transplantation. Liver Transpl 2006; 12(2):226-230.
81. Meine MH, Zanotelli ML, Neumann J et al. Randomized clinical assay for hepatic grafts preservation with University of Wisconsin or histidine-tryptophan-ketoglutarate solutions in liver transplantation. Transplant Proc 2006; 38(6):1872-1875.
82. Garcia-Gil FA, Arenas J, Guemes A et al. Preservation of the liver graft with Celsior solution. Transplant Proc 2006; 38(8):2385-2388.
83. Cavallari A, Cillo U, Nardo B et al. A multicenter pilot prospective study comparing Celsior and University of Wisconsin preserving solutions for use in liver transplantation. Liver Transpl 2003; 9(8):814-821.
84. Nardo B, Bertelli R, Montalti R et al. Preliminary results of a clinical randomized study comparing Celsior and HTK solutions in liver preservation for transplantation. Transplant Proc 2005; 37(1):320-322.
85. Moray G, Sevmis S, Karakayali FY et al. Comparison of histidine-tryptophan-ketoglutarate and University of Wisconsin in living-donor liver transplantation. Transplant Proc 2006; 38(10):3572-3575.
86. Lange R, Erhard J, Rauen U et al. Hepatocellular injury during preservation of human livers with UW and HTK solution. Transplant Proc 1997; 29(1-2):400-402.
87. Testa G, Malago M, Nadalin S et al. Histidine-tryptophan-ketoglutarate versus University of Wisconsin solution in living donor liver transplantation: results of a prospective study. Liver Transpl 2003; 9(8):822-826.
88. Khan AW, Fuller BJ, Shah SR et al. A prospective randomized trial of N-acetyl cysteine administration during cold preservation of the donor liver for transplantation. Ann Hepatol 2005; 4(2):121-126.
89. Polyak MM, Arrington BO, Kapur S et al. Donor treatment with phentolamine mesylate improves machine preservation dynamics and early renal allograft function. Transplantation 2000; 69(1):184-186.
90. Wang M, Stepkowski SM, Kahan BD. Donor pretreatment with anti-T-cell receptor monoclonal antibodies prevents graft-vs-host disease in brequinar-treated small bowel allograft recipients. Transplant Proc 1995; 27(1):383.
91. de Bruin RW, Saat RE, Heineman E et al. Effects of donor pretreatment with antilymphocyte serum and cyclosporin on rejection and graft-versus-host disease after small bowel transplantation in immunosuppressed and nonimmunosuppressed rats. Transpl Int 1993; 6(1):22-25.
92. Shaffer D, Ubhi CS, Simpson MA et al. Prevention of graft-versus-host disease following small bowel transplantation with polyclonal and monoclonal antilymphocyte serum. The effect of timing and route of administration. Transplantation 1991; 52(6):948-952.
93. Shaffer D, Maki T, DeMichele SJ et al. Studies in small bowel transplantation. Prevention of graft-versus-host disease with preservation of allograft function by donor pretreatment with antilymphocyte serum. Transplantation 1988; 45(2):262-269.

94. Mueller TH, Kienle K, Beham A et al. Caspase 3 inhibition improves survival and reduces early graft injury after ischemia and reperfusion in rat liver transplantation. Transplantation 2004; 78(9):1267-1273.

95. Chen CZ, Gallagher RC, Ardery P et al. Retrograde flush and cold storage for twenty-two to twenty-five hours lung preservation with and without prostaglandin E1. J Heart Lung Transplant 1997; 16(6):658-666.

96. Higgins RS, Letsou GV, Sanchez JA et al. Improved ultrastructural lung preservation with prostaglandin E1 as donor pretreatment in a primate model of heart-lung transplantation. J Thorac Cardiovasc Surg 1993; 105(6):965-971.

97. Vainikka T, Heikkila L, Kukkonen S et al. Donor lung pretreatment with prostaglandin E(1) does not improve lung graft preservation. Eur Surg Res 1999; 31(5):429-436.

98. Baskin-Bey ES, Washburn K, Feng S et al. Clinical Trial of the Pan-Caspase Inhibitor, IDN-6556, in Human Liver Preservation Injury. Am J Transplant 2007; 7(1):218-225.

99. Rivera-Chavez FA, Toledo-Pereyra LH, Dean RE et al. Exogenous and endogenous nitric oxide but not iNOS inhibition improves function and survival of ischemically injured livers. J Invest Surg 2001; 14:267-273.

100. Serafin A, Rosello-Catafau J, Prats N et al. Ischemic preconditioning increases the tolerance of Fatty liver to hepatic ischemia-reperfusion injury in the rat. Am J Pathol 2002; 161(2):587-601.

101. Brookes PS, Salinas EP, Darley-Usmar K et al. Concentration-dependent effects of nitric oxide on mitochondrial permeability transition and cytochrome c release. J Biol Chem 2000; 275(27):20474-20479.

102. Whiteman M, Chua YL, Zhang D et al. Nitric oxide protects against mitochondrial permeabilization induced by glutathione depletion: role of S-nitrosylation? Biochem Biophys Res Commun 2006; 339(1):255-262.

103. Ben Mosbah I, Massip-Salcedo M, Fernandez-Monteiro I et al. Addition of adenosine monophosphate-activated protein kinase activators to University of Wisconsin solution: a way of protecting rat steatotic livers. Liver Transpl 2007; 13(3):410-425.

104. Lemaster JJ, Thurman RG. Hypoxia and reperfusion injury to liver. Prog Liver Dis 1993; 11:85-114.

105. Arnault I, Bao YM, Sebagh M et al. Beneficial effect of pentoxifylline on microvesicular steatotic livers submitted to a prolonged cold ischemia. Transplantation 2003; 76(1):77-83.

106. Qing DK, Dong JH, Han BL et al. Cold preservation of pig liver grafts with warm ischemia and pentoxifylline-University of Wisconsin solution (UW). Arch Med Res 2006; 37(4):449-455.

107. Selzner N, Rudiger H, Graf R et al. Protective strategies against ischemic injury of the liver. Gastroenterology 2003; 125(3):917-936.

108. Toledo-Pereyra LH. The role of allopurinol and oxygen free radical scavengers in liver preservation. Basic Life Sci 1988; 49:1047-1052.

109. Nguyen WD, Kim DH, Alam HB et al. Polyethylene glycol-superoxide dismutase inhibits lipid peroxidation in hepatic ischemia/reperfusion injury. Crit Care 1999; 3(5):127-130.

110. He SQ, Zhang YH, Venugopal SK et al. Delivery of antioxidative enzyme genes protects against ischemia/reperfusion-induced liver injury in mice. Liver Transpl 2006; 12(12):1869-1879.

111. McNally SJ, Harrison EM, Ross JA et al. Curcumin induces heme oxygenase-1 in hepatocytes and is protective in simulated cold preservation and warm reperfusion injury. Transplantation 2006; 81(4):623-626.

112. Chen C, Johnston TD, Wu G et al. Curcumin has potent liver preservation properties in an isolated perfusion model. Transplantation 2006; 82(7):931-937.

113. van der Plaats A, t Hart NA, Morariu AM et al. Effect of University of Wisconsin organ-preservation solution on haemorheology. Transpl Int 2004; 17(5):227-233.

114. FrancoGou R, Mosbah IB, Serafin A et al. New preservation strategies for preventing liver grafts against cold ischemia reperfusion injury. J Gastroenterol Hepatol 2007; 22(7):1120-1126.

115. Ben Mosbah I, Rosello-Catafau J, FrancoGou R et al. Preservation of steatotic livers in IGL-1 solution. Liver Transpl 2006; 12(8):1215-1223.

116. Hata K, Tolba RH, Wei L et al. Impact of polysol, a newly developed preservation solution, on cold storage of steatotic rat livers. Liver Transpl 2007; 13(1):114-121.

117. Bessems M, Doorschodt BM, van Vliet AK et al. Improved rat liver preservation by hypothermic continuous machine perfusion using polysol, a new, enriched preservation solution. Liver Transpl 2005; 11(5):539-546.

118. Bessems M, Doorschodt BM, Dinant S et al. Machine perfusion preservation of the pig liver using a new preservation solution, polysol. Transplant Proc 2006; 38(5):1238-1242.

119. Guarrera JV, Arrington B, Donovan M. Early results of phase 1 trial of hypothermic machine preservation in human liver transplantation [abstract]. Transplantation. 2006; 82((1 Suppl 2): abstract 2586):P4-IV.

120. Guarrera JV, Estevez J, Boykin J et al. Hypothermic machine perfusion of liver grafts for transplantation: technical development in human discard and miniature swine models. Transplant Proc 2005; 37(1):323-325.

121. Upadhya AG, Harvey RP, Howard TK et al. Evidence of a role for matrix metalloproteinases in cold preservation injury of the liver in humans and in the rat. Hepatology 1997; 26(4):922-928.
122. Moore C, Shen XD, Gao F et al. Fibronectin-alpha4beta1 integrin interactions regulate metalloproteinase-9 expression in steatotic liver ischemia and reperfusion injury. Am J Pathol 2007; 170(2):567-577.
123. Amersi F, Shen XD, Moore C et al. Fibronectin-alpha 4 beta 1 integrin-mediated blockade protects genetically fat Zucker rat livers from ischemia/reperfusion injury. Am J Pathol 2003; 162(4):1229-1239.
124. Fondevila C, Shen XD, Moore C et al. Cyclic RGD peptides with high affinity for alpha5beta1 integrin protect genetically fat Zucker rat livers from cold ischemia/reperfusion injury. Transplant Proc 2005; 37(4):1679-1681.
125. Inuo H, Eguchi S, Yanaga K et al. Protective effects of a hibernation-inducer on hepatocyte injury induced by hypothermic preservation. J Hepatobiliary Pancreat Surg 2007; 14(5):509-513.
126. Navarro-Sabate A, Peralta C, Calvo MN et al. Mediators of rat ischemic hepatic preconditioning after cold preservation identified by microarray analysis. Liver Transpl 2006; 12(11):1615-1625.
127. Koneru B, Shareef A, Dikdan G et al. The ischemic preconditioning paradox in deceased donor liver transplantation-evidence from a prospective randomized single blind clinical trial. Am J Transplant 2007; 7(12):2788-2796.
128. Azoulay D, Del Gaudio M, Andreani P et al. Effects of 10 minutes of ischemic preconditioning of the cadaveric liver on the graft's preservation and function: the ying and the yang. Ann Surg 2005; 242(1):133-139.
129. Amador A, Grande L, Marti J et al. Ischemic preconditioning in deceased donor liver transplantation: a prospective randomized clinical trial. Am J Transplant 2007; 7(9):2180-2189.
130. Jassem W, Fuggle SV, Cerundolo L et al. Ischemic preconditioning of cadaver donor livers protects allografts following transplantation. Transplantation 2006; 81(2):169-174.
131. Cescon M, Grazi GL, Grassi A et al. Effect of ischemic preconditioning in whole liver transplantation from deceased donors. A pilot study. Liver Transpl 2006; 12(4):628-635.
132. Schreinemachers M, Doorschodt BM, van Gulik TM. Machine perfusion preservation of the liver: a worthwhile clinical activity? Current Opinion in Organ Transplantation 2007; 12(3):224-230.
133. Bessems M, Doorschodt BM, Kolkert JL et al. Preservation of steatotic livers: a comparison between cold storage and machine perfusion preservation. Liver Transpl 2007; 13(4):497-504.
134. Bessems M, Doorschodt BM, van Marle J et al. Improved machine perfusion preservation of the nonheart-beating donor rat liver using Polysol: a new machine perfusion preservation solution. Liver Transpl 2005; 11(11):1379-1388.
135. Dutkowski P, Graf R, Clavien PA. Rescue of the cold preserved rat liver by hypothermic oxygenated machine perfusion. Am J Transplant 2006; 6(5 Pt 1):903-912.
136. Dutkowski P, Furrer K, Tian Y et al. Novel short-term hypothermic oxygenated perfusion (HOPE) system prevents injury in rat liver graft from nonheart beating donor. Ann Surg 2006; 244(6):968-976; discussion 976-967.
137. van der Plaats A, Maathuis MH, Na TH et al. The Groningen hypothermic liver perfusion pump: functional evaluation of a new machine perfusion system. Ann Biomed Eng 2006; 34(12):1924-1934.

CHAPTER 10

Small Bowel Preservation

Debra McKeehen,* Luis H. Toledo-Pereyra and Roberta E. Sonnino

Introduction

Successful preservation of the small bowel has presented unique challenges because of the hollow nature of the organ, its increased ability to develop intestinal edema under minimal conditions of physiological stress and its unique immune and bacterial environment. In addition to the fundamental preservation principles established by studying other organs, the small bowel also has special metabolic needs that must be addressed to preserve the mucosal barrier and achieve normal metabolism and function.[1] Despite early developments in transplantation techniques and preservation of the small bowel during the late-1950s through the 1970s, few significant changes occurred in the 1980s and 90s.[2-24] In the last decade and a half, however, transplantation of the small bowel has been reintroduced as a potential treatment for end-stage small bowel disease (ESSBD), mainly represented by the short bowel syndrome.[25] Presently, more than 100 clinical small bowel transplants are performed in the United States annually. (UNOS, 2008) In the clinical arena, recipient survival greater than 3 years after small bowel transplantation is 65% and graft survival is approximately 50%. The longest survival is obtained when small bowel transplants are associated with liver and other organs in a cluster operation.[26] This chapter will review the historical background and the preservation of the small bowel for transplantation.

Historical Background

Preservation Using Hypothermia and Hyperbaria

Early experimental work in small bowel preservation utilizing simple hypothermic storage (5°C) was pioneered by Lillehei and colleagues in 1959.[2] They preserved canine grafts for 4 hours and achieved satisfactory nutritional support with long-term survival in some animals after autotransplantation. A few years later, the same group reported their results of small bowel auto- and allotransplantation in dogs[3] and, in 1967,[4] extended these results to the clinical application of these techniques in patients with end-stage small bowel disease.

In this same era, hyperbaric storage was tested for small bowel preservation. Lyons and associates[27] used hypothermia (4°C) along with hyperbaric oxygenation (7.9 atm) of the tissue to preserve 20 cm portions of the ileum for 24-48 hours. Adequate intestinal absorption of carbohydrates was observed, however, the organs were not further tested by subsequent transplantation. Further studies by Lillehei's group[28] utilized hyperbaric, hypothermic oxygenation (4°C, 7.9 atm) with the addition of chlorpromazine to preserve 25 cm ileum segments for 48 hours. Chlorpromazine was added to inhibit metabolism. Using these methods, 4-day graft survival was observed after autotransplantation into the recipient's neck. Additional functional studies by Lillehei and colleagues[29] demonstrated good preservation for up to 48 hours when hypothermic hyperbaria and chlorpromazine were used in combination. However, no assessment of viability was made by subsequent transplantation of the preserved small bowel. In contrast, several years later, Ruiz and

*Corresponding Author: Debra McKeehen—VA Medical Center, Minneapolis, MN.
Email: debramckeehen@va.gov

Organ Preservation for Transplantation, Third Edition, edited by Luis H. Toledo-Pereyra.
©2010 Landes Bioscience.

associates[30] tested whole blood as a perfusate, together with hypothermia and hyperbaria, for up to 48 hours and were able to obtain acceptable in vitro viability measurements for preservation up to only 12 hours.

Later work by Raju and colleagues[31] reported results of allotransplantation of canine small bowels after flushing with lactated Ringer's solution, followed by for 12 and 24 hours simple hypothermic storage. All recipients of grafts transplanted after 12 hours storage and 67% of the animals receiving grafts stored 24 hours survived beyond 5 days. Donor pretreatment with antibiotics and extensive intra-luminal irrigation of the graft were cited by the authors as important technical features in this successful preservation method.

Initial Perfusion Attempts

Initial attempts at using perfusion for experimental preservation of small bowels were also made in the 1960s. Austen and McLaughlin[32] compared the effects of normothermic pulsatile and nonpulsatile perfusion of sections of canine ileum (5 feet), using whole blood as the perfusate. With normothermic pulsatile perfusion, the segments remained viable for 18 hours, as compared to only 6 hours, using nonpulsatile normothermic perfusion. Iijima and Salerno[33] used heparinized blood diluted with heparinized Ringer's lactate as the perfusate for normothermic perfusion to successfully preserve entire canine small bowel autografts for 5 hours.

Other experimental functional studies using normothermic nonpulsatile perfusion of entire small bowel yielded similar results. The perfusates tested were heparinized bovine blood for 3-hour preservation of rat intestine[34] and erythrocyte polyvinylpirrolidone (PVP) and low molecular weight dextran for up to 5 hours preservation of canine small bowel.[35]

Subsequent studies incorporated hypothermic (4°C) conditions into small bowel preserved by perfusion, in an attempt to improve graft viability. Using homologous serum as the perfusate, Alican and coworkers,[36] were unable to extend preservation times to 24 hours after hypothermia (9°C) and allotransplantation. Toledo-Pereyra and Najarian[37] tested hypothermic pulsatile perfusion with either plasmanate or cryoprecipitated plasma (CPP). In addition, the effects of chlorpromazine, methylprednisolone and anti-dog antilymphocyte globulin (ALG) were tested. After 6-hours preservation and allotransplantation, the best survival results were observed in grafts perfused with chlorpromazine, methylprednisolone and ALG, in combination. Toledo-Pereyra and associates[38] also demonstrated that with the addition of allopurinol, a zanthine oxidase blocker, to the CPP, preservation times of 24 hours could be achieved with successful allotransplantation. Allopurinol may have contributed to improved preservation by increasing the nucleotide pool and conserving energy.

Twenty-four hour small bowel preservation using hypothermic storage versus hypothermic pulsatile perfusion was studied by Toledo-Pereyra and Najarian.[39] CPP was used as a perfusate for hypothermic pulsatile perfusion and Collins (C4) solution was used for hypothermic storage of entire canine small bowel allografts. The best graft survival results were obtained using hypothermic pulsatile perfusion with CPP for 24 hours. Acceptable results were also obtained using a combination of hypothermic storage for 6 hours followed by perfusion for 18 hours.

Entire canine small bowel allografts were also successfully preserved for 24 hours by Toledo-Pereyra and colleagues[40] on a MOX-100 perfusion apparatus with either CPP or human plasmanate as the perfusate. In fact, small bowel perfused with CPP or plasmanate and subsequently transplanted demonstrated better survival than freshly transplanted, nonpreserved allografts. In later work, more extended perfusion times using hypothermic pulsatile perfusion with CPP were attempted by Toledo-Pereyra and associates.[41] Few changes were seen in the small bowel during the first 48 hours of preservation; however, longer periods of 48-72 hours resulted in damage to the small bowel. Allotransplantation survival results correlated with the in vitro observations.

Assessment of Absorptive Capacity after Preservation

One problem encountered in small bowel transplantation has been a reduction in the absorptive capacity of the intestine immediately after transplantation. Cohen and associates in 1969[5] focused on the absorptive properties of jejunal canine auto- and homografts after transplantation. Numerous known markers of intestinal absorption were also studied by Harmel's group in the late 1980's.[42] Toledo-Pereyra and colleagues studied the effects of 24-hour preservation by hypothermic pulsatile perfusion on the ability of the entire small bowel to absorb carbohydrates and vitamins.[43] A significant impairment in the intestinal absorption of D-xylose and vitamin A was observed after allotransplantation. Recovery of these functions was only seen in long-term survivors that usually lived more than 5 weeks. No impairment was seen, however, in the vitamin B_{12} absorption. Prolonged survivors showed low normal limits of vitamin B_{12}. Guttman and his coworkers, in Montreal, studied absorption and the effect of short term preservation and transplantation on the physiologic properties of intestinal smooth muscle and nerves, key elements in bowel absorptive function. They showed that intrinsic noradrenergic nerves and excitatory innervation were intact after transplantation. Only extrinsic inhibitory innervation was lost.[44,45]

Role of Endotoxin and Bacteria in Preserved Small Bowel

A major obstacle to the development of safe and reliable small bowel transplantation is the release of bacteria and endotoxins, especially if the organ is damaged in the process of donor death, graft harvesting or during preservation. Toledo-Pereyra and Najarian had previously observed[39] that small bowel grafts preserved by hypothermic pulsatile perfusion were sometimes destroyed soon after allotransplantation by hemorrhagic necrosis. They subsequently evaluated the effect of intraluminal bacteria and endotoxin on long-term survival of small bowel allografts previously preserved by hypothermic perfusion for 12 hours.[46] Preparation of the donor small bowel with antibiotics led to improved survival after allotransplantation.

Attempts at Clinical Preservation

In 1979, several preservation methods were compared by Toledo-Pereyra and Najarian[47] using human small bowel in vitro. Results demonstrated that small bowel preserved by hypothermic pulsatile perfusion with CPP or human plasmanate, showed minimal damage after 24 hours. Hypothermic storage with Collins (C3) solution for 12 hours resulted in edema, damage to the mucosa, serosa and mesentery. Small bowel exposed to combination preservation with both methods, hypothermic storage for 6 hours and perfusion for 18 hours, demonstrated histology similar to perfused grafts.

Prevention and Amelioration of Ischemia-Reperfusion Injury

The era of the 1980s was characterized by an emphasis on the prevention and amelioration of ischemia-reperfusion injury. Several researchers extensively studied the role of oxygen free radicals (OFR), leukotriene B_4, (LTB_4,), platelet-activating factor (PAF), prostacyclin (PGI_2), thromboxane A_2 (TxA_2,) and complement during warm ischemia and their contribution to cell damage. On the other hand, the use of superoxide dismutase, catalase, dimethyl sulfoxide, allopurinol, mannitol, desferroxamine and other oxygen free radical scavengers, or various antagonists (against PAF, LTB_4, TxA_2,), administered to the recipient during or immediately before the reperfusion, decreased considerably the tissue damage seen after ischemia-reperfusion injury.[9-13] The high concentrations of xanthine dehydrogenase and xanthine oxidase in the small bowel of small animals, not as prominent in humans, have made this organ particularly suitable for the study of free radical injury.

More Work on Ischemia-Reperfusion Injury (1990s)

In the 1990s, special attention was given to cytokines (interleukin-I, interleukin-6 and tumor necrosis factor) and their role in reperfusion injury, particularly, in liver models.[48] The endothelium-derived relaxing factor (later known as nitric oxide: NO), has also been found to be important in small bowel preservation for transplantation.[49,50]

Further studies on the potential importance of nitric oxide in association with ischemic preconditioning prior to hypothermic storage, to protect small bowel transplants from reperfusion injury, were conducted by Hotter and colleagues.[51,52] The preconditioning phenomenon is defined as one or more brief periods of ischemia with intermittent reperfusion that results in protection from subsequent and sustained ischemia/reperfusion. Hotter et al concluded that preconditioning induces protection of intestinal grafts by nitric oxide from cold preservation and reperfusion injury.

The production of the potent vasoconstrictor endothelin and the expression of adhesion molecules (integrins, selectins and the immunoglobulin superfamily) in preservation reperfusion models have been proposed as the main factors of the injury seen in ischemia during and after reperfusion.[53] In the 1990s, the global mechanism of the complex preservation-reperfusion injury was conceptualized as a conjunction of phenomena where the adhesion molecules, neutrophils, cytokines and OFR play an important role. Rapamycin, glutamine, lazaroid u74389g and superoxide dismutase, substances which may function as free radical scavengers or antioxidants, were also used in this era for experimental small bowel preservation with promising results.[54-57] In addition, Sisley and colleagues found that donor depletion of neutrophils was beneficial for cell function and morphology after small bowel transplant revascularization.[58]

The role of heat shock protein (HSP) induction in protecting small bowel grafts from preservation and reperfusion injury has been explored by Tarumi and colleagues.[59] They concluded that HSP protected against small intestinal warm ischemia-reperfusion injury by inhibiting the synthesis of inflammatory cytokines and the activation of neutrophils and by accelerating the synthesis of anti-inflammatory cytokines.

More recently, Schaefer and colleagues tested the effects of perioperative glycine donor pretreatment for improving smooth muscle dysfunction as a result of ischemia/reperfusion injury.[60] Glycine (1 mg/g body weight) was infused (0.1 mL/g/hr) for 2 hours into donor rats prior to harvesting of small bowel grafts. They found that treatment with glycine attenuated the proinflammatory cascade and improved smooth muscle dysfunction.

Use of Immunosuppressive Drugs for Graft Pretreatment

Several immunosuppressive drugs have been tested as graft pretreatment in an attempt to improve graft survival. Several reports, especially on large animals, were published about the results of small bowel transplants treated with cyclosporine alone or associated with methylprednisolone.[15,16]

Search for Improved Preservation Solutions

In the late 1980s, researchers at the University of Wisconsin developed their preservation solution (UW), originally studied for pancreas preservation and later demonstrated to have protective function in kidney and liver.[17-19] It was also utilized for heart and lung short-term preservation with preliminary acceptable results.[61] For small bowel preservation, UW solution never reached the level of protection previously observed for preservation of the pancreas.[62,63] However, experimentally UW demonstrated moderately good results[19-24] and is now routinely used for clinical transplantation.

Sonnino and colleagues had previously shown that 24 hours of hypothermic storage was possible using rat small bowel syngeneic grafts. However, when hypothermic storage using UW solution was extended to 48 hours, disappointing results were obtained.[64] Although these grafts appeared histologically well-preserved prior to implantation, dramatic reperfusion injury was evident upon revascularization. This was evidenced by a loss of villi and crypts and inflammatory cells in all layers. The bowel was abnormal grossly, microscopically and ultrastructurally. This injury was irreversible with persistently abnormal histology for up to 1 week in the majority (5 of 6) of the transplants. They concluded that UW solution alone might allow for satisfactory preservation of intestinal grafts for 48 hours only in isolated cases and was therefore not adequate for predictable, satisfactory 48 hr clinical preservation. The same group showed that while a proposed new flushing solution, Carolina Rinse, did not prevent reperfusion injury after cold storage of small bowel grafts in University of Wisconsin solution, it did reduce myeloperoxidase activity (an indicator of free radical activity) in small bowel grafts.[65,66]

Later work by Sonnino and associates[67,68] studied secretory PLA_2 ($sPLA_2$) secretion during intestinal preservation in the same rat model. PLA_2 is a small enzyme that is synthesized and released from cells, including neutrophils and vascular endothelium, in response to stimuli such as ischemia. The researchers found that $sPLA_2$ is actively secreted during intestinal preservation and believe that this secretory response damages the graft, thereby limiting the useful duration of graft storage.[69,70] Experiments with the addition of PX-13, a specific $sPLA_2$ inhibitor, showed very promising protection of grafts preserved in cold storage with UW for up to 48 hours. Structural and physiologic studies suggested that $sPLA_2$ inhibition may drastically improve the survival of these grafts.[71] Later studies[72] also showed that Intestinal Fatty Acid Binding Protein (I-FABP) is another mediator of ischemic injury in preserved bowel and may play a role in graft perfusion and preservation, together with $sPLA_2$.

In 1994, Toledo-Pereyra and associates[23] compared Euro-Collins (EC), UW and albumin-dextran-adenosine-allopurinol-verapamil (ADAAV) solutions to flush small bowel rat allografts prior to 24 hour hypothermic storage. Although some improvement was observed in early survival with the ADAAV solution, no long term survival was seen in any of the groups studied. None of the preservation solutions gave consistent results with this length of preservation.

Experimental work by Li and coworkers[73] used a porcine autograft model to investigate the effect of various solutions for hypothermic storage of small bowel. Lactated Ringer's (LR), Euro-Collins, HC-A (a modified Ross, hyperosmolar citrate containing adenine) and WMO-1 (containing ATP and Ca^{2+}) solutions were compared for 12, 18 and 24 hours. The results indicated that EC and WMO-1 solutions were superior to LR and HC-A solutions for small bowel preservation in the pig model for up to 18 hours hypothermic storage.

A comparison of UW and Celsior, a cardioplegic and heart storage solution was conducted by Minor, et al[74] using small bowel segments from rats. The segments were flushed with either UW or Celsior solutions and hypothermically stored at 4°C, for 18 hours. After storage, grafts flushed with Celsior exhibited better post-ischemic recovery than those in the UW treated group, in terms of vascular perfusion characteristics, enzyme release and carbohydrate absorption.

In 2001, DeRoover and colleagues[75] investigated preservation of rat small bowel for up to 72 hours, using a perfluorocarbon emulsion added to UW solution. Perfluorocarbons were used because they are biologically inert organic liquids that have a high oxygen carrying capacity. The beneficial effect of the perfluorocarbon emulsion, however, was limited to 24-48 hours preservation.

The ability of a supplemental buffering agent, histidine and carbohydrate substrate, glucose, to facilitate glycolytic ATP production during hypothermic storage of small bowel was evaluated by Salehi and colleagues.[76] A modified UW solution, including lactobionate (100mM), raffinose (30 mM), KOH (100 mM), NaOH (15 mM), KH_2PO_4 (25 mM), $MgSO_4$ (5 mM) and adenosine (5 mM) was used, with the addition of either 90mM histidine or 20mM glucose and 40U/L insulin. Their results indicated that the histidine-buffered solution was superior with respect to enhancing the capacity of glycolysis for anaerobic energy production. The greatest improvement in adenylates occurred only when supplemental carbohydrate substrate was supplied to the preservation solution.

Intraluminal Hypothermic Perfusion

Bacterial infection has remained a major obstacle to successful intestinal transplantation. The mucosal epithelium is susceptible to even brief periods of ischemia. After transplantation, bacterial translocation across the mucosal epithelium may result in sepsis.[77] Churchill and colleagues at the University of Alberta were the first to pursue hypothermic luminal perfusion with UW solution for small bowel preservation as a means of better preservation of the mucosa.[78] Rat small bowel was flushed intravascularly and intraluminally with UW solution and then separated into groups with either no storage, 1 hour continuous intraluminal perfusion (2-4°C) and hypothermic storage (4°C), or 24 hours of continuous intraluminal hypothermic (2-4°C) perfusion. Energetics, lipid peroxidation and histology were assessed during the 24 hour storage period at (4°C). The data indicated that intraluminal perfusion improved tissue energetics, however mucosal integrity was superior only in the group with the 1 hour of luminal perfusion.

In further work by Churchill and associates,[79] rat small bowel grafts were vascularly flushed with UW solution, followed by intra-luminal flushing with an amino acid-based (AA) solution prior to either 1 hour continuous intraluminal perfusion (2-4°C) and hypothermic storage (4°C) or 24 hours of continuous intraluminal hypothermic (2-4°C) perfusion . Their data showed that 24 hours of continuous intraluminal perfusion provided improved tissue energetics. However, the group of small bowel grafts with a brief 1 hour perfusion prior to hypothermic storage demonstrated markedly superior histology as compared to the group perfused for 24 hours. In a parallel study, Sonnino et al demonstrated the importance of controlling flushing pressure for intraluminal perfusion of grafts, irrespective of the type of solution used.[80]

Effects of Donor Nutritional Status

Salehi and Churchill recently studied the effects of short-term donor fasting on the energetics and histology of rat small bowel grafts.[81] Rats were fasted (12-14 hours) or nonfasted prior to small bowel harvest, vascular flushing with UW solution and luminal flushing with either UW or AA solution, followed by hypothermic storage at 4°C for 24 hours. The best results with respect to energetics, oxidative stress and mucosal integrity were observed after luminal flushing with the AA solution. Increased oxidative stress, resulting in mucosal injury, was observed in the fasted grafts, as compared to the nonfasted small bowel grafts.

Reduction of Oxidative and Energetic Stress

Churchill and associates hypothesized that oxidative catabolism facilitated by the AA storage solution that they had developed and tested, promotes oxidative stress.[82] They tested the additions of Trolox, an antioxidant, water-soluble derivative of Vitamin E, and superoxide dismutase/catalase to the luminal perfusion with AA to reduce oxidative stress prior to hypothermic storage for up to 24 hours. The addition of superoxide dismutase and catalase did not result in a consistent reduction in oxidative stress nor conservation of energetic parameters. In contrast, Trolox increased graft energetics, reduced oxidative damage and improved the maintenance of the key functional aspects of the mucosa during hypothermic storage.

Subsequent studies by Churchill and coworkers[83] analyzed the relationship between energetic/oxidative stress responses and the fundamental kinase signaling events leading to necrosis or apoptosis of rat small bowel grafts during up to 12 hours of hypothermic storage with UW or AA solutions. They found that pre-apoptotic signaling (c-jun N-terminal kinase and P38) was abrogated and cytoprotective signals (extracellular signal-regulated kinase) were upregulated in the grafts stored in AA solution. JNK and P38, on the other hand, were strongly upregulated in the UW preserved grafts after 1 and 12 hour storage.

Recent Research Work Using Human Small Bowel

Although many preservation solutions have been studied, including the clinical gold standard, UW solution, none has been able to maintain graft integrity for hypothermic storage periods comparable to other commonly transplanted organs. The intraluminal administration of AA solution, developed by Churchill and colleagues, promises to provide better protection from reperfusion injury than the standardly used UW solution. Cellular energetics, permeability and histologic injury after 24 hours hypothermic storage of human small bowel segments were analyzed. All segments received vascular flushes with UW solution and subsequently had either no luminal flush, a luminal flush with UW or a luminal flush with AA. Only the segments flushed with AA solution showed significantly less morphologic injury and better mucosal barrier function.[84] DeRoover and colleagues[85] developed a model to compare preservation solutions for human small bowel grafts. In preliminary work, they compared UW and Celsior preservation solutions using grafts harvested from four multiorgan donors. After dissection, a 50 cm ileal segment was immediately flushed with Celsior. After the perfusion of the abdominal organs with UW, a second segment of adjacent ileum was harvested. The two intestinal grafts were then divided into segments before immersion in either Celsior or UW for 0, 6, 12, or 24 hour hypothermic storage (4°C). The results of this comparison showed severe histological alterations of graft mucosa after short periods of preservation by either UW or Celsior solutions.

Recent studies by Balaz and coworkers[76] were aimed at determining whether there was a difference between jejunal or ileal segments of the human small bowel in their susceptibility to ischemic injury. Intestinal specimens were acquired from multi-organ donors after perfusion with HTK solution. The harvested intestine was divided into jejunal and ileal segments and specimens were taken from each section and stored for 1, 6, 9, 12 and 24 hours. No difference was noted between jejunal and ileal grafts in their susceptibility to ischemic injury due to hypothermic storage for 24 hours in HTK. A significant difference was observed in the histological results only after 12-hours of cold ischemia for both segments.

The effect of early ischemia-reperfusion damage on intraoperative changes of metabolic, hemodynamic and coagulative parameters has been recently analyzed by Siniscalchi and colleagues.[87] They used Park's classification to correlate histological damage observed in graft biopsies taken at the end of the transplant procedure with intraoperative changes after reperfusion. Patients receiving small bowel grafts that exhibited mucosal damage (Park's Grades 2-8) had significant hemodynamic, metabolic and coagulative disorders, as compared to small bowel exhibiting normal mucosa and minimal damage (Park's Grades 0-1). It is important to note that all grafts were preserved by two-layer cold storage with UW solution and there was no significant difference in ischemic time between the two groups, pointing to the potential importance of detecting postoperative intestinal dysfunction for achieving ultimate graft function.

Work by Mangus and coworkers,[88] compared HTK and UW solutions for clinical small bowel transplantation of 57 grafts in 54 patients (15 pediatric and 42 adults, ranging in age from 14-66 years old). Thirty-seven allografts were preserved in HTK solution and UW solution was used in 22 allografts. Recipient and donor demographics and warm and cold ischemia times (ranging from 8.5-14 hrs and 23.13-47 hours, respectively) were similar between the two study groups. The median cold ischemia time for each group was 8 hours. Primary outcomes were graft and patient survival, early graft function and episodes of rejection. No differences were noted between these two solutions when initial graft function, appearance of the bowel on initial endoscopy and number of rejection episodes were compared. Despite the short preservation times, this study represented the first published report comparing HTK and UW solutions at a large volume transplant center.

Clinical Small Bowel Preservation—Current Status

The special metabolic properties of the small bowel have presented many challenges for achieving successful preservation of mucosal architecture, metabolic status and function. In addition, successful small bowel transplantation faces the obstacles of bacterial infection and graft rejection. The loss of mucosal barrier function and subsequent bacterial translocation, a dramatic increase in inflammation and the potential risk of a life-threatening infection can result from small bowel injury during hypothermic storage.[1] Although longer periods have been achieved experimentally, the current realistic limit of hypothermic storage in the clinical setting using UW solution is about 12 hours.

References

1. Dr. Thomas Churchill. Personal observations. www.ualberta.ca/~smri/Surgery%20 Research%20Web%20Page/DrChurchill.htm, accessed, 2009.
2. Lillehei RC, Gott K, Miller FA. The physiologic response of the small bowel of the dog to ischemia, including in vitro preservation of bowel with successful replacement and survival. Ann Surg 1959; 150:543-60.
3. Lillehei RC, Goldberg S, Gott B et al. The present status or small bowel transplantation. Am J Surg 1963; 105:58-72.
4. Lillehei RC, Idezuki Y, Feemster JA et al. Transplantation of stomach, intestine and pancreas: experimental and clinical observations. Surgery 1967; 62:721-41.

5. Cohen WB, Hardy MA, Quint J et al. Absorptive function in canine jejunal autografts and allografts. Surgery 1969; 65:440-46.
6. Toledo-Pereyra LH, Najarian JS. Small bowel preservation: a comparison of perfusion and nonperfusion systems. Arch Surg 1973; 107:875-77.
7. Toledo-Pereyra LH. Small bowel preservation. In: Toledo-Pereyra LH, ed. Basic Concepts of Organ Procurement, Perfusion and Preservation for Transplantation. New York: Academic Press, 1982:317-31.
8. Toledo-Pereyra LH. Small bowel preservation: evolution of methods and ideas and current concepts. Transplant Proc 1992; 24:1083-84.
9. Southard JH, Marsh DC, McAnulty JF et al. Oxygen-derived free radical damage in organ preservation: activity or superoxide dismutase and xanthine oxidase. Surgery 1987; 101:566-70.
10. Grace PA. Ischemia-reperfusion injury. Br J Surg 1994; 81:637-47.
11. Banda MA, Granger DN. Mechanism and protection from ischemic ischemia injury. Transplant Proc 1996; 28:2595-2597.
12. Toledo-Pereyra LH, Granger DN. Small bowel ischemia and reperfusion injury: pathophysiological mechanisms. In: Das DK, ed. Pathophysiology of Reperfusion Injury. Boca Raton: CIIC Press, 1993:137-47.
13. Parks DA, Bulkley GB, Granger DN. Role of oxygen free radicals in shock, ischemia and organ preservation. Surgery 1983; 94:428-34.
14. Calne RY, White DJG. The use of cyclosporin A in clinical organ grafting. Ann Surg 1982; 196:330-37.
15. Raju S, Didlake RH, Cayirli M et al. Experimental small bowel transplantation utilizing cyclosporine. Transplantation 1984; 38:561-66.
16. Diliz-Perez HS, McClure J, Bedetti C et al. Successful small bowel transplantation in dogs with cyclosporine and prednisone. Transplantation 1984; 37:127-32.
17. Whalberg JA, Southard JH, Belzer FO. Development of a cold storage solution for pancreas preservation. Cryobiology 1986; 23:477-81.
18. Belzer FO, D'Alessarrdro AM, Hoffmann RM et al. The use of UW in clinical transplantation. Ann Surg 1992; 215:579-85.
19. Schweizer E, Gassel AM, Deltz E et al. A comparison of preservation solutions for small bowel transplantation in the rat. Transplantation 1994; 57:1406-08.
20. Rodriguez-Quilantan FJ, Toledo-Pereyra LH, Suzuki S. Total heterotopic small bowel transplantation in the rat: technical considerations in the procurement, preservation and transplantation of the graft. Gac Med Mex 1993; 129:131-37.
21. Rodriguez-Quilantan FJ, Toledo-Pereyra LH, Suzuki S. Role of neutrophils following small bowel transplantation in the rat. Transplant Proc 1993; 25:3209.
22. Fabian MA, Bollinger RR, Wyble CW et al. Evaluation of solutions for small intestinal preservation. Biochemical changes as a function of storage time. Transplantation 1991; 52:794-99.
23. Rodriguez-Quilantan FJ, Toledo-Pereyra LH, Suzuki S. Twenty-four-hours total small bowel hypothermic storage preservation and transplantation in the rat: a study of various preservation solutions. J Invest Surg 1994; 2439-51.
24. Taguchi T et al. Evaluation of UW solution for preservation of small intestinal transplants in the rat. Transplantation 1992; 53:1202-05.
25. Shanbhogue LKR, Molenaar JC. Short bowel syndrome: metabolic and surgical management. Br J Surg 1994; 81:486-99.
26. Gondolesi G, Fauda M. Technical refinements in small bowel transplantation. Current Opinion in Organ Transplantation 2008; 13(3):259-265.
27. Lyons GW, Manax WG, Largiader F et al. Intestinal absorption of ileum preserved by hypothermia and hyperbaric oxygen. Surgery 1967; 62:721-741.
28. Eyal Z, Manax WG, Bloch JH et al. Successful in vitro preservation of the small bowel, including maintenance of mucosal integrity with chlorpromazine, hypothermia, and hyperbaric oxygenation. Surgery 1963; 57:259-268.
29. Lillehei RC, Idezuki Y, Feemster JA et al. Transplantation of stomach, intestine and pancreas: experimental and clinical observations. Surgery 1967; 62:721-741.
30. Ruiz JO, Schultz LS, Hendrickx J et al. Isolated intestinal perfusion: a method for assessing preservation methods and viability before transplantation. Trans Am Soc Artif Intern Organs 1971; 17:42-8.
31. Raju S, Fujiwara H, Lewin JR et al. Twelve-hour and twenty-four hour preservation of small bowel allografts by simple hypothermia. Transplantation 1988; 45:290-293.
32. Austen WG, McLaughlin ED. In vitro small bowel perfusion. Surg Forum 1965; 16:359-61.
33. Iijima K, Salerno RA. Survival of small intestine following excision, perfusion and autotransplantation. Ann Surg 1967; 166:968-975.

34. Kavin H, Levin NW, Stanley MM. Isolated perfused rat small bowel-technic, studies of viability, glucose absorption. J Appl Physiol 1967; 22:604-611.
35. Hohenleitner FJ, Senior JR. Metabolism of canine small intestine vascularly perfused in vitro. J Appl Physiol 1969; 26(1):119-28.
36. Alican F, Hardy JD, Cayirli M et al. Intestinal transplantation: laboratory experience and report of a clinical case. Am J Surg 1971; 121(2):150-9.
37. Toledo-Pereyra LH, Najarian JS. Preservation of the small intestine for allotransplantation. Surg Gynecol Obstet 1973; 137(3):445-50.
38. Toledo-Pereyra LH, Simmons RL, Najarian JS. Comparative effects of chlorpromazine, methylprednisolone and allopurinol during small bowel preservation. Am J Surg 1973; 126(5):631-4.
39. Toledo-Pereyra LH, Najarian JS. Small bowel preservation. Comparison of perfusion and nonperfusion systems. Arch Surg 1973; 107(6):875-7.
40. Toledo-Pereyra LH, Simmons RL, Najarian JS. Prolonged survival of canine orthotopic small intestinal allografts preserved for 24 hours by hypothermic bloodless perfusion. Surgery 1974; 75(3):368-76.
41. Toledo-Pereyra LH, Simmons RL, Najarian JS. Two-to-three day intestinal preservation utilizing hypothermic pulsatile perfusion. Ann Surg 1974; 179(4):454-9.
42. Teitelbaum DH, Sonnino RE, Dunaway DJ et al. Rat jejunal absorptive function after intestinal transplantation: effects of extrinsic denervation. Dig Dis Sci 1993; 38 (6):1099-1104.
43. Toledo-Pereyra LH, Simmons RL, Najarian JS. Absorption of carbohydrates and vitamins in the preserved and transplanted small intestine. Am J Surg 1975; 129(2):192-7.
44. Taguchi T, Zorychta E, Sonnino RE et al. Small intestinal transplantation in the rat: Effect on physiologic properties of smooth muscle and nerves. J Pediatr Surg 1989; 24(12):1258-1263.
45. Taguchi T, Zorychta E, Sonnino RE et al. Function of smooth muscle and nerve after small intestinal transplantation in rat: Effect of storing donor bowel in Eurocollins. J Pediatr Surg 1989; 24(7):634-638.
46. Toledo-Pereyra LH, Raij L, Simmons RL et al. Role of exdotoxin and bacteria in long-term survival of preserved small-bowel allografts. Surgery 1974; 76(3):474-81.
47. Toledo-Pereyra LH, Najarian JS. Human small bowel preservation: assessment of viability during storage. Bol Asoc Med P R 1979; 71(9):336-41.
48. Toledo-Pereyra LH, Suzuki S. Neutrophils, cytokines and adhesion molecules in hepatic ischemic and reperfusion injury. Role of ischemic reperfusion injury mediators. I Am Coll Surg 1994; 174:758-62.
49. Mueller AR, Platz KP et al. The effects of administration of nitric oxide inhibitors during small bowel preservation and reperfusion. Transplantation 1994; 58:1309-16.
50. Villarreal D, Grisham MB, Granger DN. Nitric oxide donors improve gut function after prolonged hypothermic ischemia. Transplantation 1995; 59:685-89.
51. Hotter G, Closa D, Prados M et al. Intestinal preconditioning is mediated by a transient increase in nitric oxide. Biochem Biophys Res Commun 1996; 222:27-32.
52. Sola A, De Oca J, Gonzalez R et al. Protective effect of ischemic preconditioning on cold preservation and reperfusion injury associated with rat intestinal transplantation. Ann Surg 2001; 234:98-106.
53. Heemann UW, Tullius SG, Azuma H et al. Adhesion molecules arid transplantation. Ann Surg 1994; 219:4-12.
54. Harward TRS, Coe G, Souba WW et al. Glutamine preserves gut glutathione levels during intestinal ischemia reperfusion. J Surg Res 1994; 56:351-55.
55. Katz SM, Sun SC, Schechner RS et al. Improved small intestinal preservation after lazaroid u74389g treatment and cold storage in University of Wisconsin preservation solution. Transplantation 1995; 59694-98.
56. Lew JI, Zhang W, Koide S et al. Glutamine improves cold-preserved small bowel graft structure and function following ischemia and reperfusion. Transplant Proc 1996; 28(5):2605-6.
57. Sun SC, Greenstein SM, Schechner RS et al. Superoxide dismutase: enhanced small intestinal preservation. J Surg Res 1992; 52:583-90.
58. Sisley MC, Desai T, Harig JM et al. Neutrophil depletion attenuates human intestinal reperfusion injury. J Surg Res 1994; 57:192-96.
59. Tarumi K, Yaghihasi A, Tsuruma T et al. Heat shock protein improves cold preserved small bowel grafts. Transplant Proc 1998; 30(7):3455-8.
60. Schaefer N, Takara K, Schuchtrup S et al. Perioperative glycine treatment attenuates ischemia/reperfusion injury and ameliorates smooth muscle dysfunction in intestinal transplantation. Transplantation 2008; 85(9):1300-10.
61. Toledo-Pereyra LH, Rodriguez-Quilantan FJ. Scientific basis and current status of organ preservation. Transplant Proc 1994; 26:309-11.
62. Alessiani M, Tzakis A, Todo S et al. Assessment of five-year experience with abdominal organ cluster transplantation. J Am Coll Surg 1995; 180:1-9.

63. Abu-Elmagd K, Todo S, Tzakis A et al. Three years clinical experience with intestinal transplantation. J Am Coll Surg 1994; 179:385-400.
64. Sonnino RE, Pritchard T, Riddle JM. Limited survival of rat small bowel transplants Preserved in University of Wisconsin Solution for 48 Hours. J Inv Surg 1993; 6(2):185-199.
65. Anveden-Hertzberg L, Li J, Riddle JM et al. Carolina Rinse solution does not prevent reperfusion injury after cold storage of small bowel grafts in University of Wisconsin solution. Transplant Proc 1994; 26(3):1482.
66. Anveden-Hertzberg L, Sidoti SA, Sonnino RE. Carolina Rinse Solution reduces myeloperoxidase activity in small bowel grafts after reperfusion. Transpl Proc 1996; 28(5):2602.
67. Sonnino RE, Wang L, Franson RC. Early secretory events during intestinal graft preservation. Transplant Proc 1998; 30(6):2643.
68. Sonnino RE, Pigatt L, Schrama A et al. Phosphlipase A_2 Secretion during intestinal graft ischemia. Digestive Diseases and Sciences 1997; 42(5):972-981.
69. Arcuni J, Wang L, Yousef K et al. Secretory event in intestinal grafts during preservation ischemia. J Surgical Research 1999; 84(2):233-239.
70. Sonnino RE, Pigatt L, Burchett S et al. Role of secretory phospholipase A_2 ($sPLA_2$) in ischemic injury to intestinal grafts during 24 hr preservation. Transplant Proc 1996; 28(5):2603-4.
71. Denney J, Ranjbar NE, Sonnino RE. Structural assessment of intestinal grafts preserved with phospholipase A_2 inhibition. Transplant Proc 1998; 30(6):2639.
72. Sonnino RE, Cross RS, Tawfik OW. Mediators of ischemia in preserved intestinal Grafts. Transplant Proc 2002; 34:975.
73. Li YS, Li JS, Li N et al. Evaluation of various solutions for small bowel graft preservation. World J. Gastroenterol 1998; 4(2):140-143.
74. Minor T, Vollmar B, Menger MD et al. Cold preservation of the small intestine with the new Celsior-solution. Transpl Int 1998; 11:32-37.
75. DeRoover A, Krafft MP, Deby-Dupont G et al. Seventy-two hours hypothermic intestinal preservation study using a new perfluorocarbon emulsion. Artif Cells Blood Substit Immobil Biotechnol 2001; 29(3):225-234.
76. Salehi P, Spratlin J, Chong T et al. Beneficial effects of supplemental buffer and substrate on energy metabolism during small bowel storage. Cryobiology 2004; 48:245-253.
77. Thomson AB, Keelan M, Thiesen A et al. Small bowel review: Part II. Can J Gastroenterol 2001; 13:446-466.
78. Zhu JZJ, Castillo EG, Salehi P et al. A novel technique of hypothermic luminal perfusion for small bowel preservation. Transplantation 2003; 76:71-76.
79. Salehi P, Zhu JZJ, Castillo EG et al. Preserving the mucosal barrier during small bowel storage. Transplantation 2003; 76:911-917.
80. Sonnino RE. Effect of flushing pressure on rat small bowel transplants. J Invest Surg 1996; 9:321-325.
81. Salehi P, Churchill TA. The influence of short-term fasting on the quality of small bowel graft preservation. Cryobiology 2005; 50:83-92.
82. Salehi P, Walker J, Madsen K et al. Control of oxidative stress in small bowel: relevance to organ preservation. Surgery 2006; 139:317-323.
83. Salehi P, Walker J, Madsen KL et al. Relationship between energetic stress and pro-apoptotic/cytoprotective kinase mechanisms in intestinal preservation. Surgery 2007; 141:795-803.
84. Olson DW, Jijon H, Madsen KL et al. Human bowel storage: The role for luminal preservation solutions. Transplantation 203; 76(4):709-14.
85. DeRoover A, De Leval L, Gilmaire J et al. Luminal contact with University of Wisconsin solution improves human small bowel preservation. Transplant Proc 2004; 36(2):274-275.
86. Balaz P, Kudla M, Lodererova A et al. Preservation injury to the human small bowel graft. Ann Transplant 2007; 12(1):15-18.
87. Siniscalchi A, Piraccini E, Miklosova Z et al. Metabolic, coagulative and hemodynamic changes during intestinal transplant: good predictors of postoperative damage? Transplantation 2007; 84(3):346-50.
88. Mangus RS, Tector J, Fridell JA et al. Comparison of Histidine-Tryptophan-Ketoglutarate Solution in Intestinal and Multvisceral Transplantation. Transplantation 2008; 86(2):298-302.

CHAPTER 11

Heart Preservation

Yoshikazu Suzuki and Francis D. Pagani*

Introduction

E xtended cardiac allograft preservation arises from the practical necessity of protecting the donor heart from the sequelae of ischemia-reperfusion and inflammatory injury. Successful preservation of the donor heart during procurement and implantation is indispensable to the success of heart transplantation for assuring early graft function and achieving both excellent short-term and long-term recipient survival.

Following early reports of the successful procurement and preservation of the heart by the method of hypothermic saline immersion during experimental canine orthotopic heart transplantation by Lower and Shumway in 1960, considerable research efforts have been focused on improving and extending the time of safe donor heart preservation. Despite these efforts, the single-flush, simple ischemic hypothermic immersion storage method is still the time-proven, practical, gold standard of heart preservation. The practical safe duration of ischemic hypothermic storage of donor hearts remains limited to approximately 4 to 6 hours at maximum, which precludes opportunities for improving allocation and outcomes by improved immunological matching. Despite these obvious limits, ischemic hypothermic storage has provided successful organ preservation permitting recipient survival rates of 90% or more at one year and approximately 50% at 10 years.[1,2]

More recently, advances in clinical and basic science research in the field of organ preservation, transplant immunology, ischemia-reperfusion injury, and endothelial and microvasculature biology have brought forward an improved understanding of the importance of preservation of the myocardium, as well as endothelium and microvasculature of the donor organ. Endothelial and microvasculature injury of the donor heart may initiate or heighten the recipient immunogeneic response to the donor organ through both antigen-dependent and nonspecific, immunologic host mechanisms. Release of allogeneic antigens through ischemic and reperfusion injury and activation of cytokine, chemokine and complement mechanisms incite the onset and progression of both acute rejection and chronic cardiac allograft vasculopathy. An improved understanding of the mechanisms of allograft injury has brought forth a number of emerging and resurgent alternative methods of heart preservation to reduce ischemia-reperfusion injury.

Background

Heart transplantation remains the gold standard for end-stage heart failure refractory to medical therapy. For appropriate candidates it provides the best long-term survival and QOL benefit. According to the latest registry reports of the International Society for Heart and Lung Transplantation (ISHLT) reflecting over 73,000 heart transplants performed worldwide, 1,5 and 10 year-survival is approximately 85%, 72% and 50%.[1,2] Following heart transplantation, approximately 90% of patients enjoy an excellent functional status without significant activity limitations, approximately 60 to 80% remain free of hospital readmission and approximately 30% of the patients return to

*Corresponding Author: Francis D. Pagani—Heart Transplant Program and Center for Circulatory Support, Section of Cardiac Surgery, 2120 Taubman Center, Box 0348, 1500 East Medical Center Drive, Ann Arbor, Michigan 48109. Email: fpagani@umich.edu

Organ Preservation for Transplantation, Third Edition, edited by Luis H. Toledo-Pereyra.
©2010 Landes Bioscience.

full-time or part-time employment. Despite these substantial advantages, patients undergoing heart transplantation continue to face the potential of death from graft failure, allograft rejection, infection, cardiac allograft vasculopathy (CAV) and malignancy.[3] The survival following heart transplantation decreases sharply during the first 6 months after transplantation, then decreases at a very linear rate of approximately 3 to 4% per year, thereafter. Over the past 3 decades, substantial improvement has been made in early survival, however, the linear mortality rate of heart transplantation due to late causes of death such as CAV, have not substantially changed over the past decade.[1-3] Within the first 30-days posttransplant, early graft failure, multiple-organ failure and non-Cytomegarovirus (non-CMV) infection accounts for approximately 40%, 14% and 13% of transplant deaths, respectively. Following 30 days posttransplant and within the first year of heart transplantation, non-CMV infection, graft failure and acute rejection account for 33%, 18% and 12% of deaths. After 5-years, CAV, late graft failure (likely due to CAV), malignancy and non-CMV infection account for 30%, 30%, 23% and 10% of deaths.[1,2] Assuring excellent early graft function can have significant beneficial effects on reducing subsequent allograft function, freedom from rejection and development of CAV and reducing the incidence of infection and malignancy by decreasing the level of immunosuppression required for occurrence of rejection.

Time Limitations with Heart Preservation

It should be emphasized that early graft failure still remains the most significant cause of death within the first 30-days posttransplant and that donor allograft ischemic time remains a powerful and significant predictor of survival.[2] Significant reductions in incidence of early allograft failure resulting in death have occurred despite current limits of allograft ischemic time. Comparison of the causes of early posttransplant deaths during 1998 to 2005 with early posttransplant deaths during 1992 to 1997, demonstrates a significant 9% reduction in primary and nonspecific early allograft failure within the first 30-days posttransplant.[2] Allograft ischemic time under current clinical practice of single-flush, simple ischemic hypothermic immersion storage method, has remained approximately $3.1 \pm 1.0(0.5-6.0)$ hours.[2] After 3 hours of cold allograft ischemic time, there is a significant and linear increase in the odds ratio of 1-year mortality and the practical limitation of cold ischemic time still remains approximately 4-6 hours.[1,2]

Limitations in the Availability of Cardiac Allografts

The number of heart transplant procedures performed each year continues to decrease, with 2945 heart transplant procedures reported to ISHLT registry in 2004 as compared with 4428 performed in 1994 (35% reduction).[4] Specifically, the UK Transplant, Euro Transplant and UNOS (United Network of Organ Sharing, USA) transplant networks have reported a 48%, 24% and 10% reduction in the number of heart transplant procedures from 1995 to 2004.[1] According to the Scientific Registry of Transplant Recipients (SRTR) in the US, 2127 heart transplant procedures were performed in 2005 among the 3223 patients on the waitlist at start of 2005 (66% transplant rate within one year) with 2833 patients newly added to the waitlist during 2005 and ending with 2980 patients on waitlist at the end of 2005. Of the number of patients on the list waiting for a heart, 21.9% were alive on waitlist one year after listing and 12.4% died waiting. In 2004, 3058 nonrecoveries of potential cardiac allografts from consented-donors occurred, of which 1690 (55.3%) were because of poor organ function.[4] The serious shortage of allograft organs emphasizes the need for the development of strategies that optimize both resuscitation of injured allograft organs during the early donor management phase and preservation of donor organs during the procurement and implantation phases.

Donor Management Strategies before Procurement: Significant Impact of Brain Death on the Outcome of Heart Transplantation

Management of a potential donor heart prior to the procurement plays a critical role in the subsequent function of the allograft posttransplantation and is an important component of the procurement and preservation process. It is apparent that additional efforts need to be directed toward more aggressive donor management to maintain the function of the donor heart and to minimize the deleterious effects of the state of brain death. Brain death is characterized by the final common pathway of the rise in intracranial pressure and herniation of the brain stem.[5] These changes lead to catecholamine surges and subsequent decreases in circulating hormones such as thyroid hormone, cortisol, insulin and anti-diuretic hormones. Brain death also manifests dramatic electrocardiographic and hemodynamic changes indicative of myocardial ischemia and arrhythmias. In potential cardiac allografts, brain death manifests pathologically with evidence of forced anaerobic metabolism, diffuse edema and myocardial ischemia and necrosis.[5,6] The importance of preprocurement donor management on subsequent posttransplant allograft function was reported by researchers from Papworth Hospital in the UK in 1995.[7] Aggressive optimization of cardiovascular performance utilizing data from pulmonary artery catheters and hormone replacement therapy with steroids, insulin, vasopressin and T3, yielded 44 (85%) additional transplantable cardiac allografts out of 52 potential cardiac allograft donors who fell initially outside the minimum acceptance criteria (35% out of total 150 cardiac allograft donor offers). The results in early and late patient survival were excellent.[7] Following the observations made by the Papworth group, a consensus guideline of donor management practices was published in 2002 that has subsequently been incorporated into the UNOS guidelines and pathway for donor cardiac allograft management.[8] A retrospective analyses of UNOS data has demonstrated that adoption of a structured donor management algorithm was associated with an overall 19.4% increase in heart transplant procedures.[9,10] Adoption of the hormone replacement therapy was associated with a 28% increase in the odds of a given donor becoming a cardiac allograft donor. Adoption of the hormone replacement therapy was also associated with 46% and 48% reduced odds of death within 30 days and cardiac allograft dysfunction.[10,11]

Evolving Concepts of the Association of Brain Death with Acute Allograft Rejection and Chronic Allograft Vasculopathy

Numerous recent clinical and basic science researches have documented the association between brain death leading to allograft ischemic injury and the subsequent incidence of acute allograft rejection and chronic CAV. Terasaki et al. reported that the outcome of renal allografts from living-related and living-unrelated sources was consistently superior to renal allografts obtained from cadavers.[12] These data suggested the importance of brain death and subsequent influence of antigen independent events on outcomes of transplanted organs. This concept was further supported by the series of superb experimental studies from Tullius et al. These investigators demonstrated antigen-independent functional, morphological and immunohistological changes of an iso-renal allograft that resembled those appearing in chronically rejecting 72-week renal allografts.[13] Takada et al. demonstrated that a 6-hour period of brain death was associated with the up-regulation of the macrophage-associated cytokines (IL-1, IL-6 and TNF-α), Th-1 cell-associated cytokines (IL-2 and TNF-γ), proinflammatory mediators and cell surface molecules (P-selectin, E-selectin and MHC class I and II antigens) and the costimulatory molecule B7, in peripheral organs in rat.[14] Wilhelm et al. demonstrated that donor hearts subjected to gradual onset brain death were rejected early and more intensely in an unmodified recipient as compared to donor allografts obtained from living donor controls.[15] Segel et al. demonstrated that coronary endothelial cells from brain dead rat hearts without hemodynamic instability were inflamed and associated with selective over-expression of IL-1β, IL-6, intercellular adhesion molecule-1 (ICAM-1) and vascular adhesion molecule-1 (VCAM-1).[16] Szabo et al. demonstrated that coronary blood flow and endothelium-dependent vasodilatation were impaired after brain death in an in-vivo animal model.[17]

Experimental data also supports the mode of brain death affecting both early and late cardiac allograft function. Shivalkar et al. demonstrated that experimentally-induced brain death in dogs elicited by a sudden increase in intracranial pressure resulted in a 1000-fold increase in the level of epinephrine after brain death (five times more than the gradual increase in intracranial pressure). Histological examination of the hearts obtained from dogs subjected to brain-death with a sudden increase in intracranial pressure demonstrated that 93% of the entire area examined demonstrated severe ischemic changes (severe myocytolysis with swollen mitochondria and necrotic nuclei) that was 4x greater than the area observed in hearts obtained from dogs subjected to brain death with a gradual increase in intracranial pressure. In a second series of their experiments, cardiac allografts obtained from dogs subjected to brain death with a sudden increase in intracranial pressure showed poor functional recovery compared to cardiac allografts obtained from non brain dead donors or cardiac allografts obtained from animals with brain death associated with a gradual increase in intracranial pressure.[18] Clinical studies in thoracic organ recipients have also supported the importance of brain death on influencing subsequent allograft function. Anyanwu et al. reported that the freedom from angiographic CAV in hearts obtained and subsequently transplanted from patients who received heart-lung transplantation (i.e., domino heart transplantation; donor cardiac allograft not exposed to brain death environment) was 99%, 83% and 77% at 1,5 and 10 years respectively and is significantly higher compared to the incidence of 55% at 5 years reported from the Cardiac Transplant Research Database of cadaveric hearts.[19] Mehra et al. reported in a clinical observational study that an explosive rise in intra-cranial pressure associated with brain death was accompanied by a more drastic recruitment of inflammatory cells, pro-inflammatory cytokines, adhesion and costimulatory molecules and was also associated with higher incidences of findings of CAV in cardiac allografts assessed by intravascular ultrasound examinations. An explosive mode of rise in intra-cranial pressure during brain death also resulted in higher adverse cardiac event rates and higher mid-term mortality.[20] These studies suggest that brain death is intimately associated with the incidence and the progress of acute allograft rejection and chronic CAV.

Ischemia-Reperfusion Injury, Inflammatory Response and Chronic CAV: Central Role of Endothelial Dysfunction

CAV is a major factor limiting long-term survival after cardiac transplantation.[1-3,21,22] CAV is an accelerated form of coronary vessel disease that is characterized by diffuse, concentric fibrous intimal hyperplasia along the entire length of the coronary vessel including both epicardial and intramural arteries and veins.[21,22] Both immunologic-dependent and non-immunologic risk factors contribute to the development of CAV causing endothelial dysfunction and injury eventually leading to progressive intimal thickening. Immunologic-dependent risk factors include the number of HLA mismatches, the number of acute rejection episodes, their duration and their time of onset posttransplantation. Immunologic-independent risk factors include hyperlipidemia, older donor age, sex, obesity, diabetes mellitus, hypertention, hyperhomocysteinemia, cytomegalovirus infection, ischemia-reperfusion injury and brain death.[21,22] Ischemia-reperfusion injury plays a significant role in endothelial injury-induced proliferation of smooth muscle cells and the pathophysiology of CAV. Numerous clinical and basic science studies have supported the original concept of the "Injury Hypothesis" first reported in 1994 by Dr. Land and the relationship between inchemia-reperfusion injury during the peri-transplant period and onset and progression of CAV.[23-27] Gaudin et al. reported in a retrospective, observational study, that the histological degree of ischemic injury present in the early posttransplant biopsy was the strongest predictor of the development of CAV.[24] Endothelial cells play an important role in maintaining normal vessel wall function by their ability to control vascular tone, to inhibit thrombus formation, leukocyte adhesion and platelet adhesion and to suppress vascular smooth muscle cell proliferation. Reactive oxygen species (i.e., free radical formation; ROS)-mediated ischemia-reperfusion injury to endothelial cells may provide the trigger for CAV through adhesion of leukocyte and platelet, upregulation of MHC class I and class II antigen expression, release of nonMHC donor antigens, growth factors and cytokines, expression of adhesion molecules on antigen-presenting cells and proliferation of vascular smooth muscle cells.[23] Thus, ROS-mediated allograft injury activates the innate immune system of both the donor and recipient. Dendritic cells (DC) are also elucidated as another major target of ischemia- reperfusion injury. Upregulation of DC-mediated T-cell alloactivity results in presentation of allogeneic peptides, MHC antigens, and costimulatory molecules. ROS-mediated intracellular appearance of nonnative oxidized proteins may induce the chaperoning activity of heat shock proteins, that may activate donor and recipient's Toll-like receptor 4 (TLR4). TLR4 causes DC to initiate the bridge to adaptive allo-immunity resulting in acute rejection. Donor TLR4-bearing vascular cells effect allo-atherogenesis via the innate immune system resulting in chronic CAV.[27]

Microvasculature Prothrombogenicity in the Pathophysiology of Cardiac Allograft Vasculopathy

There are other evidences to support that ischemia-reperfusion injury and the inflammatory response at the time of cardiac allograft procurement, preservation and implantation are strongly related to the development and the severity of chronic CAV. CAV was formally considered as a process solely representing chronic rejection that was mediated by the incompatibility of the recipient's immune cells to the cells within donor's allograft. In a study of 80 human cardiac allografts with cellular, humoral and vascular specific immuno-cytochemical antibodies, Faulk and Labarrere found a correlation of clinical severity of the cardiac allograft rejection with the vascular damage and not with the degree of cellular or humoral-mediated injury.[28] These authors contended that failure of a transplanted cardiac allografts were caused by failure of the microvessels within the allografts to remain open.[28,29] Failure of the microvasculature of the donor allografts occured because the vessels became prothrombogenic as demonstrated by the presence of microvascular fibrin. There was an approximately 50% chance of microvasculature fibrin deposition seen in transplanted hearts.[30] The presence of microvascular fibrin was associated with microvascular occulusion and myocardial cellular injury reflected with the detection of cardiomyocyte specific molecules in the circulation such as troponins T and I.[30-32] The allografts that showed myocardial fibrin deposition within the first three months after transplantation developed significantly more CAV and the CAV was more severe and progressesed more rapidly than in allografts not having any myocardial fibrin deposits. Patients with increased myocardial cell injury immediately after transplantation persisted in showing a prothrombogenic microvasculature with persistent levels of detectable troponin I in circulation during the first year after transplantation. These patients were at significantly higher risk for developing CAV during follow-up.[33]

Three risk factors that contribute to prothrombogenicity in allograft vessels are: (1) loss of anticoagulation (loss of vascular antithrombin), (2) loss of fibrinolytic activity (loss of vascular tissue plasminogen activator) and (3) microvascular endothelial activation (upregulation of endothelial ICAM-1, MHC class II, and HLA-DR and expression of tissue factor).[29] The normal microvasculature of the heart is thrombo-resistant, largely because of the presence of components of the anticoagulant pathways (the protein C-protein S anticoagulant pathway and the heparan sulfate proteoglycan-antithrombin anticoagulant pathway), the fibrinolytic pathway within the vascular structures and also the expression of prostacyclin (PGI2) and nitric oxide.[34] Thrombomodulin was present on the endothelial cell plasma membrane of the entire microvasculature. Antithrombin was present on smooth muscle cells of arteries and arterioles, on arterial intima and in endothelial cells of veins and venules. The capillary network within the heart did not show any immunohistochemically detectable antithrombin.[32,35] Thrombomodulin expression within the cardiac microvasculature after transplantation was downregulated by cellular rejection episodes, but this was not the case with antithrombin.[32,35] It was the status of vascular antithrombin after transplantation that was directly associated with the deposition of microvasculature fibrin and subsequent allograft outcome.[36] One possible mechanism for the loss of vascular antithrombin was cellular rejection, because macrophages and T-lymphocytes could release growth factors, cytokines, or heparinase that could affect the expression or availability of cellular or extracellular heparan sulfate proteoglycan molecules, which were responsible for the vascular antithrombin binding. However, this possibility seemed unlikely as the loss of vascular antithrombin and not the number of severity of rejection episodes that was associated with subsequent development of CAV.[32,35,36] The recovery of vascular antithrombin binding in allografts, which was associated with development of unusual capillary antithrombin binding, was also associated with a lower rate, delayed development, less severe and slower progression of CAD than in allografts that initially lost and maintained loss of vascular antithrombin binding during the first three month after transplantation.[32,36-38]

The arterial tree within the heart normally contains the principal activator of the fibrinolytic cascade, namely tissue plasminogen activator, which is primarily associated with smooth muscle cells. Another characteristic of a thrombo-resistant microvasculature in the normal heart is the absence of endothelial activation markers such as ICAM-1, MHC class II HLA-DR and tissue factor within the arterial and arteriolar endothelium. Microvascular fibrin deposition was also associated with both depletion of tissue plasminogen activator from arteriolar smooth muscle cells and expression of endothelial activation markers, which were associated with the subsequent development of CAV and decreased graft survival.[31,32,34,35,39-41] All of the deposition of microvascular fibrin, the depletion of vascular (arteries and veins) antithrombin, the depletion of tissue plasminogen activator from the arteriolar smooth muscle cells and the expression of arterial and arteriolar endothelial activation markers detected in the first weeks after transplantation, were independently and cumulatively related to the development and the severity of transplant CAV and subsequent allograft failure.[29,32,34] An increasing number of non-immunologic risk factors within the allograft microvasculature in the first weeks after transplantation were associated with increased risk for developing CAV and graft failure.[29,32,34] In contrast to this, cellular rejection episodes during the first three months after transplantation were not significantly associated with the subsequent development of CAV.[29,32,34] The development of microvascular pro-thrombogenicity early after transplantation was particularly relevant because it was associated with the long-term evolution of CAV. Once CAV develops, regimens of new and varied immunosuppressive agents failed to improve allograft survival.[29,42] Therefore, efforts to attenuate prothrombogenic microvasculature changes should be initiated at the time of allograft procurement, preservation and implantation.[29,32,34]

The close linkage between clotting and inflammation is readily demonstrable in human physiologic responses to a variety of potentially injurious stimuli such as bleeding, burns and sepsis. The same pro-inflammatory stimuli that activate the human clotting cascade activate the phagocytic effector cells, including neutrophils, monocytes and macrophages. Thus, increased expression of endothelial activation markers within the allograft microvasculature early after transplantation, even in the absence of acute cellular rejection, could be caused by the release of pro-inflammaotry cytokines such as TNF-α or IL-6. Endothelial cell injury can be mediated by immune cells, antibodies and perhaps ischemia/reperfusion injury. The precise triggering mechanisms of these events are still unknown, although the fact that they occur immediately after transplantation suggests that events in the peri-transplantation period such as ischemia/reperfusion and systemic inflammatory response could be the initiating events. Ischemia-reperfusion injury is associated with neutrophil recruitment, which could lead to cleavage of heparan sulfate molecules through the release of enzymes such as neutrophil elastase or heparanase. Ischemia-reperfusion injury can also induce depletion of vascular tissue plasminogen activator following the generation of microvasculature thrombin because it is well recognized that thrombin can induce the release of cellular tissue plasminogen activator. Ischemia-reperfusion injury can also induce expression of endothelial activation markers. Endothelial activation, which involves expression of endothelial tissue factor, could promote microvascular activation of the coagulation cascade with the final deposition of microvascular fibrin.[34]

Methods of Myocardial Preservation

Heart preservation is the practical necessity of the need for transportation and allocation of the cardiac allograft. This process exposes the cardiac allograft to an obligatory phase of ischemia and reperfusion injury. Successful and complete salvage of ischemic myocardial tissue can only be achieved by early reperfusion at a time when tissue is still in a reversible state of injury.[43] If early reperfusion is not feasible, attempts should be made to slow the rate of evolution of the ischemic process. Ischemia is defined as the pathological state of a shortage of oxygen-dependent high-energy phosphate, adenosine triphosphate (ATP), viability secondary to an inadequate blood supply to an organ. ATP is required to maintain cellular function and viability (structural, ionic and metabolic homeostasis). Under normal aerobic conditions, cardiomyocytes produces its energy substrate, ATP, through free fatty acid β-oxidation (70%) in mitochondria and glucose oxidation.[44] During ischemia, cardiomyocytes are forced to alternatively utilize efficient, but limited anaerobic glycolysis to produce ATP, which results in acidosis and the accumulation of lactate and pyruvate leading to decreased contractility. Ischemia leads to the reduction of ATP reserve by 65% at 15 minutes and by 90% at 40 minutes even at the cost of cessation of its contractile function for its survival.[44]

Ischemic Hypothermic Storage

To maintain cellular viability and during ischemia, the explanted cardiac allograft needs to be protected by reducing energy demand and by avoiding the depletion of ATP reserves. The most common method in clinical practice to obtain satisfactory myocardial preservation is by ischemic hypothermic immersion storage, with the supplemental use of cardioplegia at the time of myocardial arrest, storage, implantation and frequently at reperfusion. Ischemic hypothermic storage has provided a simple, effective method for cardiac allograft preservation during transplantation.[1]

Hypothermia

One of the most fundamental aspects of myocardial preservation during transplantation is the utilization of hypothermia. The efficacy of hypothermia alone is so profound that it was used as the sole method for myocardial preservation for many years.[45] The effects are obviously attributable to the ability to reduce metabolic activity and oxygen demand of the heart and also to reduce enzymatic activity and degradation rate of viable cellular components.[46] For every decrease in the temperature by 10°C, both metabolic activity and enzymatic activity of the heart are reduced by approximately half.[47] During ischemia, hypothermia slows the depletion of ATP that is essential for cell viability and delays the occurrence of enzymatic and mechanical disruption of cell membrane and intracellular structures.

Despite these profound effects on myocardial protection during ischemia, hypothermia alone is not a completely perfect means of protecting the cardiac allograft.[47,48] Anaerobic metabolism continues even at a slow rate resulting in accumulation of protons, lactate and breakdown products of adenine nucleotides (hypoxanthine) that contribute to the generation of oxygen free radicals at the time of reperfusion. The activity of ATP-generating reactions and ionic pumps are also impaired and slowed down. The loss of transmembrane ionic gradients, that is, sodium and chloride flow into the cell down a concentration gradient, increases intracellular osmolarity and promotes cellular swelling. The increased intracellular sodium concentration reverses Na/Ca antiport on the condition of the energy depletion and hypothermia, which leads to Ca^{2+} influx. Calcium overload leads to rigor contracture and the lethal hypercontracture at reperfusion in cardiomyocytes and also causes cellular damages by activation of calcium dependent phospholipases, proteases, endonucleases and adenosine triphosphatases (uncoupling oxidative phosphorylation).[49] Hypothermia is also associated with increased vascular permeability, interstitial edema and endothelial dysfunction.[49,50] The beneficial effects of supplementing hypothermia with cardioplegia during the various stages of heart procurement and implantation have been documented and, thus this procedure has become the standard of practice in clinical transplantation today.[1,51-53]

Cardioplegia and Composition of Preservation Solutions

The addition of cardioplegia to hypothermic preservation of the heart has improved upon the overall level of myocardial protection.[46,54] The principal importance of the cardioplegia solution is its ability to (1) produce rapid mechanical arrest in diastole, avoiding energy depletion from prolonged ventricular fibrillation and (2) cool the myocardium quickly and evenly. These objectives are most frequently obtained by cardioplegic solutions that are hyperkalemic leading to a depolarized electromechanical arrest. Hyperkalemic solutions are the most reliable, widespread and time-proven methods of cardioplegia. However, hyperkalemic solutions can cause transmembrane ion fluxes and promote ATP dependent ion exchanger activation, that leads to ATP depletion, even during hypothermic ischemia.[55] In addition, hyperkalemic solution can impair vasodilatory function of the endothelium through NO-independent hyperpolarizing vasodilatory factors.[56] Endothelium-dependent vascular relaxation is known to be mediated by three different endothelium derived relaxing factors (EDRFs): prostacyclin (PGI2), nitric oxide (NO) and an unidentified endothelium-derived hyperpolarizing factor (EDHF). Results of numerous studies regarding the effect of hyperkalemic heart preservation solutions on endothelial function are conflicting. However, He et al. demonstrated in a series of studies that hyperkalemic cardioplegia solutions have minimal impact on the NO-related endothelium dependent vaso-relaxation after exposure for 1 or 2 hours.[56-58] When the effect of ischemia-reperfusion injury was excluded by using oxygenated crystalloid cardioplegia and prostacyclin release was blocked by indomethacin, the NO-related endothelium dependent vasorelaxation was well preserved in either porcine epicardial coronary arteries or neonatal rabbit aorta.[57,58] With exclusion of the effect of ischemia-reperfusion injury and influences of PGI2 and NO, EDHF-mediated vaso-relaxation were impaired by hyperkalemic cardioplegia as well as St Thomas' Hospital and University of Wisconsin solutions either in porcine or human coronary arteries.[59-62] Other agents that were reported for cardioplegia additives were calcium desensitizing agents (magnesium),[63] myosin inhibitors (2,3-butanedione monoxime),[64,65] hyperpolarizing agents such as (adenosine),[66] K-ATP channel agonist (pinacidil),[67] polarizing agents such as (procaine[68] and lidocaine)[69] and sodium channel blocker, (tetrodotoxin).[70,71] The introduction of magnesium into cardioplegia helps to achieve immediate heart arrest and the enrichment of magnesium may counteract the unfavorable effect of hypocalcemia on salcolemmal membrane by preventing calcium influx.[56] In addition to the protective effect on myocardium, magnesium has been proven to be a potent vasodilator through both endothelium-dependent and independent mechanisms.[56,72-75] Magnesium preserves EDHF-mediated vasorelaxation and hyperpolarization and restores EDHF function impaired by hyperkalemia.[76]

There are a number of cardioplegia and storage solutions available for donor heart procurement and preservation today, with the clinical use of these solutions widely varying according to institutional preference.[77,78] Broadly, cardioplegia solutions can be classified into extracellular solutions defined as those containing 70 mEq/L or greater of sodium and intracellular solutions defined as those containing less than 70 mEq/L of sodium (Table 1).[78] With an improved understanding of the mechanisms that underlie ischemia-reperfusion injury, a number of components have been added to supplant this basic premise (myocardial arrest) of cardioplegia solutions. These additives include components that: (1) attenuate or prevent oxygen-derived free radical generation (superoxide dismutase, catalase, desferoxamine, allopurinol, mannitol, glutathione); (2) increase intravascular oncotic pressure and prevent interstitial and intracellular swelling (lactobionate, hydroxyethyl starch, mannitol); (3) provide substrate for cellular energy production (aspartate, glutamate, glucose, pyruvate); (4) stabilize cellular membranes (hydrocortisone, procaine, magnesium); (5) buffer against acidosis (bicarbonate, phosphate, histidine, tris-hydroxymethylaminomethane (THAM)); and (6) inhibit cellular influx of calcium (verapamil).

Extracellular solutions (i.e., St. Thomas' Hospital solution, Celsior) contain a high sodium ion concentration approximating plasma concentrations and a sufficient potassium ion concentration (15 to 20 mmoles/L) to obtain myocardial arrest. In addition, St. Thomas' Hospital solution contains magnesium ion (16 mmol/L) and procaine (1 mmole/L) to promote myocardial arrest and stabilize cellular membranes. Celsior solution contains magnesium ion (13 mmol/L), low calcium ion (0.25 mmol/L) and additives of impermeants (mannitol, lactobionate), buffer (histidine), free radical scavengers and antioxidants (glutathione, histidine, mannitol) and energy substrate (glutamate). Intracellular solutions (i.e., University of Wisconsin (UW), Euro-Collins solution) have a high potassium ion concentration (117 to 125 mmoles/L) and low sodium ion content, mimicking the ionic milieu of cells. In addition to this basic electrolyte composition, UW solution contains additives to attenuate oxygen-derived free radical generation (glutathione and allopurinol), additives to limit interstitial and intracellular swelling (hydroxyethyl starch (pentastarch), lactobionate and raffinose), along with adenosine (vasodilator, ATP precursor, calcium-channel inhibition), dexamethasone (anti-inflammation and membrane stabilization) and insulin. Euro-Collins solution contains added glucose for substrate and oncotic benefit. Bretschneider's HTK solution contains a low sodium ion concentration and sufficient potassium ion (approximately 9 mmoles/L) to obtain myocardial arrest, with added histidine (118 mmoles/L), that serves as an oncotic and buffering agent and antioxidant, in addition to tryptophane, alpha-ketoglutarate (substrate) and mannitol (oncotic agent, oxygen-derived free radical scavenger). The Stanford solution contains a low sodium ion concentration and a sufficient potassium ion concentration to obtain mechanical arrest (27 mmoles/L). Its primary additives are glucose and mannitol. Other cardioplegia formulations include CP-8,[79] Cardiosol (polyethylene glycol-based)[80] and NIH solution.[81] Demmy et al. reported in 1997 that 167 different heart preservation solutions were utilized in a survey of donor heart preservation solutions at 143 active UNOS heart transplant centers in 37 states and the District of Columbia. Of these formulations, were commonly cited solutions, 49.7% were intracellular solutions and the remainder was extracellular.[78] Significant variations in solutions utilized among heart transplant centers were observed among major regions of U.S. Moreover, logistic regression analyses of 9401 patients who underwent heart transplantation from October, 1987 to December, 1992 demonstrated a reduction in the adjusted one month mortality odds ratio for grafts preserved with intracellular solutions rather than extracellular solutions, although center selection bias could possibly account for these observed differences.[78] In comparison with extracellular solutions, intracellular solutions theoretically have the potential advantage of minimizing the sodium influx through ion gradients and subsequent cell swelling and calcium overload. However, this potential benefit of intracellular solutions is supposed to be equated by extracellular solutions when the latter are supplemented with impermeants.[49] No single optimal technique for organ preservation has been proved within the clinically relevant time frame of 4 to 6 hours.[77] Numerous clinical trials, rather, substantiated satisfactory donor heart preservation using a number of different preservation solutions of both intracellular and extracellular compostion, as long as the donor heart ischemic time is not markedly prolonged.[1,2,82-88] Michel and Ferrera et al. evaluated efficacy of various cardioplegia solutions utilizing an isolated nonworking rat heart model and biopsy specimens. A comparison of extracelluar solutions, such as Celsior solution, St. Thomas' Hospital solutions, the modified University of Wisconsin solution (UW-1), Lyon Preservation solution (LYPS) and simple isotonic saline solution wuth intracellular solutions such as standard University of Wisconsin solution (UW), Bretschneider HTK solution, Stanford solution and Euro-Collins solution (EC) demonstrated that after 8 hours cold storage at $4\degree$C, extracellular type solutions provided better preservation than did intracellular type solutions.[89] Simple isotonic saline solutions and EC were ineffective in providing adequate myocardial preservation at $4\degree$C for 8 hours. Thus, the ionic basis of the solution seems to be of primary importance in improving the quality of the preservation. However, UW and UW-1 (the inversion of the sodium-potassium ratio in UM) provided equivalent preservation of cardiac function. These authors concluded that preservation quality might be attributed to a low-concentration of calcium ion less than 0.1mmol/L to prevent calcium paradox and the effect of limiting ischemia-reperfusion injury.[89]

Pharmacologic Modifications to Cardioplegic Solutions: Attenuating Ischemia-Reperfusion Injury during Myocardial Preservation

Attenuation of Oxygen-Derived Free Radical Injury

Superoxide Dismutase and Catalase

Since the observations of the phenomenon of hypoxic-reoxygenation injury to the myocardium (i.e., oxygen paradox) by Hearse et al., the role of oxygen-derived free radicals in the initiation of cellular damage reoxygenation of ischemic myocardium has been well recognized in numerous clinical and experimental models.[90,91] These observations have led to inclusion of many specific additives into current clinical cardioplegic solutions that scavenge or prevent formation of these reactive oxygen-derived metabolites. There are several important sources of oxygen-derived free radicals. These include generation of superoxide anion species during electron transport at the internal mitochondrial membrane from the autoxidation of ubizuinones,[92,93] and nicotinamide-adenine dinucleotide (NADH) dehydrogenase.[91,94] Normally, mitochondrial enzymes of the electron transport chain reduce oxygen to water without the production of significant quantities of oxygen-derived free radical metabolites. This generally accounts for 95% of tissue oxygen consumption.[95] However, in the absence of oxygen (i.e., myocardial ischemia), electron passage through the mitochondrial electron transport system ceases.[96] Following reperfusion and restoration of tissue oxygen levels, the presence of diminished cellular levels of adenosine diphosphate (ADP) prevents rapid utilization of oxygen by the mitochondrial electron transport system. Oxygen then undergoes incomplete reduction with the formation of the superoxide anion ($\cdot O2^-$). Superoxide anion may then act as a reducing agent donating its electrons, or as an oxidizing agent, forming hydrogen peroxide (H_2O_2). Superoxide anion may, in addition, react with membrane phospholipids to form alkoxy radicals that can lead to changes in membrane fluidity and permeability and intracellular calcium accumulation. Tissue protection against the toxic superoxide anion radical is accomplished through the dismutase reaction that is catalyzed by the enzyme superoxide dismutase (SOD). This enzyme catalyzes the formation of H_2O_2 from superoxide anion and increases the rate of intracellular dismutation by a factor of 10^9. The major danger of H_2O_2 accumulation, as a product of the dismutase reaction, is the production of the hydroxyl radical ($\cdot OH$).[97] The $\cdot OH$ is a very reactive and unstable oxidizing species that reacts

Table 1. Composition of selected preservation solutions used in clinical heart transplantation

Names of Types of Solutions	UW: University Wisconsin	Bretschneider's-HTK-4	Stanford	St. Thomas	St. Thomas 2	Euro-Collins	Celsior
Intracellular(I) / Extracellular (E) type	UW-1 (I)	I	I	E	E	I	E
pH	7.4	7.1-7.2	7.8-8.4	5.5-7.8	7.3-7.8	7.2-7.4	7.3
mOsm/L	320	310	440	290-326	324	300-375	320-360
Na(mmol/L)	20.0-35.0	15.0	20.0	120.0-144.0	120.0	10.0-15.0	100.0
K (mmol/L)	125.0	9.0	27.0	16.0-20.0	16.0	115.0-117.0	15.0
Cl (mmol/L)		50.0	30.0	160-203	160-203	10.0-15.0	28.0-42.0
Mg (mmol/L)	5.0	4.0		16.0	16.0	0-80.0	13.0
Ca (mmol/L)		0.015		1.2-2.4	1.2		0.245-0.250
Phosphate (mmol/L)	25.0					102.0	
Sulfate (mmol/L)	5.0					8.0	
Procaine (mmol/L)				1.0	1.0		
Histidine (mmol/L)		180.0					30.0
Histidine-HCL (mmol/L)		18.0					
Tryptophane (mmol/L)		2.0					
α-Ketoglutarate (mmol/L)		1.0					
Glutamate (mmol/L)			20.0				20.0
Bicarbonate (mmol/L)			20.0	10.0	10.0	10.0-15.0	
Albumin (gm/L)						214	

continued on next page

Table 1. Continued

Names of Types of Solutions	I	E	I	E	E	I	E
Glucose (mg/L)			45.0			139 (mmol/L)	
Mannitol (mg/L)			11.4			198.0	60.0
Lidocaine (mg/L)		30.0 (mmol)					
Acetate (mg/L)							
Gluconate (mg/L)							
Hydroxyethyl Starch (mg/L)	50.0	50.0					
Lactobionate (mmol/L)	100.0	100.0					80.0
Raffinose (mmol/L)	30.0	35.0					
Allopurinol (mmol/L)	1.0	1.0					
Adenosine (mmol/L)	5.0	5.0					
Glutathione (mmol/L)	3.0	3.0					3.0
Dexamethasone (mg/L)	16.0						
Insulin (units/L)	100.0						
Penicillin (U/L)	20,000						

with a variety of organic compounds and biological membranes. H_2O_2 accumulation is prevented by reaction with catalase or glutathione peroxidase. Therefore, significant concentrations of $\cdot OH$ or superoxide anion do not exist under physiological conditions, as they are scavenged by the actions of SOD, catalase and glutathione peroxidase. There were numerous experimental studies that demonstrated, for the most part, the benefits of SOD plus catalase in attenuating ischemia-reperfusion myocardial injury.[98-100] In a heterotopic transplant rat heart model, Sun et al. demonstrated that following a 3.5 hour hypothermic storage period, the administration of SOD and catalase into the recipient 5 min before reperfusion resulted in improved graft performance.[98] These authors found an increased mean myocardial blood flow, decreased end-diastolic pressure and increased maximim dP/dt as a function of left ventricular volume with SOD and catalase as compared to saline control hearts. Jurmann et al. demonstrated that following 3 hours of hypothermic ischemic storage, hearts reperfused with SOD and catalase had significant increases in developed left ventricular pressure and increased coronary blood flow following 15 minutes of reperfusion in an ex vivo heart-lung preparation[99] Gharagozloo et al. demonstrated that the addition of SOD and catalase preserved left and right ventricular compliance and myocardial oxygen consumption and reduced lactate production as compared to control animals in an ex-vivo preservation model in sheep.[100]

Allopurinol

The role of allopurinol in attenuating reperfusion injury arises from its ability to inhibit the enzyme xanthine oxidase (XO). Normally, in the presence of adequate tissue oxygenation, hypoxanthine is catabolized by xanthine dehydrogenase to xanthine and ultimately, uric acid. However, in the presence of ischemia, calcium-dependent proteolysis of cellular xanthine dehydrogenase occurs and results in the production of XO.[89] Samples of ischemic myocardium biopsied 30 minutes after coronary occlusion showed a greater than 300% increase in the activity of XO.[101] XO utilizes molecular oxygen instead of NAD^+ as its electron acceptor and catalyzes the production of superoxide anion and H_2O_2. Therefore, at the time of reperfusion, superoxide anion and H_2O_2 are produced by the enzyme in concentrations that are deleterious to the cells. During ischemia, ATP is degraded to its nucleoside precursors: adenosine, inosine and hypoxanthine. As ischemia develops, the cell is no longer able to maintain proper ion gradients across its membranes and calcium-activated protease converts the xanthine dehydrogenase to XO. When oxygen is reintroduced into the system, a burst of superoxide anion and lipid peroxidation follow. Allopurinol becomes a noncompetitive inhibitor of XO function when given in higher doses, so the formation of oxygen-derived free radicals are attenuated during the reperfusion. Bando et al. demonstrated that significant generation of oxygen-derived free radicals occured during reperfusion and not during perfusion or storage of the heart with modified Euro-Collins solution.[102] Administration of allopurinol during the period of reperfusion significantly reduced the generation of oxygen-derived free radicals.

Desferoxamine

The prevention of $\cdot OH$ production can be achieved by scavenging its substrates (superoxide anion or H_2O_2) with SOD and catalase, or by reducing the amount of the metal catalyst essential for its formation, with the iron chelator, desferoxime. Menasche et al. demonstrated that desferoxamine given as an additive to single-dose cardioplegia solution at the end of arrest and to the reperfusate during the initial phase of reoxygenation, improved postreperfusion ventricular pressure development, maximal rate of rise of ventricular pressure, left ventricular compliance and coronary flow.[103]

Leukocyte Depletion

Neutrophils play an important role in the pathogenesis of ischemia-reperfusion injury to the heart (myocardial infarction and contractile dysfunction) by generating oxygen free radicals, releasing proteases and inducing a localized and generalized inflammatory response. Neutrophil-mediated ischemia-repefusion injury involves specific interactions among neutrophils, cardiomyocytes and coronary artery and venous endothelial cells in the early moments of reperfusion. Neutrophils are one of the largest sources of oxygen-derived free radical generation during ischemia-reperfusion of the myocardium. Neutrophils release more than 20 different proteolytic enzymes such as acid hydrolases, the serine protease, elastase contained in azurophilic granules and metalloproteinases, collagenese and gelatinase. Upon reperfusion and reoxygenation, the activation of neutrophils initiates a "respiratory burst" with a sudden and large increase in oxygen consumption. This leads to the generation of oxygen metabolites and toxic free radicals that are released into the external tissue environment and cause tissue injury, cell death and also upregulation and release of cytokines (TNF-α, IL-8, IL-1, IL-6 and macrophage inflammatory protein-2 (MIP-2)), complement (C5a) and lipid mediators (LTB4, TxA2 and platelet activating factor (PAF)).[44,104-106] Greater than 90% of the oxygen consumed by neutrophils during this period can be accounted for by superoxide anion production through the action of a NAD(P)H-oxidase enzyme, located in the cell membrane.[104,107,108] Following the formation of H_2O_2 by the dismutase reaction, H_2O_2 can react with neutrophil myeloperoxidase found in azurophilic granules to produce the toxic species, hypochlorous acid and N-substituted chloramines (R-NHCl).[104,109] The toxic effects of leukocytes are not only related to their ability to generate toxic oxygen metabolites, but also to their ability to participate in capillary plugging and contribute to the "no-reflow" phenomenon at the time of reperfusion.[110-113] Neutrophils are recruited to the reperfused myocardium during the early minutes of reperfusion by proinflammatory cytokines, complement and chemotactic factors (IL-8) released by the myocardium during ischemia. The first interaction is a tethering or "rolling" of neutrophils along the endothelial surface, which is mediated by P-selectin and E-selectin on the endothelium and a sialylated glycoprotein on the neutrophil, most likely sialyl Lewisx or the sialomucin P-selectin glycoprotein ligand-1 (PSGL-1).[106,114] Two to four hours later, PAF and LTB4 in the microcirculation can increase the surface expression of CD11/CD18 on neutrophils, while IL-1 and TNF-α increase ICAM-1 expression on the endothelium. Firmer adherence of neutrophils to the endothelium then ensues. The initial loose adherence step is obligatory for later firm adherence mediated by the CD11/CD18 complex on neutrophils and ICAM-1 endothelial cells and is critical in the pathogenesis of myocardiocyte necrosis, microvascular injury and apoptosis.[106]

Recently, the cellular mechanism underlying endothelial activation has been elucidated with the study of nuclear factor κ-B (NF-κB). Oxidative stress activates the tyrosine phospholrylation of IκBα, an inhibitor of NF-κB that binds to NF-κB in the cytoplasm; such phosphorylation dissociates IκBα from NF-κB. The translocation of functional NF-κB to the nucleus, with binding to the target genes results in transcriptional activation of those genes.[56,115] Targeting on the signaling pathway of endothelial cell activation may ameliorate ischemia-reperfusion injury induced endothelium impairment. Better recovery of coronary vascular response to serotonin and bradykinin in porcine coronary vessels was obtained by adding desferoxamine or manganese superoxide dismutase to the cardioplegic solution to reduce the oxygen-derived free radicals.[56,116] Better recovery of coronary vascular response to acetylcholine in neonatetal lamb heart was also obtained by adding leukocyte CD18 (ligand for ICAM-1) antibody before cardioplegic ischemic arrest.[56,117] Moreover, transfection of NF-κB decoy oligonucleotides into isolated heart blocked ICAM-1 upregulation and inhibited increase in neutrophil adhesion.[56,118]

Two lines of evidence support the hypothesis that leukocytes contribute to myocardial ischemia-reperfusion injury. First, neutrophils activated in vivo with tetradecanoyl phorbol acetate induce cardiovascular dysfunction, which can be prevented by treatment with SOD and catalase.[114] Second, leukocyte depletion enhances recovery of ischemic-reperfused myocardium.[119-121] Canine hearts immersed in modified Collins solution were transplanted heterotopically and orthotopically to evaluate the effect of terminal warm-blood cardioplegia with or without leukocyte depletion.[119] Preload recruitable stroke work was measured before harvesting (control) and after transplantation, as a load-insensitive index of myocardial function. The heterotopic hearts were divided into four groups according to the preservation method. Hearts were preserved for 3 hours in group 1 and for 24 hours in groups 2,3 and 4. Terminal warm-blood cardioplegia with leukocyte depletion was used in group 4. Among the heterotopic transplants, the percentage of baseline recovery of preload recruitable stroke work was highest in group 4 (91 ± 6% of baseline). To evaluate the effect of leukocyte-depleted terminal blood cardioplegia on prolonged preservation, 41 canine hearts were stored in modified Euro-Collin's solution and transplanted heterotopically.[120] Leukocyte-depleted terminal blood cardioplegia preserved the percentage of preload recruitable stroke work and diastolic compliance after transplantation compared with preharvesting values. Coronary blood flow was not different from controls. There was no significant detection of malondialdehyde production during terminal blood cardioplegia and 10 minutes after aortic unclamping. ATP content increased to the preharvesting value in animals treated with leukocyte-depleted terminal warm cardioplegia. In a randomized, double-blinded clinical trial in transplanted human hearts, Pearl et al. demonstrated that the administration of leukocyte-depleted terminal warm blood cardioplegia reduced creatinine phosphokinase release in the coronary effluent, reduced thromboxane B2 release from the heart and the perioperative duration of inotropic support.[121] All hearts functioned adequately in the control and treatment groups. Leukocyte-depleted reperfusion decreased biochemical evidence of reperfusion injury, but did not significantly enhance myocardial function as compared to controls. This may have been attributed to the short ischemic duration (3 hours) of the study.

Myocardial Cellular Protection from "Stone Heart" (Cardiomyocyte Hypercontracture): Irreversible Ischemia-Reperfusion Injury

In ischemic myocardium, contracture develops by means of a rigor-type mechanism. A force generating cross bridge cycling is initiated when cytosolic ATP is reduced to a low (<100 µmol/L) but not zero level.[122,123] In ischemia, this window of low cytosolic ATP concentrations is open only during brief period, because cellular ATP reserves are quickly exhausted. The myofibrillar shortening then stays fixed, as all cross bridges between actin and myosin remain in an attached state. When energy depletion is rapidly relieved, ischemic rigor contracture is usually reversible.[122]

After prolonged ischemia myocardial cells may develop severe contracture, which can lead to cytoskeletal defects that consequently increase the fragility of cardiomyocytes upon reperfusion. Substantial contracture is accompanied by a specific form of tissue necrosis, the so-called "contraction band necrosis".[122,124] The histologic picture is characterized by coexistence of supercontracted sarcomeres, overextension of spaces in between and sarcolemmal disruptions, all in the same cells resulting from strong and inhomogeneous mechanical forces leading to cardiomyocyte necrosis.

Reperfusion-induced myocardial cell contracture can have two different causes: (1) Ca^{2+} overload-induced hypercontracture and (2) rigor-type contracture. Ca^{2+} overload-induced contracture results from rapid re-energization of contractile cells with a persistent Ca^{2+} overload during ischemia. High cytosolic Ca^{2+} plus energy leads to uncontrolled activation of the contractile machinery and calcium-dependent enzymes such as calpains, endonucleases, ATPases and phospholipases, leading to cardiomyocyte necrosis. When hyper contracture affects the entire heart, as may occur immediately after reperfusion following global prolonged severe ischemia, the rapidly developing lethal structural damage has been termed as the "stone heart" phenomenon, histologically characterized by hypercontracted myofibrils and ruptured cellular membranes.[122]

Ischemic cells become energy-depleted and subsequently develop a Ca^{2+} overload of the cytosol. Cardiomyocytes in ischemia accumulate H^+ from anaerobic glycolysis and Na^+ by means of (1) the Na^+/H^+ exchanger, (2) the $Na^+/HCO3-$ symporter, (3) other routes. With reduction of the Na^+ gradient and membrane depolarization, the sarcolemmal Na^+/Ca^{2+} exchanger is turned into its "reverse mode", which finally leads to cytosolic accumulation of Ca^{2+}. If the ability of mitochondria to resume ATP synthesis is not critically impaired during the ischemic period, reoxigenation leads to a rapid recovery of energy production. Resynthesis of ATP can enable cardiomyocytes to recover promptly from the loss of cytosolic cation balance as energy recovery reactivates $N^+–K^+–ATPase$ and restores the Na^+ gradient and the membrane potential. The "forward mode" of the Na^+/Ca^{2+} exchanger thus extrudes excess cytosolic Ca^{2+}. However, it also reactivates the contractile machinery with resynthesis of ATP that had been fixed in ischemic rigor contracture. The latter effect is normally faster than the former, which leads to an uncontrolled Ca^{2+} dependent contraction.[122]

In detailed studies of early reperfusion by Piper et al, Ca^{2+} level declines rapidly within a minute but starts to oscillate during the first five minutes of reperfusion caused by the cyclic uptake and release of Ca^{2+} by the sarcoplasmic reticulum. During this period, the transsarcolemmal Na^+ gradient is still reduced and the Na^+/Ca^{2+} exchanger still operates in reverse mode. It is during the oscillations when further uncontrolled extensive myofibrillar hypercontracture develops.[122,125,126] Strategies to prevent this type of injury are directed experimentally at cytosolic Ca^{2+} control or myofibrillar Ca^{2+} sensitivity limited at the initial time of reperfusion. Agents used for the protocols to reduce the Ca^{2+} sensitivity of myofibrils are (1) phosphatase 2,3-butanedione monoxime,[127,128] (2) cGMP-mediated effectors (nitric oxide, atrial natriuretic peptides)[129-132] and (3) constant cytosolic acidosis (simultaneous proton transport inhibition of both the Na^+/H^+ exchanger and the $Na^+/HCO3-$ symporter).[125,133] Agents used for the protocols to reduce salcoplasmic reticulum-dependent Ca^{2+} oscillations are (1) Ca^{2+} –ATPase inhibitor on salcoplasmic reticulum, (2) Ca^{2+} release inhibitor on salcoplasmic reticulum and (3) the anesthetic halothane or intracellular acidosis.[122,125,126,133]

In contrast to Ca^{2+} overload-induced hypercontracture, rigor-type contracture occurs during reoxigenation as well as during ischemia if reenergization of the ischemic cardiomyocytes by mitochondrial ATP production proceeds very slowly, independent of Ca^{2+} overload.[122] In comparison to during ischemia, upon re-oxygenation cardiomyocytes may spend much more time at the window of low cytosolic ATP suitable to induce reperfusion- induced rigor-type contracture. Thus, cell shortening can be much more pronounced in reperfusion-induced rigor-type contracture than observed in ischemia-induced rigor contracture. It may be prevented by strategies improving early mitochondrial reactivation for energy production. Agents used for these protocols are (1) mitochondrial oxidative energy production substrates (succinate) and (2) protection of mitochondria from compulsory calcium uptake.[122]

Modifying Substrate Availability: Substrate Enhancement

L-glutamate and L-aspartate

Substrate enhancement with the amino acids, L-glutamate and L-aspartate, when added to both cardioplegia and reperfusate solutions after periods of ischemic arrest, have been shown to attenuate ischemia-reperfusion injury and improve myocardial performance.[134-137] The rationale for the use of L-aspartate and L-glutamate as additives to cardioplegic solutions is based on the alterations in cellular metabolism that occur during ischemia and reperfusion. During ischemia, the myocardium progresses from aerobic metabolism to anaerobic glycolysis, accompanied by a depletion of ATP and reduction in tissue L-glutamate and L-aspartate levels.[138-140] The protective effects of L-glutamate appear to arise from its ability to form important Krebs cycle intermediates that are depleted during ischemia.[141] The importance of L-glutamate is highlighted by the observation that it is the only amino acid extracted by the normal myocardium. Further, the myocardium of patients with coronary artery disease extracts more L-glutamate than does the myocardium of individuals with normal coronary arteries.[142] Clinical studies have also shown that the heart extracts glutamate and exports alanine in response to oxygen deprivation.[143] During periods of ischemia, succinic acid, a Krebs cycle intermediate, becomes depleted in the myocardium, leading to derangements in mitochondrial metabolism and decreased ATP production. Periods of anoxia have been reported to stimulate the metabolic conversion of L-glutamate to succinate.[144] During ischemia, glutamate enters into a transamination reaction with pyruvate to form alanine and alpha-ketoglutarate, a Krebs cycle intermediate. Alpha-ketoglutarate can be oxidized in the mitochondria by the alpha-ketoglutarate dehydrogenase and succinyl coenzyme A synthetase reactions to form succinate and GTP.[145] Similarly, aspartate is transaminated in the presence of alpha-ketoglutarate to form glutamate and oxaloacetate. The oxaloacetate enters the Krebs cycle and is reduced to malate, accompanied by oxidation of NADH. Oxidized nicotinamide-adenine dinucleotide (NAD) can enter the cytoplasm and promote ATP production by means of glycolysis. Further support of the transamination hypothesis comes from studies demonstrating that the protective effect of L-glutamate can be blocked by adding aminooxyacetic acid, an inhibitor of mitochondrial transaminases.[146] Haan et al. reported on a study to determine whether substrate enhancement with

L-glutamtate during periods of cold storage would improve ventricular function in transplanted hearts. Following excision of donor hearts and hypothermic storage for 3 hours, hearts were reperfused using a Langendorff preparation. Hearts were treated in one of five ways: recieved no L-glutamate, L-glutamate before ischemia, L-glutamate during storage, during reperfusion, or during both storage and reperfusion. Only hearts supplemented with L-glutamate during reperfusion or during storage and reperfusion demonstrated improved functional recovery.[147] Choong et al. reported on the effects of supplementing oxygenated St. Thomas' cardioplegia solution with L-aspartate and/or D-glucose for long-term preservation of excised rat hearts following arrest and preservation for 20 hours of hypothermic low-flow perfusion.[148] Following normothermic reperfusion with a crystalloid perfusate, hearts treated with L-aspartate or L-aspartate plus D-glucose, but not D-glucose alone, demonstrated significant improvement in aortic flow rates, increased levels of myocardial ATP, a reduction in cellular uptake of sodium and calcium and a reduction in ultrastructural damage.[148] In an earlier study, Choong et al. demonstrated in a rat model that following 10 hours of hypothermic storage, hearts preserved with St. Thomas' solution enriched with L-aspartate or L-aspartate and L-glutamate, but not L-glutamate alone, recoverd 93% of aortic flow rates, 95% of coronary flow and 88% of developed aortic pressure compared to baseline.[149] Tixier et al. demonstrated that administration of warm substrate-enriched (L-glutamate) blood cardioplegia for induction of myocardial arrest during heart procurement restored the levels of high-energy substrates that were depleted in an experimental model of brain-death.[150]

L-arginine, Nitroglycerine, Sodium Nitroprusside (Nitric Oxide Donors)

Numerous studies reported that the L-arginine-nitric oxide (NO) pathway plays an important role in ischemia- reperfusion injury.[151] In the heart, NO is synthesized from oxygen and L-arginine by coronary endothelium,[152] cardiac myocytes[153] and endocardial cells.[154] It has a host of physiologic effects, including autoregulatory modulation of coronary blood flow,[155,156] superoxide scavenger and inhibition of neutrophil-endothelial interaction[157,158] and platelet aggregation. NO is shown to have a beneficial effect by increasing post-ischemic blood flow and decreasing the "no-reflow" phenomenon,[159] decreasing leukocyte adhesion and expression of cell adhesion molecules[160] and quenching free radicals, particularly superoxide radicals.[161-164] The benefit of supplementation of NO donors, L-arginine or nitroglycerin in cardioplegia on post-ischemic ventricular performance and endothelial function has been well established.[165-167] However, some authors report deleterious effects of NO during cardiac ischemia and reprefusion probably due to the inhibition of mitochondrial respiration or formation of peroxynitrite, a precursor of a highly oxidant species, by the reaction of NO with superoxide.[151,168,169]

Specific Channel Agonists or Antagonists

Adenosine

Endogenously occurring adenosine is formed by the actions of the membrane bound and cytosolic enzyme 5'-nucleotidase, which dephosphorylates adenosine monophosphate (AMP) to adenosine.[170] Formation of adenosine can also occur from the hydrolysis of S-adenosylhomocysteine (SAH) to adenosine and homocysteine by the actions of the enzyme, SAH hydrolase. However, this enzymatic reaction favors the formation of SAH, thereby forming a large intracellular pool of adenosine that is bound to SAH. There are three pathways by which adenosine breakdown is facilitated. These include phosphorylation by adenosine kinase to AMP, degradation to inosine by adenosine deaminase and washout into the circulation. Phosphorylation by adenosine kinase to AMP is the enzymatically preferred reaction and this pathway becomes an important route for purine salvage by the myocardium. Other pathways of purine salvage include breakdown of adenosine to inosine, which is further metabolized to hypoxanthine, phosphorylated to inosine monophosphate (IMP) and ultimately aminated to AMP. Adenosine can also be converted to adenine and then subsequently ribosylated to AMP.

Its cardiovascular actions are mediated primarily by activating membrane receptors coupled to G-proteins that differentially couple to adenylate cyclase. These cardiovascular actions include vasodilatory effect proposed as a central mechanism of autoregulation of coronary blood flow, negative inotropic, chronotropic and dromotropic effects, anti-neutrophil effects and inhibitory effects on the inflammatory response and cardioprotective effects against ischemia-reperfusion injury during all three windows of treatment opportunity (pretreatment, ischemia and reperfusion).[171] There are at least four receptor subtypes: A1, A2A, A2B, A3 receptors. The A1 and A3 subtypes are coupled to inhibitory G-proteins (Go, Gi) that inhibit adenylate cyclase. The A1 and A3 receptor subtypes have been well established to be the trigger and the mediator of ischemic preconditioning, most importantly being related to the opening of mitochondrial membrane ATP-sensitive potassium channels (K-ATP) through the protein kinase C (PKC) signaling pathway. They are coupled to kinases including PKC with resultant sarcolemmal K-ATP channels' opening and transmembrane hyperpolarization and tyrosine kinases with resultant phosphorylation and desensitization of those receptors after prolonged exposure to agonists. The A1 receptor activation is also the mediator of conductant cardiomyocyte membrane hyperpolarization and bradycardia by the same mechanisms involving opening of K-ATP channel. Activation of the A1 receptor subtype in cardiomyocytes may reduce the generation of reactive oxygen species, promote glycolysis and improve anaerobic metabolism and suppress the L-type calcium channels during ischemia-reperfusion.[171-176] The A2 receptor subtype is coupled to stimulatory Gs proteins that stimulate adenylate cyclase, leading to an increase in cAMP. Stimulation of the A2A receptor cause vasodilatation in vascular smooth muscle and may be associated with positive inotropy, mediated by increased cAMP levels. While A1 receptor subtype mediated cardioprotective effects of adenosine predominate during pretreatment and ischemia, cardioprotective effects of adenosine during early reperfusion are the result of activation of A2A receptor subtype augmented by simultaneous activation of A2B receptor subtype. Activation of the A2B receptor subtype has been shown to be involved in the cardioprotection at the time of reperfusion by the mechanism of ischemia postconditioning.[171,172] Adenosine has been shown to inhibit neutrophil-endothelial cell interactions during ischemia-reperfusion. Activation of the A2 receptor subtype may directly inhibit the generation of reactive oxygen species, superoxide, by PAF stimulated netrophils during ischemia-reperfusion resulting in the inhibition of neutrophil-endothelial adhesion, interaction and subsequent endothelial dysfunction, although this "neutrophil theory" has been challenged by the evidences that suggest this cardioprotection of adenosine is a direct effect of stimulating A2 receptor subtypes on isolated cardiomyocytes rather than neutrophils.[171,172,177-179] Although there has been no direct evidence showing that A2 receptor subtypes activation can modulate Ca^{2+} dynamics at reperfusion, adenosine enhances NO production through endothelial NO synthase (eNOS) in vascular smooth muscle cells and cardiomyocytes by activating A2A receptor subtype. The protective effect of adenosine against mitochondrial oxidant damage is prevented by blockade of NO synthesis and inhibition of PKG that has shown to attenuate Ca^{2+} induced contracture.[172,180-182]

The phosphatidylinositol-3-kinase (PI3-kinase)/Akt pathway is a cell survival mechanism that attenuates apoptosis and regulates glycogen synthesis and glucose transport. Akt is activated subsequent to the production of phosphatidylinositol-3, 4, 5, -triphosphate (PIP3) by PI3-kinase. Akt is a serine/threonine kinase that mediates several functions through the phosphorylation and inactivation of the pro-apoptotic kinase, glycogen synthase kinase-3 (GSK-3 α/β). It also plays an essential role in cardioprotection against ischemia-reperfusion injury.[172,183,184] Hausenloy and Yellon proposed that pharmacological manipulation and activation of the antiapoptotic PI3-kinase/Akt and extracellular signal regulated kinase (ERK) pathways early in reperfusion prevent reperfusion-induced injury, thereby salvaging myocardium and limiting infarct size.[185] Adenosine A2 receptor agonists activate the PI3-kinase/Akt pathway evidenced by the blockage of this cardioprotective effect by the PI3-kinase inhibitors.[186] Adenosine induced NO generation through A2A receptor subtype is partially mediated by activation of the PI3-kinase/Akt pathway, which may contribute to the protective effect of adenosine against mitochondrial oxidant damage. The PI3-kinase/Akt pathway is likely to be a critical target of A2 receptor activation at reperfusion that may mediate cardioprotection.[172]

ERK is a member of the mitogen-activated protein kinase (MAPK) family and plays an important role in the regulation of cell proliferation, differentiation and survival.[172] ERK influences cardioprotection against ischemia-reperfusion by reducing apoptosis[187] and infarct size.[188] Both ischemic preconditioning and opioid induced preconditioning promoted phospholylation of cytosolic ERK during reperfusion in rat hearts and the selective MAPK/ERK kinases inhibitor blocked the cardioprotective effect of preconditioning, implying that ERK activation during reperfusion is critical if preconditioning is to protect against ischemia-reperfusion injury.[172,189] Experimental evidences showed ERK might serve as a downstream target of A2 receptor activation and might be an important element of the signaling pathway of the reperfusion injury salvage kinase (RISK)-pathway leading to cardioprotection from A2 receptor activation at reperfusion.[172,185,186,190]

Mitochondria have recently emerged as a central regulator of cell death either by necrosis or apoptosis in a variety of disease states.[191,192] Under physiological conditions, the mitochondrial inner membrane is relatively impermeable to metabolites and ions. However, under stress conditions that produce reactive oxygen species and elevated cytosolic Ca^{2+}, a nonselective, high-conductance pore known as the mitochondrial permeability transition pore (MPTP) opens in the mitochondrial inner membrane allowing free passage of molecules up to 1.5 kDa, including protons, Ca^{2+} and water leading to release of proapoptotic cytochrome C with caspase activation and cessation of ATP production.[193,194] Some investigators have proposed that MPTP opening is the lethal event in reperfusion injury and cardioprotective interventions against reperfusion injury should be targeted to the prevention of MPTP opening.[194,195] Experimental evidence in isolated rat cardiomyocytes showed that adenosine A2 receptor activation might protect the heart at reperfusion by preventing MPTP opening and the subsequent loss of mitochondrial membrane potential caused by oxidative stress through the mechanism of NO generation by endothelial NO synthase via A2A receptor activation.[172,181]

Prevention of hypercontracture occurring in the first minutes of reperfusion is believed to be critical for cardioprotection against reperfusion injury.[182] Experimental evidence in rabbit hearts showed that delaying the onset of administration of adenosine A2 receptor agonist to just 10 minutes after reperfusion completely abolished its cardioprotective effect to reduce infarct size and that treatment of adenosine A2 receptor agonist starting at reperfusion and lasting for 30 or 40 minutes were not protective compared to lasting for 70 minutes reduced infarction.[196] Therefore, it is underscored that the proper timing and duration of the treatment is extremely important to produce the cardioprotective effect of adenosine receptors activation against ischemia-reperfusion injury.

Na-H Exchange Porter Antagonist: Cariporide

The cardioprotective effect of Na-H exchanger inhibitors was demonstrated when administered perioperatively to the patients undergoing coronary artery bypass graft surgery and when added into blood cardioplegia at the time of the induction of ischemic arrest. The decreased accumulation of intracellular Na^+ and subsequent Ca^{2+} overload contributed to the reduced post-ischemic infarct size and tissue edema due to the supplementation of the selective NHE type-1 isoform inhibitor, cariporide. Cariporide also showed the preservation effect on endothelial function in a canine model. However, evidences showed that optimal benefits were seen only when the treatment was administered prior to the onset of ischemia.[56,197-200]

K-ATP Channel Opener: Pinacidil, Nicorandil, Aprikalim/mK-ATP Channel Opener

Diazoxide, BMS-180448 (Cardiac Selective)

ATP sensitive potassium channel openers when added to hyperkalemic cardioplegia, reduced the NA^+-Ca^{2+} exchange outward current elevated by hyperkalemia and might attenuate calcium overload, leading to improved contractile function after cardioplegic ischemic arrest in rat heart.[201] It also showed endothelial protective effect when added to hyperkalemic cardioplegia by preserving EDHF mediated coronary relaxation.[202] BMS-180448 is a cardiac selective mK-ATP channel opener, mimics the phenomenon of ischemic preconditioning and does not cause systemic hypotension even with intravenous administration. Intravenous administration of BMS-180448 prior to coronary ligation showed to reduce myocardial infarct size by approximately 50% in a dog infarct model.[203] In isolated rat working hearts model, pretreatment of BMS-180448 in combination with the Na/H exchange inhibitor, cariporide resulted in additive cardioprotection with significantly recovered cardiac function from prolonged (6 hours) deep hypothermic storage in an extracellular-based cardioplegic solution, but BMS-180448 only did not improve significant recovery of cardiac function.[204] Furthermore, in a porcine brain-dead heart transplant model, addition of BMS-180448 with the Na/H exchange inhibitor, cariporide showed no significant additional benefit.[205] The process of brain death could possibly activate the mK-ATP channel through endogenous ischemic preconditioning and there was an experimental evidence that showed cardioplegic solutions that contain high concentrations of potassium and magnesium might mediate their cardioprotective effect in part through the activation of mK-ATP channels.[206,207] Therefore, exogenous administration of mK-ATP channel opener may have limited benefit in a setting where the mK-ATP channel is already activated by myocardial ischemia from brain death and cardioplegia.[208]

Miscellaneous

Brief episodes of ischemia-reperfusion protect the myocardium from damage induced by subsequent more prolonged ischemia. When first described by Murry et al[209] such ischemic pre-conditioning was elicited by brief coronary occlusion and the endpoint was reduced infarct size. Since then, a variety of preconditioning stimuli have been uncovered including hypoxia, rapid cardiac pacing, thermal stress, stretch and various pharmacological agents.[210] Receptor-dependent or receptor-independent endogenous or exogenous triggers of preconditioning that would be expected to reduce ischemia-reperfusion cardiac injury when added to cardioplegia solutions include: (1) (receptor-dependent triggers of preconditioning) adenosine, bradykinin, acetylcholine, opioids (morphine, δ1-opioid), prostaglandins, norepinephrine, angiotensin and endothelin; and (2) (receptor-independent triggers of preconditioning) free radicals[211] and volatile anesthetic agents (generate small amounts of reactive oxygen species that trigger the preconditioning secondary messenger pathways).[211,212]

Apoptotic cell death, which was shown to contribute to the myocyte death sustained during ischemia-reperfusion injury, might be accelerated during the reperfusion period. The pro-survival PI3K-Akt and Erk 1/2 kinase cascades were activated in response to ischemia-reperfusion injury and initiated myocardial protection through their anti-apoptotic actions. Growth factors, which activate these pro-survival PI3K-Akt and Erk 1/2 kinase cascades during the first few minutes of reperfusion following a lethal ischemic insult, were expected to attenuate reperfuion-induced cell death. Activation of these pro-survival kinase cascades prior to ischemia was associated with the profound cardio-protection induced by the phenomenon of ischemic preconditioning, suggesting perhaps that agents which activate these signaling pathways, should be able to provide protec-tion at time of reperfusion and also precondition the myocardium.[213] Growth-factor-mediated up-regulation of these pro-survival kinase cascades at reperfusion was demonstrated to protect the heart against reperfusion injury. Furthermore, other agents such as HMG-Co-A reductase inhibitors and G-protein-coupled receptor ligands such as bradykinin were also shown to initiate cardio-protection at the time of reperfusion by activating these pro-survival kinase cascades.[213]

Alternative Strategies to Hypothermic Ischemic Storage

Other alternatives to simple hypothermic ischemic storage include continuous or intermit-tent extracorporeal hypothermic perfusion or microperfusion, beating continuous extracor-poreal normothermic perfusion and coronary oxygen persufflation. Proctor and Parker first demonstrated in 1968 the ability of a mechanical perfusion pump to preserve canine hearts for 72 hours ex vivo. Advances in extracorporeal hypothermic perfusion have permitted clinical application of this technique in 1981.[214] The superiority of perfusion versus cold storage for hypothermic heart preservation has been shown in numerous studies. However, this technique is considerably more complicated and except for a few scattered clinical reports of its use, has not been routinely utilized for preservation of thoracic organs.[214] Extracorporeal hypothermic perfusion of donor heart provides a continuous supply of oxygen and substrates while removing metabolic byproducts. Its greatest potential utility is its ability to provide a longer duration of donor heart preservation, as compared to simple hypothermic ischemic storage. The efficacy of continuous hypothermic perfusion is dependent on a number of variables including perfusion pressure and composition of the preservation solution.[215,216] Wicomb and colleagues successfully performed allo-transplantation of baboon hearts following 24 hours or 48 hours of preservation using hypothermic perfusion.[217,218] Experimental animals survived until rejection or elective euthanasia. All animals demonstrated normal hemodynamic function on cardiac catheterization and normal or only minor ischemic changes on myocardial histologic examination.[217,218] Guerraty et al reported on the ability to perform orthotopic canine transplantation following 24 hours of continuous hypothermic, low-pressure (22cm H2O) perfusion preservation of the donor heart.[219] The hypothermic perfusion was compared to a group of control animals receiving a heart removed by the same cardiectomy technique, but transplanted immediately. Following transplantation, all grafts from both experimental groups supported the recipient circulation after weaning from cardiopulmonary bypass. There was no difference in survival or graft function when studied at 5 to 10 days following operation.[219]

Microperfusion is the modified technique of the administration of the flush solution with a very low flow and perfusion pressure to alleviate the deleterious effect of hypothermia toward tissue edema in continuous hypothermic perfusion method. Wicomb et al described the technique of hypothermic microperfusion.[220] Donor hearts were perfused at 3-6 ml/g/24 hour using a modified UW solution. Following 24 hours of preservation, donor hearts functioned at a level 121% above baseline. In an experimental rat model comparing hypothermic preservation by ischemic storage vs low-flow perfusion, Larese et al demonstrated that the ratio of ATP to inorganic phosphate (a measure of cellular acidosis), as assessed by 31P-NMR spectroscopy, was significantly greater (less cellular acidosis) following 15 hours of low-flow hypothermic perfusion as compared to hypothermic ischemic storage.[221] Ferrera et al reported in a comparison of continuous hypothermic microperfusion with simple cold ischemic storage, that microperfusion proved superior to simple cold storage after 24 hours of preservation in an ex vivo pig model.[222] These investigators noted better ventricular compliance and higher mean left ventricular developed pressure in a 24-hour preservation with continuous hypothermic microperfusion than that with a simple cold immersion storage, whereas in a 6-hour preservation, no functional difference was observed between the continuous hypothermic microperfusion and the simple cold storage. They also noted higher levels of ATP and total adenine nucleotides in the biopsy specimens in a 6-hour preservation with continuous hypothermic microperfusion than that with a simple cold storage. In a later comparison of continuous perfusion, microperfusion, or simple storage, Ferrera et al again demonstrated the superior performance of perfusion techniques compared to simple storage.[223] However, there was no significant difference between continuous perfusion and microperfusion techniques. Zhu et al reported on the technique of hypothermic intermittent perfusion.[224] During storage, hearts were perfused intermittently at a perfusion pressure of 60 mmHg for 3 min at 25°C with an oxygenated cardioplegia solution. Hearts underwent perfusion every 4, 6, 8, 10, 11 hours, or at 10 and 17 hours of storage, only. The best preservation was obtained with intermittent perfusion at 10 and 17 hours of storage and every 10 hours. Karck et al compared simple storage to intermittent perfusion or continuous perfusion with either St. Thomas' solution or UW solution and found in the group receiving St. Thomas's solution, better preservation with intermittent perfusion every 90 min compared to simple storage or continuous perfusion.[225] In contrast, hearts preserved with UW solution demonstrated better performance following simple storage compared to all other groups. These authors postulated that intermittent or continuous infusion of UW solution was injurious to the myocardium as a result of potassium overloading.

Beating, warm, oxygenated and nutrient-rich, leukocyte and platelet depleted donor blood perfusion method was aimed toward zero ischemia and was designed to maintain and improve organ function. This method is expected to allow continuous clinical evaluation and to provide the increased amount of time that an organ can remain outside of the body. This method is hoped to enable the utilization of marginal donor organs that are currently consented but discarded due to questionable viability and function by resuscitating the organ outside of the body and by reassessing fully functional, biochemical and metabolic parameters. There is a commercially available portable organ transport device called the Organ Care System by TransMedics, Inc. (Boston, MA).[226] This is a sterile, portable organ transport system which consists of oxygenated pulsatile pump circuit system with improved biocompatibility, nutrient supplier, sample drawer, robust platform and wireless monitor. This device was clinically applied in Germany and UK in 2006 and both cases were reported successful. Currently PROTECT (the Prospective multi-center European Trial to Evaluate the Safety and Performance of the Organ Care System for Heart Transplants) clinical trial is underway in four institutions in Germany and UK.[226]

Conclusion

Cardiac allograft procurement and preservation, that is, protection from the sequelae of ischemia-reperfusion and inflammatory injury is the time-limited practical necessity for transportation and allocation of the donor heart. Successful preservation of the donor heart during procurement and implantation is indispensable to the success of heart transplantation. Current practice of the single-flush, simple ischemic hypothermic immersion storage method with the aid of cardioplegia is the time-proven, gold standard of heart preservation. It allows approximately 3 hours (4 to 6 hours at maximum) of safe, reliable ischemic storage period resulting in recipient survival rates of 90% or more at one year and approximately 50% at 10 years.

Recent substantial progresses in understanding of the mechanisms of allograft injury have brought forth a number of emerging and resurgent alternative methods of heart preservation. These include continuous or intermittent extracorporeal hypothermic perfusion or microperfusion, beating continuous extracorporeal normothermic perfusion and coronary oxygen persufflation. These alternatives are to reduce ischemia-reperfusion injury and inflammation, to further improve both short-term and long-term outcomes and to prolong storage time in order to improve donor pool and immunological matching.

Among them, beating, warm, oxygenated and nutrient-rich, leukocyte and platelet depleted donor blood perfusion method aims at ischemia. It is designed to allow continuous clinical evaluation and to provide the increased amount of time that an organ can remain outside of the body. This method is hoped to enable the utilization of marginal donor organs that are currently consented but discarded due to questionable and function by resuscitating the organ outside of the body and by reassessing fully functional, biochemical and metabolic parameters. Currently PROTECT clinical trial is underway in four institutions in Germany and UK.

References

1. Taylor DO, Edwards LB, Boucek MM et al. Registry of the International Society for Heart and Lung Transplantation: twenty-third official adult heart tranplantation report—2006. J Heart Lung Transplant 2006; 25:869-879.
2. Taylor DO, Edwards LB, Boucek MM et al. The registry of the International Society for Heart and Lung Transplantation: twenty-first official adult heart tranplant report—2004. J Heart Lung Transplant 2004; 23:796-803.
3. Kirklin JK, Naftel DC, Bourge RC et al. Evolving trends in risk profiles and causes of death after heart transplantation: a ten-year multi-institutional study. J Thorac Cardiovasc Surg 2003; 125:881-890.
4. The 2005 Organ Procurement and Transplantation Network (OPTN)/Scientific Registry of Transplant Recipients (SRTR) Annual Report HYPERLINK "http://www.ustransplant.org/annual_ reports/current/ data_tables.htm" http://www.ustransplant.org/annual_reports/current/data_tables.htm
5. Novitzky D, Wicomb WN, Cooper DKC et al. Electrocardiographic, hemodynamic and endocrine changes occurring during experimental brain death in the Chacma baboon. J Heart Transplant 1984; 4:13-19.
6. Novitzky D, Horak A, Cooper DK et al. Electrocardiographic and histopathologic changes developing during experimental brain death in the baboon. Transplant Proc 1989; 21:2567-2569.
7. Wheeldon DR, Potter CD, Oduro A et al. Transforming the "unacceptable" donor: outcomes from the adoption of standardized donor management technique. J Heart Lung Transplant 1995; 4:734-742.
8. Zaroff JG, Rosengard BR, Armstrong WF et al. Consensus Conference Report. Maximizing use of organs recovered from the cadaver donor: cardiac recommendations. Circ 2002; 106:836-841.
9. Rosendale JD, Chabalewski FL, McBride MA et al. Increased transplanted organs from the use of a standardized donor management protocol. Am J Transpl 2002; 2:761-768.
10. Rosendale JD, Kauffman HM, McBride MA et al. Hormonal resuscitation yields more transplanted hearts, with improved early function. Transplantation 2003; 75:1336-1341.

11. Rosendale JD, Kaufffman HM, McBride MA et al. Aggressive pharmacological donor management results in more transplanted organs. Transplantation 2003; 75:482-487.
12. Terasaki PI, Cecka JM, Gjertson DW et al. High survival rates of kidney transplants from spousal and living unrelated donors. N Engl J Med 1995; 333:333-336.
13. Tillius SG, Hemann U, Hancock WW et al. Long-term kidney isografts develop functional and morphologic changes that mimic those of chronic allograft rejection. Ann Surg 1994; 220:425-435.
14. Takada M, Nadeau KC, Hancock WW et al. Effects of explosive brain death on cytokine activation of peripheral organs in the rat. Transplantation 1998; 65(12):1533-1542.
15. Wilhelm MJ, Pratschke J, Beato F et al. Activation of the heart by donor brain death accelerates acute rejection after transplantation. Circulation 2000; 102:2426-2433.
16. Segel LD, vonHaag DW, Zhang J et al. Selective overexpression of inflammatory molecules in hearts from brain-dead rats. J Heart Lung Transplant 2001; 21:804-811.
17. Szabo G, Buhmann V, Bahrle S et al. Brain death impairs coronary endothelial function. Transplantation 2002; 73:1846-1848.
18. Shivalkar B, Van Loon J, Wiland W et al. Variable effects of explosive or gradual increase of intracranial pressure on myocardial structure and function. Circulation 1993; 87:230-239.
19. Anyanwu AC, Banner NR, Mitchell AG et al. Low incidence and severity of transplant-associated coronary artery disease in heart transplants from live donors. J Heart Lung Transplant 2003; 22:281-286.
20. Mehra MR, Uber PA, Ventura HO et al. The impact of mode of donor brain death on cardiac allograft vasculopathy. J Am Coll Cardiol 2004; 43:806-810.
21. Ramzy D, Rao V, Brahm J et al. Cardiac allograft vasculopathy: a review. Can J Surg 2005; 48:319-327.
22. Rahmani M, Cruz RP, Granville DJ et al. Allograft vasculopathy versus atherosclerosis. Circ Res 2006; 99:801-815.
23. Valantine HA. Cardiac allograft vasculopathy: central role of endothelial injury leading to transplant "atheroma". Transplantation 2003; 76:891-899.
24. Gaudin PB, Rayburn BK, Hutchins GM et al. Peritransplant injury to the myocardium associated with the development of accelerated arteriosclerosis in heart transplant recipients. Am J Surg Pathol 1994; 18:338-346.
25. Land W, Schneeberger H, Schleibner S et al. The beneficial effect of human recombinant superoxide dismutase on acute and chronic rejection events in recipients of cadaveric renal transplants. Transplantation 1994; 57:211-217.
26. Land W, Messmer K. The impact of ischemia/reperfusion injury on specific and nonspecific, early and late chronic events after organ transplantation. Transplant Rev 1996; 10:108-127.
27. Land WG. The role of postischemic reperfusion injury and other nonantigen-dependent inflammatory pathways in transplantation. Transplantation 2005; 79:505-514.
28. Faulk WP, Labarrere CA, Pitts D et al. Vascular lesions in biopsy specimens devoid of cellular infiltrates: qualitative and quantitative immunocytochemical studies of human cardiac allografts. J Heart Lung Transplant 1993; 12:219-229.
29. Labarrere CA, Nelson DR, Spear KL. Non-immunologic vascular failure of the transplanted heart. J Heart Lung Transplant 2003; 22:236-240.
30. Laberrere CA, Nelson DR, Faulk WP. Myocardial fibrin deposits in the first month after transplantation predict subsequent coronary artery disease and graft failure in cardiac allograft recipients. Am J Med 1998; 105:207-213.
31. Labarrere CA. Relationship of fibrin deposition in microvasculature to outcomes in cardiac transplantation. Curr Opin Cardiol 1999; 14:133-139.
32. Labarrere CA, Nelson DR, Park JW. Pathologic markers of allograft arteriopathy: insight into the pathophysiology of cardiac allograft chronic rejection. Curr Opin Cardiol 2001; 16:110-117.
33. Labarrere CA, Nelson DR, Cox CJ et al. Cardiac-specific troponin I levels and risk of coronary artery disease and graft failure following heart transplantation. JAMA 2000; 284:457-464.
34. Labarrere CA, Deng MC. Microvascular prothrombogenicity and tranplant coronary artery disease. Transplant Immunology 2002; 9:243-249.
35. Labarrere CA. Anticoagulation factors as predictors of transplant-associated coronary artery disease. J Heart Lung Transplant 2000; 19:623-633.
36. Labarrere CA, Troy RJ, Nelson DR et al. Vascular antithrombin and clinical outcome in hearttransplant patients. Am J Cardiol 2001; 87:425-431.
37. Torry RJ, Labarrere CA, Torry DS et al. Vascular endothelial growth factor expression in transplanted human hearts. Transplantation 1995; 60:1451-1457.
38. Torry RJ, Bai L, Miller SJ et al. Increased vascular endothelial growth factor expression in human hearts with microvascular fibrin. J Mol Cell Cardiol 2001; 33:175-184.
39. Labarrere CA, Pitts D, Nelson DR et al. Vascular tissue plasminogen activator and the development of coronary artery disease in heart-transplant recipients. N Engl J Med 1995; 333:1111-1116.

40. Labarrere CA, Nelson DR, Faulk WP. Endothelial activation and development of coronary artery disease in transplanted human hearts. JAMA 1997; 278:1169-1175.
41. Labarrere CA, Nelson DR, Miller JM et al. Value of serum-soluble intercellular adhesion molecule-1 for the noninvasive risk assessment of transplant coronary artery disease, posttransplant ischemic events and cardiac graft failure. Circulation 2000; 102:1549-1555.
42. Mehra MR, Ventura HO, Chambers RB et al. The prognostic impact of immunosuppression and cellular rejection on cardiac allograft vasculopathy: time for a reappraisal. J Heart Lung Transplant 1997; 16:743-751.
43. Hearse DJ. Myocardial protection during ischemia and reperfusion. Mol Cell Biochem 1998; 186:177-184.
44. Hoffman JW Jr, Gilbert TB, Poston RS et al. Myocardial reperfusion injury: etiology, mechanisms and therapies. J Extra Corpor Technol 2004; 36:391-411.
45. Lower RR, Stofer RC, Hurley EJ et al. Successful homotransplantation of the canine heart after anoxic preservation for seven hours. Am J Surg 1962; 104:302-306.
46. Bigelow WG, Mustard WT, Evans JG. Some physiological concepts of hypothermia and their application to cardiac surgery. J Thorac Surg 1954; 28:463-480.
47. Greenberg JJ, Edmunds LH Jr, Brown RB. Myocardial metabolism and postarrest function in the cold and chemically arrested heart. Surgery 1960; 48:31-42.
48. Rosenfeldt FL, Hearse DJ, Cankovic-Darracott S et al. The additive protective effects of hypothermia and chemical cardioplegia during ischemic cardiac arrest in the dog. J Thorac Cardiovasc Surg 1980; 79:29-38.
49. Parolari A, Rubini P, Cannata A et al. Endothelial damage during myocardial preservation and storage. Ann Thorac Surg 2002; 73:682-690.
50. Hansen TN, Dawson PE, Brockbank GM. Effects of hypothermia upon endothelial cells: mechanism and clinical importance. Cryobioloby 1994; 31:101-106.
51. Watson DC, Reitz BA, Baumgartner WA et al. Distant heart procurement for transplantation. Surgery 1979; 86:56-59.
52. Hardesty RL, Griffith BP, Deeb GM et al. Improved cardiac function using cardioplegia during procurement and transplantation. Transplantation Proc 1983; 15:1253-1255.
53. Swanson D, Dufek JH, Kahn DR. Myocardial preservation for transplantation. Transplant Proc 1979; 11:1478-1479.
54. Conti VR, Bertranou EG, Blackstone EH et al. Cold cardioplegia versus hypothermia for myocardial protection. Randomized clinical study. J Thorac Cardiovasc Surg 1978; 76:577-589.
55. Cohen NM, Damiano RJ Jr, Wechsler AS. Is there an alternative to potassium arrest? Ann Thorac Surg 1995; 60:858-863.
56. Yang Q, He GW. Effect of cardioplegic and organ preservation solutions and their components on coronary endothelium-derived relaxing factors. Ann Thorac Surg 2005; 80:757-767.
57. He GW, Yang Q, Wilson GJ et al. Tolerance of epicardial coronary endothelium and smooth muscle to hyperkalemia. Ann Thorac Surg 1994; 57:682-688.
58. He GW, Yang Q, Rebeyka IM et al. Effects of hyperkalemia on neonatal endothelium and smooth muscle. J Heart Lung Transplant 1995; 14:92-101.
59. He GW. Hyperkalemia exposure impairs EDHF-mediated endothelial function in the human coronary artery. Ann Thorac Surg 1997; 63:84-87.
60. He GW, Yang CQ, Yang JA. Depolarizing cardiac arrest and endothelium-derived hyperpolarizing factor-mediated hyperpolarization and relaxation in coronary arteries: the effect and mechanism. J Thorac Cardiovasc Surg 1997; 113:932-941.
61. Ge ZD, He GW. Altered endothelium-derived hyperpolarizing factor-mediated endothelial function in coronary microarteries by St Thomas' Hospital solution. J Thorac Cardiovasc Surg 1999; 118:173-180.
62. Ge ZD, He GW. Comparison of University of Wisconsin and St Thomas' Hospital solutions on endothelium-derived hyperpolarizing factor-mediated function in coronary micro-arteries. Transplantation 2000; 70:22-31.
63. Wakabayashi A, Tsunehiro N, Guilmette JE. Experimental evaluaiton of magnesium cardioplegia. J Thorac Cardiovasc Surg 1982; 84:685-688.
64. Stowe DF, Boban M, Graf BM et al. Contraction uncoupling with butanedione monoxime versus low calcium or high potassium solutions on flow and contractile function of isolated hearts after prolonged hypothermic perfusion. Circulation 1994; 89:2412-2420.
65. Stringham JC, Paulsen KL, Southard JH et al. Forty-hour preservation of the rabbit heart: optimal osmolarity, (Mg2+) and pH of a modified UW solution. Ann Thorac Surg 1994; 58:7-13.
66. Hudspeth DA, Nakanishi K, Vinten-Johansen J et al. Adenosine in blood cardioplegia prevents postischemic dysfunction in ischemically injured hearts. Ann Thorac Surg 1994; 58:1637-1644.

67. Lawton JS, Harrington GC, Allen CT et al. Myocardial protection with pinacidil cardioplegia in the blood-perfused heart. Ann Thorac Surg 1996; 61:1680-1688.
68. Harlan BJ, Ross D, Macmanus Q et al. Cardioplegic solutions for myocardial preservation: analysis of hypothermic arrest, potassium arrest and procaine arrest. Circulation 1978; 58:I114-I118.
69. Hearse DJ, O'Brien K, Braimbridge MV. Protection of the myocardium during ischemic arrest. Dose-response curves for procaine and lignocaine in cardioplegic solutions. J Thorac Cardovasc Surg 1981; 81:873-879.
70. Chambers DJ, Hearse DJ. Developments in cardioprotection: "polarized" arrest as an alternative to "depolarized" arrest. Ann Thorac Surg 1999; 68:1960-1966.
71. Hearse DJ. Polarization and myocardial protection. Curr Opin Cardiol 1999; 14:495-500.
72. Hearse DJ, Stewart DA, Braimbridge MV. Myocardial protection during ischemic cardiac arrest. The importance of magnesium in cardioplegic infusates. J Thorac Cardiovasc Surg 1978; 75:877-885.
73. Shakerinia T, Ali IM, Sullivan JA. Magnesium in cardioplegia: is it necessary? Can J Surg 1996; 39:397-400.
74. Yang ZW, Gebrewold A, Nowakowski M et al. Mg(2+)-induced endothelium-dependent relaxation of blood vessels and blood pressure lowering: role of NO. Am J Physiol 2000; 278:R628-R639.
75. Longo M, Jain V, Vedernikov YP et al. Endothelium dependence and gestational regulation of inhibition of vascular tone by magnesium sulfate in rat aorta. Am J Obstet Gynecol 2001; 184:971-978.
76. Yang Q, Liu YC, Zou W et al. Protective effect of magnesium on the endothelial function mediated by endothelium-derived hyperpolarizing factor in coronary arteries during cardioplegic arrest in a porcine model. J Thorac Cardiovasc Surg 2002; 124:361-370.
77. Wheeldon D, Sharples L, Wallwork J et al. Donor heart preservation survey. J Heart Lung Transplant 1992; 11:986-993.
78. Demmy TL, Biddle JS, Bennett LE et al. Organ preservation solutions in heart transplantation—patterns of usage and related survival. Transplantation 1997; 63:262-269.
79. Wang T, Batty PR, Hicks GL Jr et al. Long-term hypothermic storage of the cardiac explant. Comparison of four solutions. J Cardiovasc Surg (Torino) 1991; 32:21-25.
80. Wicomb WN, Hill DJ, Collins GM. Twenty-four-hour ice storage of rabbit heart. J Heart Lung Transplant 1994; 13:891-894.
81. Mollhoff T, Sukehiro S, VanAken H et al. Long-term preservation of baboon hearts. Effects of hypothermic ischemic and cardioplegic arrest on high-energy phosphate content. Circulation 1990; 82(Suppl 4):IV264-268.
82. Reichenspurner H, Russ C, Uberfuhr P et al. Myocardial preservation using HTK solution for heart transplantation. A multicenter study. Eur J Cardiothorac Surg 1993; 7:414-419.
83. Reichenspurner H, Russ C, Wagner F et al. Comparison of UW versus HTK solution for myocardial protection in heart transplantation. Transpl Int 1994; 7(Suppl 1):S481-484.
84. Garlicki M. May preservation solution affect the incidence of graft vasculopathy in transplanted heart? Ann Transplant 2003; 8:19-24.
85. Demertzis S, Wippermann J, Schaper J et al. University of Wisconsin versus St. Thomas' Hospital solution for human donor heart preservation. Ann Thorac Surg 1993; 55:1131-1137.
86. Jeevanandam V, Barr ML, Auteri JS et al. University of Wisconsin solution versus crystalloid cardioplegia for human donor heart preservation. A randomized blinded prospective clinical trial. J Thorac Cardiovasc Surg 1992; 103:194-198.
87. Wildhirt SM, Weis M, Schulze C et al. Effects of Celsior and University of Wisconsin preservation solutions on hemodynamics and endothelial function after cardiac transplantation in humans: a single-center, prospective, randomized trial. Transpl Int 2000; 13(Suppl 1):S203-211.
88. Stringham JC, Love RB, Welter D et al. Impact of University of Wisconsin solution on clinical heart transplantation. A comparison with Stanford solution for extended preservation. Circulation 1998; 98(Suppl II):II157-161.
89. Michel P, Vial R, Rodriguez C et al. A comparative study of the most widely used solutions for cardiac graft preservation during hypothermia. J Heart Lung Transplant 2002; 21:1030-1039.
90. Hearse DJ, Humphrey SM, Chain EB. Abrupt reoxygenation of the anoxic potassium-arrested perfused rat heart: a study of myocardial enzyme release. J Mol Cell Cardiol 1973; 5:395-407.
91. McCord JM. Oxygen-derived free radicals in postishemic tissue injury. N Engl J Med 1985; 312:159-163.
92. Fridovich I. Quantitative aspects of the production of superoxide anion radical by milk xanthine oxidase. J Biol Chem 1970; 245:4053-4057.
93. Cadenas E, Boveris A, Ragan CI et al. Production of superoxide radicals and hydrogen peroxide by NADH-ubiquinone reductase and ubiquinol-cytochrome c reductase from beef heart mitochondria. Arch Biochem Biophys 1977; 180:248-257.

94. Nilsson R, Pick FM, Bray RC. EPR studies on reduction of oxygen to superoxide by some biochemical systems. Biochem Biophys Acta 1969; 192:145-148.
95. Fridovich I. The biology of oxygen radicals. Science 1978; 201:875-880.
96. Meerson FZ, Kagan VE, Lozlov YP et al. The role of lipid peroxidation in pathogenesis of ischemic damage and the antioxidant protection of the heart. Basic Res Cardiol 1982; 77:465-485.
97. Graf E, Mahoney JR, Bryant RG et al. Iron-catalyzed hydroxyl radical formation. Stringent requirement for free iron co-ordination site. J Biol Chem 1984; 259:3620-3624.
98. Sun SC, Appleyard R, Masetti P et al. Improved recovery of heart transplants by combined use of oxygen-derived free radical scavengers and energy enhancement. J Thorac Cardiovasc Surg 1992; 104:830-837.
99. Jurmann MJ, Schaefers HJ, Dammenhayn L et al. Oxygen-derived free radical scavengers for amelioration of reperfusion damage in heart transplantation. J Thorac Cardiovasc Surg 1988; 95:368-377.
100. Gharagozloo F, Melendez FJ, Hein RA et al. The effect of superoxide dismutase and catalase on the extended preservation of the ex vivo heart for transplantation. J Thorac Cardiovasc Surg 1988; 95:1008-1013.
101. Chambers DE, Parks DA, Patterson G et al. Xanthine oxidase as a source of free-radical damage in myocardial ischemia. J Moll Cell Cardiol 1985; 17:145-152.
102. Bando K, Tago M, Teramoto S. Prevention of free radical-induced myocardial injury by allopurinol. Experimental study in cardiac preservation and transplantation. J Thorac Cardiovasc Surg 1988; 95:465-473.
103. Menasche P, Grousset C, Mouas C et al. A promising approach for improving the recovery of heart transplants. Prevention of free radical injury through iron chelation by deferoxamine. J Thorac Cardiovasc Surg 1990; 100:13-21.
104. Lucchesi BR, Mullane KM. Leukocytes and ischemia-induced myocardial injury. Ann Rev Pharmacol Toxicol 1986; 26:201-224.
105. Babior BM. The respiratory burst of phagocytes. J Clin Invest 1984; 73:599-601.
106. Jordan JE, Zhao ZQ, Vinten-Johansen J. The role of neutrophil in myocardial ischemia-reperfusion injury. Cardiovasc Res 1999; 43:860-878.
107. Roos DJ, Hamon-Muller WT, Weening RJ. Effect of cytochalasin-B on the oxidative metabolism of human peripheral blood granulocytes. Biochem Biophys Res Commun 1976; 68:43-50.
108. McPhail LC, DeChatelet LR, Shirley PS. Further characterization of NADPH oxidase activity of human poylmorphonuclear leukocytes. J Clin Invest 1976; 58:774-780.
109. Fantone JC, Ward PA. Role of oxygen-derived free radicals and metabolites in leukocyte-dependent inflammatory reactions. Am J Pathol 1982; 107:394-418.
110. Engler RL, Dahlgren MD, Pererson MA et al. Accumulation of polymorphonuclear leukocytes during 3-h experimental myocardial ischemia. Am J Physiol 1986; 251:H93-H100.
111. Engler RL, Schmid-Schonbein GW, Pavelec RS. Leukocyte capillary plugging in myocardial ischemia and reperfusion in the dog. Am J Path 1983; 111:98-111.
112. Ambrosio G, Tritto I. Reperfusion injury: experimental evidence and clinical implications. Am Heart J 1999; 138:S69-S75.
113. Rezkalla SH, Kloner RA. No-reflow phenomenon. Circulation 2002; 105:656-662.
114. Rowe GT, Eaton LR, Hess ML. Neutrophil-derived, oxygen free radical-mediated cardiovascular dysfunction. J Moll Cell Cardiol 1984; 16:1075-1079.
115. Boyle EM Jr, Canty TG Jr, Morgan EN et al. Treating myocardial ischemia-reperfusion injury by targeting endothelial cell transcription. Ann Thorac Surg 2001; 68:1949-1953.
116. Sellke FW, Shafique T, Ely DL et al. Coronary endothelial injury after cardiopulmonary bypass and ischemic cardioplegia is mediated by oxygen-derived free radicals. Circulation 1993; 88:II395-400.
117. Kawata H, Aoki M, Hickey PR et al. Effect of sntibody to leukocyte adhesion molcule CD18 on recovery of neonatal lamb hearts after 2 hours of cold ischemia. Circulation 1992; 86:II364-370.
118. Kupatt C, Habazettl H, Goedecke A et al. Tumor necrosis factor-alpha contributes to ischemia- and reperfusion-induced endothelial activation in isolated hearts. Circ Res 1999; 84(4):392-400.
119. Fukushima N, Shirakura R, Nakata S et al. Effects of terminal cardioplegia with leukocyte-depleted blood on heart grafts preserved for 24 hours. J Heart Lung Transplant 1992; 11:676-682.
120. Fukushima N, Shirakura R, Nakata S et al. Study of efficacies of leukocyte-depleted terminal blood cardioplegia in 24-hour preserved hearts. Ann Thorac Surg 1994; 58:1651-1656.
121. Pearl JM, Drinkwater DC, Laks H et al. Leukocyte-depleted reperfusion of transplanted human hearts: a randomized, double-blind clinical trial. J Heart Lung Transplant 1992; 11:1082-1092.
122. Piper HM, Meuter K, Schafer C. Cellular mechanisms of ischemia-reperfusion injury. Ann thorac Surg 2003; 75:S644-648.
123. Nichols CG, Lederer WJ. The role of ATP in energy-deprivation contractures in unloaded rat ventricular myocytes. Can J Physiol Pharmacol 1990; 68(2):183-194.
124. Ganote CE. Contraction band necrosis and irreversible myocardial injury. J Mol Cell Cardiol 1983; 15:67-73.

125. Ladilov YV, Siegmund B, Piper HM. Protection of recxygenated cardiomyocytes against hypercontracture by inhibition of Na+/H+ exchange. Am J Physiol 1995; 268:H1531-1539.
126. Siegmund B, Schlack W, Ladilov YV et al. Halothane protects cardiomyocytes against reoxygenation-induced hypercontracture. Circulation 1997; 96:4372-4379.
127. Siegmund B, Klietz T, Schwartz P et al. Temporary contractile blockage prevents hypercontracture in anoxic-reoxygenated cardiomyocytes. Am J Physiol 1991; 260:H426-435.
128. Gercia-Dorado D, Theroux P, Duran JM et al. Selective inhibition of the contractile apparatus. A new approach to modification of infarct size, infarct composition and infarct geometry during coronary occlusion and reperfusion. Circulation 1992; 85:1160-1174.
129. Padilla F, Gercia-Dorado D, Agullo L et al. L-Arginine administration prevents reperfusion-induced cardiomyocyte hypercontracture and reduces infarct size in the pig. Cardiovasc Res 2000; 46:412-420.
130. Schluter KD, Weber M, Schraven E et al. NO donor SIN-1 protects against reoxygenation-induced cardiomyocyte injury by a dual action. Am J Physiol 1994; 267:H1461-1466.
131. Hempel A, Friedrich M, Schluter KD et al. ANP protects against reoxygenation-induced hypercontracture in adult cardiomyocytes. Am J Physiol 1997; 273:H244-249.
132. Padilla F, Gercia-Dorado D, Agullo L et al. Intravenous administration of the natriuretic peptide urodilatin at low doses during coronary reperfusion limits infarct size in anesthetized pigs. Cardiovasc Res 2001; 51:592-600.
133. Schafer C, Ladilov YV, Siegmund B et al. Importance of bicarbonate transport for protection of cardiomyocytes against reoxygenation injury. Am J Physiol Heart Circ Physiol 2000; 278:H1457-1463.
134. Lazar HL, Buckberg GD, Manganaro AM et al. Reversal of ischemic damage with amino acid substrate enhancement during reperfusion. Surgery 1980; 88:702-709.
135. Lazar HL, Buckberg GD, Manganaro AM et al. Myocardial energy replenishment and reversal of ischemic damage by substrate enchancement of secondary blood cardioplegia with amino acid reperfusion. J Thorac Cardiovasc Surg 1980; 80:350-359.
136. Gailis L, Benmouyal E. Endogenous alanine, glutamate, aspartate and glutamine in the perfused quinea pig heart: effect of substrates and cardioactive agents. Can J Biochem 1973; 51:11-20.
137. Rosenkranz ER, Okamoto F, Buckberg GD et al. Safety of prolonged aortic clamping with blood cardioplegia in energy-depleted hearts after ischemic and reperfusion injury. J Thorac Cardiovasc Surg 1986; 91:428-435.
138. Matsuoka S, Jarmakani JM, Young HH et al. The effect of glutamate on hypoxic newborn rabbit heart. J Mol Cell Cardiol 1986; 18:897-906.
139. Pozefsy T, Felig P, Tobin JD et al. Amino acid balance across tissues of the forearm in post-absorptive man; effects of insulin at two dose levels. J Clin Invest 1969; 48:2273-2282.
140. Taegtmeyer H. Metabolic responses to cardiac hypoxia: increased production of succinate by rabbit papillary muscles. Circ Res 1978; 43:808-815.
141. Digerness SB, Reddy WJ. The malate-aspartate shuttle in heart mitochondria. J Mol Cell Cardiol 1976; 8:779-785.
142. Mudge GH, Mills RM, Taegtmeyer H et al. Alterations of myocardial amino acid metabolism in chronic ischemic heart disease. J Clin Invest 1976; 58:1185-1192.
143. Thomassen AR, Nielsen TT, Baggin JP et al. Myocardial exchange of glutamate, alanine and citrate in controls and patients with coronary artery disease. Clin Sci 1983; 64:33-40.
144. Sanborn T, Gavin W, Berkowithz S et al. Augmented conversion of aspartate and glutamate to succinate during anoxia in rabbit hearts. Am J Physiol 1979; 237:H535-541.
145. Hunter FE Jr. Anaerobic phosphorylation due to a coupled oxidation-reduction between alpha-ketoglutaric acid and oxalacetic acid. J Biol Chem 1949; 177:361-372.
146. Bush LR, Warren S, Mesh CL et al. Comparative effects of aspartate and glutamate during myocardial ischemia. Pharmacology 1981; 23:297-304.
147. Haan CK, Lazar HL, Rivers S et al. Improved myocardial preservation during cold storage using substrate enhancement. Ann Thorac Surg 1990; 50:80-85.
148. Choong YS, Gavin JB, Buckman J. Long-term preservation of explanted hearts perfused with L-aspartate- enriched cardioplegic solution. Improved function, metabolism and ultrastructure. J Thorac Cardiovasc Surg 1992; 103:210-218.
149. Choong YS, Gavin JB. L-aspartate improves the functional recovery of explanted hearts stored in St. Thomas' Hospital cardioplegic solution at 4°C. J Thorac Cardiovasc Surg 1990; 99:510-517.
150. Tixier D, Matheis G, Buckberg GD et al. Donor hearts with impaired hemodynamics. Benefit of warm substrate- enriched blood cardioplegic solution for induction of cardioplegia during cardiac harvesting. J Thorac Cardiovasc Surg 1991; 102:207-213.
151. Szabo G, Bahrle S, Batkai S et al. L-Arginine: effect on reperfusion injury after heart transplantation. World J Surg 1998; 22:791-798.

152. Cocks TM, Angus JA, Campbell JH et al. Release and properties of endothelium-derived relaxing factor (EDRF) from endothelial cells in culture. J Cell Physiol 1985; 123:310-320.
153. Schultz R, Nava E, Moncada S. Introduction and potential biological relevance of a Ca2+-independent nitric oxide synthase in the myocardium. Br J Pharmacol 1992; 105:575-580.
154. Schultz R, Smith JA, Lewis MJ et al. Nitric oxide synthase in cultured endocardial cells of the pig. Br J Pharmacol 1991; 104:21-24.
155. Kelm M, Schrader J. Control of coronary vascular tone by nitric oxide. Circ Res 1990; 66:1561-1575.
156. Pohl U, Lamontagne D. Impaired tissue perfusion after inhibition of endothelium-derived nitric oxide. Basic Res Cardiol 1991; 86(Suppl. 2):97-105.
157. Lefer AM, Tsao PS, Lefer DJ et al. Role of endothelial dysfunction in the pathogenesis of reperfusion injury after myocardial ischemia. FASEB J 1991; 5:2029-2034.
158. Kurose I, Wolf R, Grisham MB et al. Modulation of ischemia- reperfusion-induced microvascular dysfunction by nitric oxide. Circ Res 1994; 74:376-382.
159. Pabla R, Buda AJ, Flynn DM et al. Intracoronary nitric oxide improves postischemic coronary blood flow and myocardial contractile function. Am J Physiol 1995; 269:H1113-1121.
160. Adams MR, Jessup W, Hailstones D et al. L-arginine reduces human monocyte adhesion to vascular endothelium and endothelial expression of cell adhesion molecules. Circulation 1997; 95:662-668.
161. Siegfried MR, Erhardt J, Rider T et al. Cardioprotection and attenuation of endothelial dysfunction by organic nitric oxide donors in myocardial ischemia-reperfusion. J Pharmacol Exp Ther 1992; 260:668-675.
162. Schulz R, Wambolt R. Inhibition of nitric oxide synthesis protects the isolated working rabbit heart from ischemia-reperfusion injury. Cardiovasc Res 1995; 30:432-439.
163. Naseem SA, Kontos MC, Rao PS et al. Sustained inhibition of nitric oxide by NG-nitro-L- arginine improves myocardial function following ischemia/reperfusion in isolated perfused rat heart. J Mol Cell Cardiol 1995; 27:419-426.
164. Depre C, Vanoverschelde JL, Goudemant JF et al. Protection against ischemic injury by nonvasoactive concentrations of nitric oxide synthase inhibitors in perfused rabbit heart. Circulation 1995; 92:1911-1918.
165. Sato H, Zhao ZQ, McGee DS et al. Supplemental L-Arginine during cardioplegic arrest and reperfusion avoids regional postischemic injury. J thorac Cardiovasc Surg 1995; 110:302-314.
166. Lefer AM Attenuation of myocardial ischemia-reperfusion injury with nitric oxide replacement therapy. Ann Thorac Surg 1995; 60:847-851.
167. McKeown PP, McClelland JS, Bone DK et al. Nitroglycerin as an adjunct to hypothermic hyperkalemic cardioplegia. Circulation 1983; 68:II107-111.
168. Kronon MT, Allen BS, Halldorsson A et al. Dose dependency of L-arginine in neonatal myocardial protection: the nitric oxide paradox. J Thorac Cardiovasc Surg 1999; 118:655-664.
169. Nossuli TO, Hayward R, Jensen D et al. Mechanism of cardioprotection by peroxynitrite I myocardial ischemia and reperfusion injury. Am J Physiol 1998; 275:H509-519.
170. Worku Y, Newby AC. The mechanism of adenosine production by poymorphonuclear leukocytes. Biochem J 1983; 214:1-6.
171. Vinten-Johansen J, Zhao ZQ, Corvera JS et al. Adenosine in myocardial protection in on-pump and off-pump cardiac surgery. Ann Thorac Surg 2003; 75:S691-699.
172. Xu Z, Mueller RA, Park S et al. Cardioprotection with adenosine A2 receptor activation at reperfusion. J Cardiovasc Pharmacol 2005; 46(6):794-802.
173. Lasley RD, Rhee JW, Van Wylen DGL et al. Adenosine A1 receptor mediated protection of the globally ischemic isolated rat heart. J Mol Cell Cardiol 1990; 22:39-47.
174. Lasley RD, Mentzer RM Jr. Adenosine improves recovery of postischemic myocardial function via an adenosine A1 receptor mechanism. Am J Physiol 1992; 263:H1460-1465.
175. ToombsCF, McGee DS, Johnston WE et al. Protection from ischemic-reperfusion injury with adenosine pretreatment is reversed by inhibition of ATP-sensitive potassium channels. Cardiovasc Res 1993; 27:623-629.
176. Narayan P, Mentzer RM Jr, Lasley RD. Adenosine A1 receptor activation reduces reactive oxygen species and attenuates stunning in ventricular myocytes. J Mol Cell Cardiol 2001; 33:121-129.
177. Constein BN, Kramer SB, Weissmann G et al. Adenosine: a physiological modulator of superoxide anion generation by human neutrophils. J Exp Med 1983; 158:1160-1177.
178. Zhao ZQ, Sato H, Williams MW et al. Adenosine A2-receptor activation inhibits neutrophil-mediated injury to coronary vascular endothelium. Am J Physiol 1996; 271:H1456-1464.
179. Jordan JE, Zhao ZQ, Sato H et al. Adenosine A2-receptor activation attenuates reperfusion injury by inhibiting neutrophil accumulation, superoxide generation and coronary endothelium adherence. J Pharmacol Exp Ther 1997; 280:301-309.
180. Xu Z, Downey JM, Cohen MV. AMP 579 reduces contracture and limits infarction in rabbit heart by activating adenosine A2 receptors. J Cardiovasc Pharmacol 2001; 38:474-481.

181. Xu Z, Park SS, Mueller RA et al. Adenosine produces nitric oxide and prevents mitochondrial oxidant damage in rat cardiomyocytes. Cardiovasc Res 2005; 65:803-812.
182. Piper HM, Abdallah Y, Schafer C. The first minutes of reperfusion: a window of opportunity for cardioprotection. Cardiovasc Res 2004; 61:365-371.
183. Armstrong SC. Protein kinase activationand myocardial ischemia/reperfusion injury. Cardiovasc Res 2004; 61:427-436.
184. Tong H, Chen W, Steenbergen C et al. Ischemic preconditioning activates phosphatidylinositol-3-kinase upstream of protein kinase C. Circ Res 2000; 87:309-315.
185. Hausenloy Dj, Yellon DM. Nw directions for protecting the heart against ischemia-reperfusion injury: targeting the Reperfusion Injury Salvage Kinase (RISK)-pathway. Cardiovasc Res 2004; 61:448-460.
186. Yang XM, Krieg T, Cui L et al. NECA and bradykinin at reperfusion reduce infarction in rabbit hearts by signaling through PI3K, ERK and NO. J Mol Cell Cardiol 2004; 36:411-421.
187. Yue TL, Wang C, Gu JL et al. Inhibition of extracellular signal-regulated kinase enhances ischemia- reoxigenation-induced apoptosis in cultured cardiac myocytes and exaggerates reperfusion injury in isolated perfused heart. Circ res 2000; 86:692-699.
188. Shimizu N, Yoshiyama M, Omura T et al. Activation of mitogen-activated protein kinases and activator protein-1 in myocardial infarction in rats. Cardiovasc Res 1998; 38:116-124.
189. Fryer RM, Pratt PF, Hsu AK et al. Differential activation of extracellular signal regulated kinase isoforms in preconditioning and opioid-induced cardioprotection. J Pharmacol Exp Ther 2001; 296:642-649.
190. Kis A, Baxter GF, Yellon DM. Limitation of myocardial reperfusion injury by AMP579, an adenosine A1/A2A receptor agonist: role of A2A receptor and Erk1/2. Cardiovasc Drugs Ther 2003; 17:415-425.
191. Kroemer G, Dallaporta B, Resche-Rigon M. The mitochondrial death/life regulator in apoptosis and necrosis. Annu Rev Physiol 1998; 60:619-642.
192. Suleiman MS, Halestrap AP, Griffiths EJ. Mitochondria: a target for myocardial protection. Pharmacol Ther 2001; 89:29-46.
193. Crompton M. The mitochondrial permeability transition poreand its role in cell death. Biochem J 1999; 341:233-249.
194. Halestrap AP, Clarke SJ, Javadov SA. Mitochonrial permeability transition pore opening during myocardial reperfusion—a target for cardioprotection. Cardiovasc Res 2004; 61:372-385.
195. Halestrap AP, Kerr PM, Javadov S et al. Elucidating the molecular mechanism of the permeability transition pore and its role in reperfusion injury of the heart. Biochem Biophys Acta 1998; 1366:79-94.
196. Xu Z, Downey JM, Cohen MV. Timing and duration of administration are crucial for antiinfarct effect of AMP579 infused at reperfusion in rabbit heart. Heart Dis 2003; 5(6):368-371.
197. Theroux P, Chaitman BR, Danchin N et al. Inhibition of the sodium-hydrogen exchanger with cariporide to prevent myocardial infarction in high-risk ischemic situations. Main results of the GUARDIAN trial. Guard during ischemia against necrosis (GUARDIAN) Investigators. Circulation 2000; 102:3032-3038.
198. Muraki S, Morris CD, Budde JM et al. Blood cardioplegia supplementation with the sodium-hydrogen exchange inhibitor cariporide to attenuate infarct size and coronary artery endothelial dysfunction after severe regional ischemia in a canine model. J thorac Cardiovasc Surg 2003; 125:155-164.
199. Cropper JR, Hicks M, Ryan JB et al. Cardioprotection by cariporide after prolonged hypothermic storage of the isolated working rat heart. J Heart Lung transplant 2003; 22:929-936.
200. Kevelaitis E, Oubenaissa A, Mouas C et al. Ischemic preconditioning with opening of mitochondrial adenosine triphosphate-sensitive potassium channels or Na/H exchange inhibition: which is the best protective strategy for heart transplants? J Thorac Cardiovasc Surg 2001; 121:155-162.
201. Li HY, Wu S, He GW et al. Aprikalim reduces the Na+ -Ca2+ exchange outward current enhanced by hyperkalemia in rat ventricular myocytes. Ann Thorac Surgt 2002; 73:1253-1260.
202. He GW. Potassium-channel opener in cardioplegia may restore coronary endothelial function. Ann Thorac Surg 1998; 66:1318-1322.
203. Grover GJ, McCullough JR, D'Alonzo AJ et al. Cardioprotective profile of the cardiac-selective ATP-sensitive potassium channel opener BMS-180448. J Cardiovasc Pharmacol 1995; 25:40-50.
204. Cropper JR, Hicks M, Ryan JB et al. Enhanced cardioprotection of the rat heart during hypothermic storage with combined Na+ -H+ exchange inhibition and ATP-dependent potassium channel activation. J Heart Lung transplant 2003; 22:1245-1253.
205. Ryan JB, Hicks M, Cropper JR et al. Sodium-hydrogen exchanger inhibition, pharmacologic ischemic preconditioning, or both for extended cardiac allograft preservation. Transplantation 2003; 76:766-771.
206. Halejcio-Delophont P, Siaghy EM, Devaux Y et al. Increase in myocardial interstitial adenosine and net lactate production in brain-dead pigs: an in vivo microdialysis study. Transplantation 1998; 66:1278-1284.
207. McCully JD, Levitsky S. The mitochondrial K(ATP) channel and cardioprotection. Ann Thorac Surg 2003; 75:S667-673.
208. Hicks M, Hing A, Gao L et al. Organ preservation. Methods Mol Biol 2006; 333:331-374.

209. Murry CE, Jennings RB, Reimer KA. Preconditioning with ischemia: a delay of lethal cell injury in ischemic myocardium. Circulation 1986; 74:112-1136.
210. Schulz R, Cohen MV, Behrends M et al. Signal transduction of ischemic preconditioning. Cardiovasc Res 2001; 52:181-198.
211. Bienengraeber MW, Weihrauch D, kersten JR et al. Cardioprotection by volatile anesthetics. Vascular Pharmacol 2005; 42:243-252.
212. Davidson SM, Duchen MR. Effects of NO on mitochondrial function in cardiomyocytes: pathophysiological relevance. Cardiovasc Res 2006; 71:10-21.
213. Hausenloy CJ, Yellon DM. New directions for protecting the heart against ischemia-reperfusion injury: targeting the Reperfusion Injury Salvage Kinase (RISK)-pathway. Cardiovasc Res 2004; 61:448-460.
214. Wicomb WN, Cooper DK, Novitzky D et al. Cardiac transplantation following storage of the donor heart by a portable hypothermic perfusion system. Ann Thorac Surg 1984; 37:243-248.
215. Chambers DJ, Takahashi A, Hearse DJ. Long-term preservation of the heart: the effect of infusion pressure during continuous hypothermic cardioplegia. J Heart Lung Transplant 1992; 11:665-675.
216. Jackson SJ, Pu LQ, Guerraty A et al. Cardiac preservation by continuous perfusion of the University of Wisconsin solution. Can J Surg 1992; 35:165-168.
217. Wicomb W, Cooper DK, Hassoulas J et al. Orthotopic transplantation of the baboon heart after 20 to 24 hours of preservation by continuous hypothermic perfusion with an oxygenated hyperosmolar solution. J Thorac Cardiovasc Surg 1982; 83:133-140.
218. Wicomb WN, Novitzky D, Cooper DK et al. Forty-eight hours hypothermic perfusion storage of pig and baboon hearts. J Surg Res 1986; 40:276-284.
219. Guerraty A, Alivizatos P, Warner M et al. Successful orthotopic canine heart transplantation after 24 hours of in vitro preservation. J Thorac Cardiovasc Surg 1981; 82:531-537.
220. Wicomb WN, Collins GM. Twenty-four hour rabbit heart storage with UW solution: effects of low flow perfusion, colloid and shelf storage. Transplantation 1989; 48:6-9.
221. Larese A, Aussedat J, Ray A et al. Hypothermic preservation of the rat heart. Comparison of immersion and low-flow perfusion methods: contribution of 31P-NMR spectroscopy. Cryobiology 1990; 27:430-438.
222. Ferrera R, Marcsek P, Larese A et al. Comparison of continuous microperfusion and cold storage for pig heart preservation. J Heart Lung Transplant 1993; 12:463-469.
223. Ferrera R, Larese A, Marcsek P et al. Comparison of different techniques of hypothermic pig heart preservation. Ann Thorac Surg 1994; 57:1233-1239.
224. Zhu Q, Yang X, Claydon MA et al. Twenty-four hour intermittent perfusion storage of the isolated rat heart: the effect of perfusion intervals on functional preservation. J Heart Lung Transplant 1994; 13:882-890.
225. Karck M, Frantzen V, Schwalb H et al. Prolonged myocardial protection with St. Thomas' Hospital solution and University of Wisconsin solution. The importance of preservation techniques. Eur J Cardiothorac Surg 1992; 6:261-266.
226. TransMedics, Inc., Boston, MA; http://www.transmedics.com/wt/home/index

CHAPTER 12

Heart and Lung Preservation

Jason A. Williams, Ashish S. Shah and William A. Baumgartner*

Introduction

Like transplantation for solid organs, the wide applicability of lung and heart-lung transplantation remains limited by organ availability. To complicate this problem, strict selection criteria for donor organs allow only 10-15% of heart donors to be eligible for concomitant lung donation. Furthermore, unlike abdominal solid organs, lungs and heart-lung blocs tolerate relatively short ischemic periods before irreversible damage occurs to these tissues, making optimal donor-recipient matching difficult and hindering graft function in organs with prolonged ischemic time.[1] Heart and lung allografts tolerate ischemic times of only 4-6 hours.[2-4]

Primary graft failure (PGF), defined as the development of alveolar infiltrates, reduction in lung compliance and impairment in alveolar gas exchange without any other definable reason for pulmonary failure, has an incidence of 10% to 25% among lung transplant recipients.[5-8] This complication is responsible for 15-20% of early lung transplant deaths and up to 75% of patients with PGF die during their hospitalization.[7,9] Risk factors for PGF have been difficult to define, but may include recipient diagnosis of pulmonary hypertension, donor female gender, donor African American race and donor age <21 or >45 years.[10] Regardless of these risk factors, most investigators believe that untoward ischemia-reperfusion injury is primarily responsible for PGF, underscoring the importance of optimal donor and recipient management, as well as optimal donor organ preservation prior to reimplantation.

Although investigators have worked diligently to improve upon the acceptable duration of organ storage, several barriers to improved clinical outcomes with prolonged ischemic times have hindered progress in this area. First, pulmonary and cardiac allografts must maintain adequate function immediately after implantation because suitable artificial replacements (e.g., dialysis) do not currently exist. Furthermore, the heart and lungs do not tolerate the depletions in intracellular adenosine triphosphate (ATP) that occur during ischemia. Finally, the alveolar capillary barrier of the lung is more susceptible to ultrastructural injury during preservation and ischemia than are other solid abdominal organs.[11,12]

Additional factors regarding pulmonary and cardiac allograft sensitivities include the damage incurred by suboptimal donor management prior to harvest and the use of older donors with more comorbidities. The myocardium and pulmonary parenchyma are very susceptible to insults incurred during inadequate management of the complex donor physiology. Furthermore, as investigators attempt to expand the donor pool, the use of older donors with more advanced coronary artery disease and pulmonary parenchymal disease has lead to diminished reserve in these organs with less capacity for recovery after extended periods of ischemia.

The use of nonheart-beating pulmonary allograft donors is one area of investigation that could potentially expand the donor pool and increase the number of available organs. Because the pulmonary tissues do not rely on perfusion for their oxygen supply, lungs may remain viable

*Corresponding Author: William A. Baumgartner—Division of Cardiac Surgery, Johns Hopkins Hospital Blalock 618, 600 North Wolfe Street, Baltimore, Maryland 21287. Tel: 410-955-5248; Fax: 410-9553809. Email: wbaumgar@csurg.jhmi.jhu.edu

Organ Preservation for Transplantation, Third Edition, edited by Luis H. Toledo-Pereyra.
©2010 Landes Bioscience.

for longer periods of time after circulation ceases than other transplantable organs.[13-15] Although still an experimental concept, very limited clinical experience using this technique has demonstrated promising results with regard to pulmonary graft function in lungs transplanted from nonheart-beating donors.[16-19] However, the pathology behind PGF and ischemia-reperfusion injury will need to be better characterized before this technique is broadly applied in the field of heart-lung and lung transplantation.

In order to understand the principles of heart-lung and lung procurement and preservation, this chapter will attempt to review the complex pathophysiology that leads to organ damage. This will be followed by a discussion of optimal management of the organs prior to and during surgical harvest and will conclude with discussions of the various storage and reperfusion techniques currently in use today.

Donor Organ Injury

Injury to the heart-lung or lung allografts can occur at any time from declaration of brain death to reimplantation of the organs. Suboptimal mismanagement of the donor, technical errors during procurement, hypothermia and ischemia-reperfusion may all contribute to suboptimal outcomes in the transplantation of these organs.

Donor Management and Procurement

Prior to harvesting the lung or heart—lung blocs, the donor must be monitored intensely for variations in hemodynamics, fluids, electrolytes, hormones and ventilatory/acid-base status. Optimization of the donor includes maintaining a mean arterial pressure of 80-90 mm Hg using intravenous vasopressors, inotropes, or afterload reducers as the situation dictates. Volume resuscitation in lung and heart-lung donors is essential for the success of the transplantation. This becomes especially important in patients with diabetes insipidus who have excessive fluid losses. However, care must be taken not to overload the lungs with excessive fluid. In these cases, vasopressin should also be used to treat the pituitary dysfunction in these patients.

Donors with devastating cerebral incidents like trauma or spontaneous hemorrhage often have electrolyte imbalances such as hypernatremia, hypokalemia, hypophosphatemia and hypomagnesemia.[20] These electrolyte abnormalities should be adequately corrected in the preharvest period and donors should be strictly monitored for the prevention and correction of accompanying acid-base disorders. Furthermore, ventilator management should be optimized in these patients, both to ensure adequate acid-base management and to prevent barotrauma to the donor lungs.

Hypothermia should be prevented using warming blankets and warm intravenous fluids as necessary. Hemoglobin should be maintained above 10 gm/dL to maximize myocardial oxygen delivery before harvesting. In addition, studies have shown the benefits of pretreatment with hormone replacement therapies such as free triiodothyronine (T3), cortisol and insulin, since a variety of hormones become depleted after brain death.[21-23] Finally, due to the increased risk of aspiration pneumonia following devastating neurological injury, broad spectrum antibiotics should be administered to donors to prevent complications of aspiration from occurring in the recipient after transplantation.

Prior to organ harvesting, most centers give donors a bolus of solumedrol, 2 grams intravenously, in order to protect the lungs during transportation and reimplantation. In addition, maintenance of allograft hypothermia and avoiding excessive exposure to air prior to immersion in storage solution may improve outcomes in the transplanted organ. It is important to recognize that donor selection and management before and during procurement are at least as important as the preservation technique employed in determining the outcome of a transplantation procedure.

Hypothermia

While hypothermia is necessary for the safe preservation of donor organs, it can also contribute to the injury incurred by these allografts.[24] Deep hypothermia inactivates membrane-bound enzymes and alters the permeability of lipid bilayers leading to tissue edema.[25,26] Inhibition of the Na^+-K^+ ATPase pump results in loss of membrane potential as sodium is permitted to enter the cell. This leads to an osmotic gradient that also results in intracellular edema.[27,28] Finally, the intracellular milieu is further altered due to perturbations in calcium homeostasis that result from hypothermic inactivation of the Ca^{2+} ATPase.[29] Recent investigations have also revealed that Tumor Necrosis Factor-α (TNF-α) expression and free radical formation can both occur in response to hypothermia alone, leading to changes in the vascular endothelium that can be detrimental to allograft function.[30-33] Techniques that avoid the necessity of deep hypothermic storage, such as continuous organ perfusion and substrate delivery, have reportedly shown promise in minimizing the deleterious effects of hypothermia seen at extremely cold temperatures.[34,35]

Ischemia-Reperfusion Injury

Our understanding of the mechanisms of ischemia-reperfusion injury is constantly evolving. While organs clearly sustain more cellular damage during prolonged periods of ischemia due to lack of oxygen and other substrates, the period of reperfusion contributes significantly to further damage sustained in transplanted organs. Oxygen free radical and leukocyte dysfunction play key roles in this phenomenon.

Oxygen Free Radicals

Oxygen-derived free radicals are highly reactive molecules that exert their cytotoxic effects via either direct attack on cells or initiation of chain reactions leading to the formation of other harmful radicals. The endothelial xanthine oxidase (XO) pathway is primarily responsible for the development of these free radicals in post-ischemic tissue.[36] During ischemia, the catabolism of ATP leads to accumulation of hypoxanthine, the purine substrate for XO. Reperfusion leads to a burst of superoxide and peroxide synthesis via this pathway. Superoxide ultimately leads to the generation of hydroxyl radical via a number of different pathways and this radical seems to account for most of the destructive processes involved with free radical injury.[37]

Free radicals lead to cellular damage through the peroxidation of membrane lipids, denaturation of proteins and disruption of cellular homeostatic mechanisms.[38] Intracellular edema and mitochondrial dysfunction result from these insults, leading to global cellular dysfunction. Furthermore, the by-products of arachidonic acid oxidation (prostaglandins, leukotrienes and thromboxanes) act as chemotactic factors for the sequestration of leukocytes in these damaged tissues.

While cells have natural defense mechanisms against oxygen free radical damage, such as superoxide dismutase, catalase and peroxidase, these defenses are overwhelmed in the face of excessive free radical production during ischemia-reperfusion injury. In an effort to prevent some of the untoward effects of ischemia-reperfusion injury, investigators are developing novel strategies that augment some of these endogenous defenses. Augmentation of peroxynitrite decomposition and enhancement of Heat shock proteins (HSP) activity have recently been shown to ameliorate some of the damage sustained by allografts due to ischemia reperfusion.[39-41]

Leukocytes

Expression of neutrophil adhesion molecules on the vascular endothelium after hypoxic-ischemic insults and generation of chemotactic factors (C3a and C5a) via activation of the complement cascade result in capillary accumulation of leukocytes and platelets during reperfusion.[42,43] These cells can cause mechanical obstruction of the microvasculature leading to diminished allograft function. In addition, bound leukocytes synergistically release cytotoxic molecules including oxygen and halide radicals, proteases and metabolites of arachidonic acid.[44,45] These substances directly mediate tissue injury and local vasospasm resulting in an increase in pulmonary vascular resistance. Efforts to attenuate some of the complement mediated damage incurred during reperfusion using complement receptor Type 1 have met promising results in experimental models of transplantation.[46] However, although a recent randomized, controlled trial of an experimental complement inhibitor demonstrated mild clinical benefit, no survival benefit or reduction in infection/rejection episodes was noted in this study.[47]

Depletion of Energy Stores

Static cold storage inevitably results in exhaustion of high energy phosphate stores. Unlike the liver and kidney, cardiopulmonary allografts do not readily tolerate extended storage depletion of ATP. Without exogenous substrate, allografts must resort to finite glycogen and lipid stores, which contribute to the limitations of safe storage times. Renal perfusion devices which allow for continuous substrate delivery during storage have dramatically increased storage duration for kidney transplantation.[48] Similar efforts in cardiopulmonary transplantation have had limited success in experimental models, but translation into clinical practice has not been achievable using current techniques.[35,49]

Manifestations of Cardiopulmonary Preservation Injury

Cardiac allografts and pulmonary allografts respond differently to the injuries previously described (Table 1). In the lungs, the alveolar-capillary membrane is quite vulnerable to injury and disturbances in the epithelial and endothelial membranes in this region of the lung lead to interstitial edema and primary graft failure.[50] Other histopathological changes that can be seen in lung allografts include alveolar edema, disruption of the alveolar-capillary membrane, hemorrhage, organelle swelling, platelet aggregation and neutrophil sequestration. Each of these can occur globally or in isolated regions with interspersed segments of normal parenchyma. Clinically, these injuries lead to impaired gas exchange, increased pulmonary vascular resistance, bronchoconstriction and/or reduced airway compliance.[51,52]

Cardiac allografts implanted during en-bloc heart-lung transplantation develop myocardial edema and ischemic contracture bands that inhibit optimal myocardial performance. Contracture bands form as a result of energy depletion and oxygen free radical mediated injury to the intracellular organelles with subsequent disturbances in calcium homeostasis. Mitochondrial damage further reduces ATP availability leading to diminished Ca^{2+} ATPase activity in the sarcoplasm. Sluggish resequestration of calcium stores into the sarcoplasmic reticulum hinders optimal diastolic relaxation and reduces

---

Table 1. Manifestations of cardiopulmonary injury

1. Pulmonary injury
   a. Histological changes
      i. Disruption of the alveolar-capillary membrane
      ii. Interstitial edema
      iii. Alveolar edema
      iv. Hemorrhage
      v. Organelle swelling
      vi. Platelet aggregation
      vii. Neutrophil sequestration
   b. Clinical manifestations
      i. Impaired gas exchange
      ii. Increased pulmonary vascular resistance
      iii. Bronchoconstricition
      iv. Reduced airway compliance
      v. Primary graft failure
2. Cardiac injury
   a. Histological changes
      i. Myocardial edema
      ii. Ischemic contracture bands
   b. Clinical manifestations
      i. Diastolic dysfunction (↓ ventricular compliance)
      ii. Systolic dysfunction (↓ contractility)

ventricular compliance by causing the development of irreversible myocardial contractures that impair cardiac function.

Heart-Lung and Lung Preservation Techniques

The literature remains filled with controversy regarding optimal cardiopulmonary organ preservation techniques despite intensive investigative efforts to quell some of the differing opinions (Table 2). The following discussion will attempt to summarize the currently accepted strategies for organ preservation while acknowledging that the optimal strategy to provide a simple, inexpensive and reproducible strategy has yet to be defined. Although simple hypothermic immersion, normothermic autoperfusion, donor core-cooling and single hypothermic pulmonary artery flush (PAF) will each be discussed, only PAF is in current clinical practice today for lung preservation.

Hypothermia

Hypothermia is the only factor that is universally considered to be essential for effective organ preservation.[28,53] This leads to a dramatic reduction in cellular metabolism and oxygen consumption, which minimizes the amount of ATP depletion and free radical production that occurs during ischemia. Unfortunately, cellular respiration does not completely cease during cold storage, so hypothermia provides only a finite amount of protection. However, other benefits of hypothermia include increased allograft tolerance of ischemic and anoxic insults, suppressed activity of hydrolytic enzymes, stabilization of lysosomal membranes and inhibition of microbial growth in the allograft. All of these factors combine to improve allograft function in the posttransplant period. Although optimal storage temperature remains a matter of some controversy, most investigators agree on temperatures between 4-10°C.[54]

Table 2. Preservation techniques for lung and heart-lung allografts

1. Simple hypothermic immersion
2. Normothermic autoperfusion
3. Donor core-cooling
4. Single hypothermic pulmonary artery flush
    a. Prostaglandin analogs
      b. Flush conditions
        i. Temperature
        ii. Volume of perfusate
        iii. Perfusion pressure
        iv. State of lung inflation
    c. Bronchial circulation flush
      i. Antegrade
      ii. Retrograde
    d. Compositon of perfusate
      i. Electrolyte composition
        1. Intracellular
        2. Extracellular
      ii. Pharmacological additives
        1. Impermenats
        2. Antioxidants
        3. Metabolic substrates
        4. Platelets activating factor inhibitors
        5. Steroids
        6. Calcium channel blockers
        7. Lazaroids

Simple Hypothermic Immersion

Early clinical experiences with lung transplantation utilized the technique of topical cooling of the allograft by immersing the atelectatic lung in cold saline slurry at 4°C.[55] Although somewhat effective in storing allografts for 4-6 hours at a time, this method is inefficient for cooling larger allografts or heart lung blocs.[56] Experimental models continue to use simple hypothermic immersion alone or in combination with other modalities to enhance organ preservation, but this technique alone is not currently used by clinical transplant centers.[57,58]

Normothermic Autoperfusion

In order to increase acceptable storage duration, investigators in the 1980's worked diligently to perfect the autoperfused working heart-lung (AWHL) circuit developed by Robicsek. The benefits of an autoperfusion system include elimination of allograft ischemia during harvest and storage, as well as the potential for serial monitoring of cardiopulmonary function during preservation. Early in the course of clinical transplantation, the AWHL preparation was successfully used for long-distance clinical cardiopulmonary preservation.[59-62] Unfortunately, the system was not able to extend the safe ischemic period past 6 hours due to early deterioration of lung function.[59,63] This deterioration resulted from pulmonary hypertension due to flow-dependent pulmonary vasoconstrictor responses in these allografts[62,64]; embolization of platelet and leukocyte aggregates[65]; and the accumulation of extravascular edema.[66,67] The technique is also cumbersome and is plagued by complications such as hypothermia, accumulation of metabolic waste products, edema and bleeding. The idea of this technique has been applied with some success to experimental systems that employ continuous hypothermic perfusion to cardiopulmonary allografts, but translation into clinical practice has been hindered by device complexity and inconsistency.[35,49]

Donor Core Cooling

The first clinical success with core-cooling of the pulmonary donor using cardiopulmonary bypass (CPB) prior to harvest was reported in 1985.[68] This brought about widespread use of this technique for cardiopulmonary organ procurement in the 1980's and 1990's.[69-71]

Operative preparation of the donor using donor core-cooling involves aortic and right atrial cannulation. The procurement team usually brings their own portable CPB apparatus and this is used to initiate extracorporeal circulation.[72] Preliminary organ dissection proceeds as the donor is gradually cooled to an esophageal temperature of 10-15°C. When the donor heart fibrillates, the left ventricle is vented or manually compressed to prevent distension. Once profound systemic hypothermia is achieved after approximately 30 minutes of CPB, the donor is exsanguinated into the bypass reservoir and organ procurement commences. The explanted organs are immersed in the donor's cold blood for maintenance of allograft hypothermia during transport.

Core-cooling provides several advantages during organ procurement. Thoracic and abdominal allografts can be uniformly and gradually cooled while tissue oxygenation is maintained by continuous extracorporeal blood flow. CPB facilitates the intraoperative management of these often labile patients in turn minimizing allograft injury due to extremes in blood pressure. It provides a relatively bloodless field for organ procurement. The use of blood as the perfusate has theoretical advantages with its buffering capacity, metabolic substrates and endogenous free radical scavengers. Finally, unlike autoperfusion, this technique potentially permits the allocation of thoracic allografts from each donor to three recipients.

Conversely, certain disadvantages of donor core-cooling must also be noted. First, allograft cooling can be more inefficient and slower compared to pulmonary flush perfusion since pulmonary perfusion on CPB is essentially limited to the bronchial artery circulation. A bypass apparatus and an individual trained in perfusion must be available at the donor hospital or must travel with the procurement team. The bloodless field that facilitates intrathoracic organ procurement is not favored by abdominal organ transplant teams because it makes identification of potential sites of hemorrhage (i.e., transected collateral vessels) more difficult. Finally, activation of the cellular and humoral limbs of the inflammatory cascade due to contact with the synthetic surfaces of the CPB unit may result in pulmonary injury.[73-75] The advent of newer, less cumbersome techniques has lead most transplant centers to abandon this practice today.

Single Hypothermic Pulmonary Artery Flush

In an attempt to extend donor organ storage times, investigators have spent years trying to develop a continuous perfusion system for heart and lung allografts.[35,76,77] Unfortunately, the theoretical advantages of continuous perfusion systems over static preservation systems (constant supply of substrate and oxygen, washout of toxic byproducts of metabolism) has been overshadowed by the progressive increase in allograft pulmonary vascular resistance and pulmonary/myocardial edema. Coupled with the complexity of the perfusion apparatuses, these complications have prevented continuous perfusion from being applied broadly in clinical practice.

However, the Stanford group created a unique alternative with the Pulmonary Artery Flush (PAF) technique, which provides excellent (6-9 hours) short-term preservation when combined with cold storage.[78,79] PAF provides rapid cooling of the allograft while producing rapid washout of toxic metabolic byproducts. Most centers, including our center at Johns Hopkins, currently utilize this technique for cardiopulmonary allograft harvesting and storage.[80,81]

Technique of Single Hypothermic Pulmonary Artery Flush

Following mobilization and isolation of the trachea and great vessels with umbilical tapes and systemic heparinization of the donor, cannulation purse-strings are placed in the aorta and pulmonary artery. A cardioplegia catheter is inserted in the ascending aorta and the main pulmonary artery is cannulated with a 20-Fr pulmonary artery cannula at the site of its ultimate division during explantation. A bolus of a prostanoid vasodilator is injected into the main pulmonary artery via a 25-gauge needle to dilate the pulmonary vascular bed prior to PAF. Some groups supplement this preflush vasodilation with a systemic infusion of prostaglandin initiated at the time of donor median sternotomy.

Inflow occlusion to the heart is achieved with double ligation of the superior vena cava and clamping or venting of the inferior vena cava near the diaphragm. With perfusate solutions prepared and procurement teams ready to proceed, the following steps occur in rapid succession. The inferior vena cava is vented proximal to its clamp and the aorta is cross-clamped. A single flush of hypothermic cardioplegic solution is administered into the aortic root at 180 mm Hg to achieve electromechanical arrest. Concurrently, a high-flow, low-pressure hypothermic flush of the lungs through the pulmonary cannula is initiated and the tip of the left atrial appendage is amputated to vent the heart. This pulmonary artery flush usually consists of 4 liters (or 60 mL/kg) of pneumoplegia solution infused antegrade by gravity. Rapid topical cooling of the heart and lungs is achieved with copious volumes of saline ice slurry poured into the pericardial well and pleural cavities. The heart is vented through the proximal end of the divided inferior vena cava and the left atrial appendage. Care should be taken to ensure that the heart remains decompressed and the effluent flush clears, which signals adequate washout.

For isolated pulmonary allografts, 250-500 mL of cold pneumoplegia is also flushed through each pulmonary vein. This is usually accomplished by cannulating each pulmonary vein orifice with a right angle cannula after explantation of the heart. Once explantation of the heart and lungs is complete, the allografts are placed in two sterile bowel bags, each filled with cold storage solution, a saline filled air tight container and a standard cooler of ice for transport.

Optimal Conditions of Pulmonary Artery Flush

Optimal preservation of cardiopulmonary allograft tissue can only be achieved if the hypothermic preservation solution thoroughly flushes all segments of the lung for homogenous cooling. This also ensures removal of vasoactive substances and potentially harmful mediators of the inflammatory response.[81] Mimicking the physiologic conditions of the lung with a high volume, low pressure flush appears to provide the best results. However, direction of the flush, perfusate temperature, infusion pressure, infusion volume and allograft ventilation are all factors that remain controversial and influence the quality of PAF.

Most centers utilize the antegrade flush described in this text. However, experimental models of PAF have shown conflicting results regarding the optimal method of perfusion. Incorporating the bronchial circulation into the PAF technique provides the theoretical advantages of achieving a more complete flush without much additional risk. In particular, investigators have demonstrated experimental improvement in allograft performance using both retrograde flush and combined pulmonary artery/bronchial artery antegrade flush.[82-87]

The ideal temperature for PAF remains a difficult question to answer because few clinicians are comfortable manipulating the temperatures currently used in clinical practice. Although the most common storage temperature used clinically is 4°C, there are advocates for temperatures ranging from 4°C to 23°C. Although hypothermia is essential for successful preservation, profound hypothermia can lead to significant vasoconstriction even in the setting of parenteral vasodilators. Experimental studies have been mixed as to which temperature is optimal in this setting. Investigators have shown equivalent results in lungs perfused at 4°C and 10°C, while others have shown some benefit in warmer temperatures up to 23°C.[88-91] However, since most surgeons have concerns about lengthening the duration of warm ischemia and since there are logistical complications when cooling organs to temperatures other than 4°C, this temperature still tends to be favored by most centers.[28]

Optimal perfusion pressure and volume also remains a debated topic within the transplant community. Most centers use a volume of 60 mL/kg, but increase this volume if the effluent has not cleared after the initial bolus.[81,92] Experimental models demonstrate that increased perfusion pressure during PAF and during reperfusion can cause deleterious effects such as reflex vasoconstriction, increased capillary permeability and increased tissue edema.[28,93] Conversely, maintaining PAF and reperfusion pressures between 20-30 mm Hg can have protective effects on the allograft leading to improved compliance, pulmonary vascular resistance, oxygen tension and overall graft function.[93-95,81] In clinical practice, infusion of perfusate by gravity (20-45 mm Hg) provides adequate pressure to perfuse the allograft while preventing barotrauma to the microvasculature and parenchyma. Lower pressures are generally inadequate for overcoming gravity, leading to insufficient perfusion of the anterior portions of the lung.

Adequate ventilation of the lungs is also important for optimal PAF. During flushing, the lungs are gently hand ventilated to expand any atelectatic regions so that optimal flushing can occur. Some centers hyperventilate their allografts during flushing because certain studies have demonstrated the beneficial release of surfactant and increased vasodilation at higher tidal volumes.[96-99] However, other centers fear the complications of barotrauma that may come with these maneuvers and choose to ventilate at half normal tidal volumes.

While current PAF techniques have expanded cardiopulmonary allograft survival to 6-9 hours of cold ischemia, the optimal methods for achieving these results remain controversial. More research in the field of lung and heart-lung preservation will be needed to answer these questions.

Composition of Perfusate and Storage Solutions

Although investigators have achieved ischemic times of up to nine hours in experimental models, the currently accepted limit for "safe" ischemic time remains 4-6 hours for heart and lung allografts. Current preservation techniques continue to use a combination of flush cardioplegia/pneumoplegia, organ preservation solution and hypothermic storage. A variety of storage solutions exist, the studies of which demonstrate conflicting results regarding the benefit of one particular solution over another (Table 3).[100,101]

Solution compositions vary according to ionic concentrations that resemble either intracellular or extracellular fluids. Theoretically, storage solutions such as University of Wisconsin (UW) and Euro-Collins (EC) solutions that have intracellular compositions (high potassium, low sodium concentrations) prevent intracellular edema by minimizing fluid shifts into the myocardial cells.[102] By mimicking the normal intracellular environment, these solutions reduce the electrochemical gradients of ions across the cell membrane and in turn minimize sodium influx and osmotic re-equilibrium with passive water entry. Furthermore, energy is conserved as the activity of the Na^+-K^+ ATPase pump is limited.[103] Unfortunately, high potassium concentrations in these solutions

Table 3. *Composition of commonly used perfusate/storage solutions*

Component	EC	UW	Celsior	Perfadex	HTK
Na$^+$ (mmol/l)	9.3	30	100	168	15
K$^+$ (mmol/l)	107	125	15	4	9
Cl$^-$ (mmol/l)	14	–	41.5	103	50
HCO$_3^-$ (mmol/l)	9.3	–	–	–	–
Mg^{2+} (mmol/l)	4.7	5	13	2	4
PO^{4-} (mmol/l)	55	25	–	37	–
Sulfate (mmol/l)	4.7	5	–	–	–
Glucose (%)	5.6	–	–	–	45
Dextran 40 (gm/l)	–	–	–	20	–
Hydroxyethyl starch (%)	–	5	–	–	–
Mannitol (%)	0.58	–	60	–	30
Lactobionate (mmol/l)	–	100	80	–	–
Raffinose (mmol/l)	–	30	–	–	–
Glutathione (mmol/l)	–	3	3	–	–
Glutamate (mmol/l)	–	–	20	–	–
Histidine (mmol/l)	–	–	30	–	180
Adenosine (mmol/l)	–	5	–	–	–
Allopurinol (mmol/l)	–	1	–	–	–
Insulin (U/l)	–	100	–	–	–
Dexamethasone (mg/l)	–	8	–	–	–

EC = Euro-Collins, UW = University of Wisconsin, HTK = Hystidine, Tryptophane, Ketoglutarate.
HTK also contains Tryptophane (2 mmol/L) and α-Ketoglutarate (1 mmol/L).

can cause direct cellular injury in the setting of hypothermia.[104-107] Furthermore, these solutions can lead to vasoconstriction with resulting inhomogeneous flushing of the allograft.[108,109]

Conversely, solutions with extracellular compositions (low potassium, high sodium concentrations), such as Stanford, St. Thomas' Hospital and Hopkins' solutions, purportedly avoid cellular damage and increased vascular resistance imposed by high potassium concentrations.[110] Unfortunately, these solutions fail to adequately preserve allografts due to the loss of endothelial integrity and the development of significant extracellular pulmonary edema.[81]

As previously discussed, pretreatment with prostaglandins can attenuate the vasoconstrictive response to intracellular solutions. In addition, most proponents of intracellular solutions feel the data demonstrating true cellular damage to the allograft from the high potassium content of these solutions are lacking. On the other hand, reports indicate improved graft function by adding impermeants, such as Dextran, to extracellular solutions, which add the advantage of avoiding intracellular edema while minimizing the potassium concentration in the solution.[111]

Solutions Currently in Use

Initially, EC was the most widely used solution for pulmonary allograft perfusion and storage.[80,112] However, studies comparing EC to UW as a preservation solution have had conflicting results, demonstrating equivalent outcomes in some series and demonstrating the benefit of either solution over the other in some series.[113-118] Another intracellular solution, Bretschneider's HTK, which is used commonly in Europe, has also been shown to have equivalent allograft preservation to UW and excellent cardioprotection in experimental models.[119,120] Therefore, the use of these solutions for optimal graft preservation remains open to individual preference.

In the past few years, the addition of certain substrates to extracellular solutions has dramatically increased their use in transplant centers around the world. Celsior has an electrolyte composition comparable to other extracellular solutions, but also contains glutamate, mannitol, lactobionate and histidine.[101] The addition of impermeants, such as mannitol and lactobionate, enables Celsior to prevent edema and cell damage without the detrimental effects of high potassium concentrations.[121] Similar benefits are seen with low potassium Dextran (LPD) solutions such as Perfadex, which contain 40 gm/L of the impermeant Dextran.

Recent interest in these solutions has prompted investigators to compare them in experimental models and clinical studies to more established solutions used over the previous two decades. Celsior demonstrated better lung preservation in experimental models when compared to both UW and low potassium EC solutions.[101,122] These results have been verified in clinical studies, as well, where Celsior was associated with improved heart and lung preservation when compared to more established solutions.[123,124] Perfadex demonstrates improved graft function in experimental transplant models, as well.[125,126] Furthermore, two separate clinical studies have demonstrated benefit in allografts preserved using this solution when compared to EC.[127,128] When Celsior and Perfadex were compared head-to-head in experimental models of lung transplantation, both demonstrated safe, effective preservation of the allograft, but Celsior provided better pulmonary preservation at extended ischemic times with slightly improved endothelial preservation.[129,130] However, these findings have yet to be verified in clinical settings.

Based on their improved performance in recent laboratory experiments and clinical studies, our group at Hopkins prefers Perfadex solution for cardiopulmonary allograft preservation.[131] We have found this solution to be equally effective at providing cardioplegia/pneumoplegia while improving overall allograft function when compared to other methods.

Pharmacological Solution Additives

The inclusion of pharmacologic additives is another area of active research and controversy in cardiopulmonary preservation. While a plethora of substances have been added to perfusate-storage solutions, the greatest potential for future routine use may lie with impermeants, antioxidants, substrates and platelet activating factor inhibitors (Table 4).

Impermeants

Impermeants are high molecular weight compounds that do not readily penetrate the cell membrane. These substances act as extracellular osmotic agents that increase the extracellular osmotic pressure, thereby minimizing the amount of hypothermia-induced cellular edema in allografts.[27,69] The most common impermeants added to perfusate-storage solutions include mannitol, lactobionate and low molecular weight Dextran. While still under investigation, these additives have certainly demonstrated benefit for allograft function in recent studies.[101,122-129]

Antioxidants

Due to the deleterious effects of oxygen-derived free radicals on allograft viability and due to the fact that pulmonary allograft recipients demonstrate compromised antioxidant status, interventions to neutralize these molecules have been the subject of much research.[132] Administration of exogenous free radical scavengers is believed to assist the recipient's intrinsic antioxidant defenses in the metabolism of the oxygen intermediates to benign molecules.[133,134] Investigators have administered scavenger molecules such as allopurinol, glutathione, superoxide dismutase (SOD),

Table 4. Perfusate-storage solution additives

1. Impermeants
   a. Low molecular weight Dextram
   b. Mannitol
   c. Lactobionate
2. Antioxidants
   a. Allopurinol
   b. Glutathione
   c. Superoxide dismutase
   d. Catalase
   e. Dimethylthiourea
   f. Tocopherol
   g. Pyrrolidine dithiocarbamate
   h. Melatonin
   i. Mannitol
   j. Dextrose
   k. Lazaroids
3. Metabolic Substrates
   a. Glucose
   b. Glutathione
   c. Adenosine
   d. L-pyruvate
   e. L-glutamate
4. Prostaglandin Analogs
5. Vasodilators
   a. Nitric Oxide
   b. 8-Br-cGMP
   c. Nitroprusside
   d. Calcium channel blockers
   e. Cyclooxygenase inhibitors
6. Platelet-activating factor inhibitors
   a. BN 52021
   b. WEB 2170
   c. TCV-309
7. Other agents
   a. Na^+/H^+ exchanger inhibitors (e.g., cariporide)
   b. Ischemic preconditioning agents (e.g., Nicorandil, Pinacidil)

catalase, dimethylthiourea, tocopherol, pyrrolidine dithiocarbamate, melatonin, mannitol and dextrose with varying success.[135-146]

One group has demonstrated that flushing the pulmonary vasculature with a second rinse solution containing antioxidants prior to reimplantation can lead to improved outcomes.[147] It is hypothesized that this method not only provides more antioxidants, but it also allows for the washout of oxygen-derived free radicals and other toxic substrates prior to reperfusion.

Another group of compounds that have received attention in the past are *lazaroids*. Lazaroids are aminosteroids that inhibit lipid peroxidation.[28] Various experimental models have demonstrated reduced lipid peroxidation and improved allograft function with the addition of lazaroids to the preservation protocol.[148-152] However, other experimental models have shown that even with reduction in lipid peroxidation, the benefit on graft function is marginal and does not warrant use in the clinical setting.[153,154] Given these conflicting findings, lazaroids continue to be a topic of intense debate and research. Like all of the antioxidants under investigation, the use of lazaroids holds promise for improved graft preservation, but translation into clinical practice has been slow and results have been inconsistent.

Metabolic Substrates

In light of the fact that storage of organs at 4°C does not completely arrest metabolic activity, it is not surprising that metabolic substrates are common additives in perfusate-storage solutions. The maintenance of myocardial high energy phosphates to prevent contracture bands during ischemia and to permit the rapid regeneration of energy stores during reperfusion in order to fuel the newly contracting heart are the potential benefits of substrate-enhanced media. Substrates may be particularly important in lung preservation given the fact that aerobic metabolism persists in the stored pulmonary allograft.

Most experience with metabolic substrates in lung transplantation has focused on glucose and glutathione, while investigation of adenosine, L-pyruvate and L-glutamate has dominated the myocardial preservation literature.[155-157] While the biochemical rationale for provision of substrates in perfusate-storage solutions appears theoretically sound, it is still unclear if exogenous substrates can be taken up and utilized by the hypothermic cells of the allograft or if their addition significantly improves outcomes.

Prostaglandin Analogs

Pulmonary vasoconstriction coincident with PAF may occur in response to hypothermic stimulus or the hyperkalemic electrolyte composition of the perfusate. Intrapulmonary shunting associated with this increased vascular tone results in inhomogeneous cooling and incomplete washout of blood components. Furthermore, if PAF infusion pressure acutely rises due to vasoconstriction, the integrity of the endothelium may be compromised resulting in exacerbation of pulmonary edema.

Prostaglandins are potent vasodilators via the cAMP pathway, making their use in cardiopulmonary transplantation quite appealing for the prevention of complications that may result from vasoconstriction.[158,159] In addition, prostaglandins have the added beneficial effects of reducing platelet aggregation, endothelial swelling, vascular permeability and leukocyte sequestration.[158,160-164] Numerous studies have demonstrated the beneficial effects of prostaglandin infusion prior to PAF on the pulmonary vasculature and allograft function as a whole.[165-168] However, other investigators have reported little or no added benefit to allografts treated with prostaglandins prior to PAF.[169-174] The conflicting reports in the literature make drawing conclusions difficult, so the use of prostaglandins in donors prior to organ procurement remains transplant center dependent.

Endogenous and Exogenous Vasodilators

In light of the fact that cardiopulmonary allografts sustain a certain amount of damage due to vasoconstriction and inhomogeneous perfusion, the use of vasodilators to allow for more uniform organ perfusion is an intense area of investigation. In addition to prostaglandins, nitric oxide (NO) and its downstream effectors are some of the most commonly studied molecules. Nitric oxide provides two potential benefits to allograft preservation. First, this molecule is a selective pulmonary vasodilator that, when inhaled, can lead to more homogenous blood flow through the organ.[175] Second, NO and its downstream effectors have free radical scavenger properties which can augment the beneficial effects of vasodilation seen in allografts treated with this molecule.

Multiple experimental studies have demonstrated the benefits of NO.[176-181] In addition to direct administration of NO, researchers have attempted to modify the allograft response to injury at sites downstream in this signaling pathway. The major intracellular action of NO is to stimulate production of cGMP. Administration of a cGMP analogue (8-Br-cGMP) has demonstrated promise in different experimental models and the addition of essential cofactors in the NO pathway has proven beneficial to pulmonary allograft function.[182-184] Nitroprusside and nitroglycerin, both of which directly activate the production of cGMP, have also been shown to ameliorate injury to pulmonary allografts.[159,185] Analogues of cAMP (the effector molecule in the prostaglandin pathway) and molecules that prevent the breakdown of cAMP have also shown great potential in alleviating injury and improving organ preservation.[186-190] Given the theoretical advantages of NO, most centers administer inhaled NO during reperfusion, even though this practice has yet to demonstrate significant improvements during clinical trials.[191]

Ischemic injury to the allograft is partially mediated by calcium influx as a result of potassium-induced opening of voltage dependant ion channels.[192] Therefore, the addition of calcium channel blockers has also been proposed to ameliorate allograft injury (e.g., verapamil in the UCLA solution).[193-195] In fact, nifedipine has been shown experimentally to preserve allograft function better than prostaglandins in a high potassium environment.[196,197]

Finally, since the cyclooxygenase (COX) metabolites, thromboxanes, produce vasoconstriction as their major downstream effect, investigators have proposed the use of COX inhibitors to reduce pulmonary vascular resistance and improve allograft function. The addition of indomethacin to both EC and UW solutions resulted in improved graft function when compared to the use of these two solutions alone.[198]

Platelet-Activating Factor Antagonists

Platelet-activating factor is a potent phospholipid released by a variety of cells in response to inflammatory and immunologic stimuli.[199-201] Increased platelet activating factor levels are associated with increased vascular permeability, cellular edema and platelet and leukocyte aggregation. These pathophysiological cellular changes are analogous to those observed secondary to ischemia-reperfusion injury in clinical lung and heart-lung transplantation.[202-204] Increased oxygen-derived free radicals due to an augmented respiratory burst in phagocytic cells and enhanced synthesis of arachidonic acid metabolites appear to be the mechanism by which platelet activating factor exerts its deleterious effects. The use of several platelet activating factor inhibitors (BN 52021, WEB 2170 and TCV-309) has ameliorated preservation injury in several experimental lung allograft models, especially when used in conjunction with EC and UW solutions.[205-211] However, like the use of other additives, translation into clinical practice has been slow and inconsistent.

Adenosine

Recent attention has also been paid to the action of adenosine receptors in modifying the immune response that leads to ischemia-reperfusion injury. The adenosine receptor on neutrophils is known to act endogenously in an anti-inflammatory capacity.[212] Research has demonstrated that administration of adenosine A2 and A3 receptor agonists (DWH-146e, ATL-146e, IB-MECA) reduces lung reperfusion injury in rabbit, porcine and feline transplant models.[212-215] Lungs pretreated with adenosine receptor agonists demonstrated improved $PaCO_2$, PaO_2, pulmonary vascular resistance, airway compliance and microvascular permeability.[212,214] Histological improvement has also been demonstrated after administration of these agents, with decreased inflammatory reactions and reduction in apoptosis.[214,215]

Adenosine agonists have additional beneficial effects on cardiac function in lung transplant recipients. Patients suffering from ischemia reperfusion injury often demonstrate some degree of cardiac dysfunction, most likely due to the release of oxygen free radicals into the bloodstream after reperfusion. In an experimental model, the administration of ATL-146e during reperfusion improved both cardiac output and arterial oxygenation in these animals.[216]

There are many unresolved issues regarding perfusate-storage solutions and the spectrum of diverse experimental models complicates reliable comparison of results in the literature. Perhaps, as some investigators have theorized, if allograft ischemic periods are extended in the future, electrolyte composition and pharmacologic additives will become increasingly more important clinically. However, despite the passionate claims of investigators worldwide, there is no compelling clinical evidence that any solution is clearly superior to the others for cardiac or pulmonary preservation when used within the current limits of safe ischemic periods.

pH Modifications

Tradition holds that acidity damages cells during storage, so most perfusion and storage solutions maintain a pH of 7.4. However, recent work has demonstrated possible benefits to allografts stored at slightly lower pH concentrations. Theoretically in cardiac myocytes, the presence of acidic environments leads to a reversible reduction in cardiac contractility which may prevent injury by reducing metabolic requirements.[28,217,218] Acidic environments also reduce the function of the Na[+]/H[+] exchanger, which has been shown to prevent intracellular enzyme degradation and reduce neutrophil activation.[219,220] Pharmacologic inhibition of these Na[+]/H[+] exchangers using cariporide has proven beneficial in experimental models of lung transplantation.[221]

Ischemic Preconditioning

Recently, investigators have taken interest in another technique to improve pulmonary allograft preservation. Ischemic preconditioning was initially described in cardiac myocytes as a method to prevent ischemia-reperfusion injury.[222] Using this technique, cells undergo an initial period of short ischemia prior to the major ischemic event. This brief period of conditioning ischemia is then protective against further insults of longer duration.

Based on the results initially seen in the myocardium, experimental models of lung transplantation have been developed to test the hypothesis that ischemic preconditioning improves pulmonary allograft function.[223-225] Not only have these studies demonstrated a benefit to ischemic precondition, but these investigations have also outlined the possible mechanisms for its protective effects. Ischemic preconditioning in the cardiac and pulmonary allograft appears to be mediated through ATP dependant K[+] channels within these cells. Pharmacologic modification of these K[+]-ATP channels using agents such as Nicorandil and Pinacidil has lead to significant functional and molecular benefit to pulmonary allografts, demonstrating a precise molecular target which can improve storage of these organs.[226-228] Although this technique has yet to be translated into clinical practice, it holds great promise for improving allograft function.

Lung Inflation during Storage

Experimental evidence suggests that maintenance of alveolar volume is critical for minimizing lung injury during ischemia.[229-231] Among other things, prevention of atelectasis prevents ischemia-reperfusion injury by stimulating the release of surfactant, which has been shown to benefit the stored pulmonary allograft.[232,233] Furthermore, inflation of the lung allows oxygen to penetrate the pulmonary tissue. The lung is uniquely capable of maintaining aerobic metabolism during storage by utilizing the oxygen supply in the alveoli.[11,234] Direct oxygenation of the pulmonary cells is possible in the lung allograft due to the short diffusion distance from the alveolar space. In deflated lungs, the parenchyma acts like a solid organ and cannot benefit from aerobic metabolism.

Conflicting reports in the literature make it difficult to determine the optimal gas concentration of the inflated lung. While providing the lung with some amount of oxygen should improve overall allograft function, excessive amounts of alveolar oxygen during storage can lead to free radical production and further damage of the allograft.[235,236] As a result, common practice is to keep alveolar oxygen concentration under 40% during storage. Our institution prefers to partially inflate the lung with 30% O_2 prior to stapling the mainstem bronchus or trachea and storing the lung at 4°C.

Preservation during Implantation

To prevent rewarming during implantation prior to perfusion, the dependent portions of the heart and lungs are immersed in cold saline and the exposed regions are covered in saline slurry or cold saline soaked gauze pads. Frequent addition of cold saline into the thoracic cavity is important for keeping the lung or heart-lung bloc cold and preventing warming via direct thermal contact with other intrathoracic structures.

Reperfusion Modification

To help prevent ischemia-reperfusion injury, certain modifications are being made to the reperfusion stage of transplantation. By diminishing the leukocyte burden, controlling reperfusion pressures and administering exogenous surfactant, investigators hope to limit the untoward injurious effects that occur during this stage of the transplant procedure.

Leukocyte Depletion

Pulmonary leukostasis may result in microvascular occlusion during restoration of cardiopulmonary blood flow. In addition, leukocytes are intimately involved in reperfusion injury through the release of oxygen derived free radicals.[237,238] By diminishing the leukocyte burden and reducing leukocyte adhesion to myocytes, pneumatocytes and endothelial cells, these effects can be reduced.[239]

Early investigation focused on leukocyte depletion using mechanical blood filters.[237] Experimental results were quite impressive, demonstrating improved post-ischemic pulmonary and cardiac function as determined by systemic arterial oxygenation, pulmonary vascular resistance, airway pressures, lung water content and histology.[237-240] More recently, monoclonal antibodies to the integrin adhesion molecules have been shown to prevent intercellular contact between leukocytes and endothelium in both the heart and lung.[242-245] Pharmacological strategies to inhibit leukocyte adhesion using agents such as pentoxifylline have also shown promise in experimental models.[246] A randomized, controlled trial using modified perfusate that included leukocyte depleted blood demonstrated significant improvements in survival and incidence of ischemia-reperfusion injury in patients receiving modified perfusate.[247]

Most centers use filtering strategies during reperfusion to mechanically reduce the leukocyte burden in the reperfused allograft.[248-251] This technique can be safely and easily achieved using commercially available filters placed in the cardiopulmonary bypass circuit. Another method is the use of filtered warm blood cardioplegia during the initial reperfusion period. Unfortunately, optimal filtration requires placing both donor and recipient on cardiopulmonary bypass, which is not always required using current transplantation techniques. Furthermore, the additional immunosuppression associated with filtration may counteract some of the benefits that recipients would otherwise enjoy. Because of the benefit shown with pharmacologic leukocyte inhibition, the use of filtration will likely diminish over the next few years in favor of this easier, less cumbersome alternative.

Controlled Reperfusion

Controlled reperfusion refers to the incremental increase of flow or perfusion pressure into the pulmonary allograft at the time of reperfusion. Theoretically, gradual increases in flow and pressure to physiologic levels over the course of 10-15 minutes may prevent shearing of edematous endothelial cells. Use of this technique has shown benefit in experimental models and it has also enjoyed success after translation into clinical practice.[252-256] Combining controlled reperfusion with a reduction in FiO_2 may further prevent many of the complications of ischemia reperfusion injury by limiting the amount of oxygen available for free radical production. Our center titrates the FiO_2 on the cardiopulmonary bypass machine to achieve systemic $PaO_2 < 100$ mm Hg and we set the ventilator on room air during reperfusion. We also give a bolus of mannitol 25 grams prior to the controlled release of the pulmonary artery clamp. Many other centers have also started to adopt this technique into lung and heart-lung transplant practice with good results.

Exogenous Surfactant

Surfactant, a phospholipid secreted by type II pneumatocytes, reduces surface tension within the alveoli. Lung preservation strategies are known to compromise alveolar surfactant secretion and function after implantation.[232,233,255] Recent experimental evidence demonstrates improved pulmonary function and allograft hemodynamics after endobronchial administration of exogenous surfactant into donor lungs.[256-259]

Conclusion

Poor tolerance of prolonged ischemic time continues to be a significant problem in the field of cardiopulmonary transplantation. Newer advancements in donor management, organ procurement techniques, perfusate/storage solutions and reimplantation techniques are providing promising results for patients undergoing cardiopulmonary transplantation. However, continued investigation into strategies that will expand storage times past 4-6 hours is vital to the expansion of transplantation as a treatment option for patients with end-stage cardiopulmonary disease.

The current strategy of a single cardioplegic and pneumoplegic flush with prostaglandin pretreatment followed by static hypothermic storage certainly provides a simple, reliable technique for satisfactory cardiopulmonary preservation over relatively short ischemic times. Nevertheless, improved techniques permitting extension of the storage interval would have many benefits, including increasing the donor pool, better allocation of organs and improved allograft performance with minimal rejection episodes. However, until conflicts in the literature can be resolved and until prevention of ischemia-reperfusion injury can be adequately addressed, optimal organ preservation will continue to be one of the most challenging aspects of lung and heart-lung transplantation.

References

1. Thabut G, Mal H, Cerrina J et al. Graft ischemic time and outcome: A multi-center analysis. Am J Respir Crit Care Med 2005; 171(7):786-91.
2. Breen T, Ali I, Novick R et al. The effect of ischemic time on survival after lung transplantation. J Heart Lung Transplant 1995; 14:38.
3. Kshettry V, Kroshus T, Burdine J et al. Does donor organ ischemia >4 hours affect long term survival after lung transplantation? J Heart Lung Transplant 1995; 14:42.
4. Veno T, Snell G, Williams T et al. Impact of graft ischemia time on outcomes after bilateral sequential single-lung transplantation. Ann Thorac Surg 1999; 67:1577-82.
5. Levine S, Angel L. Primary graft failure: who is at risk? Chest 2003; 124:1190-2.
6. Meyers B, de la Morena M, Sweet S et al. Primary graft dysfunction and other selected complications of lung transplantation: a single-center experience of 983 patients. Thorac Cardiovasc Surg 2005; 129(6):1421-9.
7. Christie J, Sager J, Kimmerl S et al. Impact of primary graft failure on outcomes following lung transplantation. Chest 2005; 127(1):161-5.
8. Christie J, Bavaria J, Palevsky H et al. Primary graft failure following lung transplantation. Chest 1998; 114:51-60.
9. International Society of Heart and Lung Transplantation 2002 Registry. Available at: www.ishlt.org.
10. Christie J, Kotloff R, Pochettino A et al. Clinical risk factors for primary graft failure following lung transplantation. Chest 2003; 124(4):1232-41.
11. Date H, Matsumura A, Manchester J et al. Changes in alveolar oxygen and carbon dioxide concentration during lung preservation. J Thorac Cardiovasc Surg 1993; 105:492-501.
12. Southard J. Advances in organ preservation. Transplant Proc 1989; 21:1195-6.
13. Egan T, Lambert C, Reddick R et al. A strategy to increase the donor pool: the use of cadaver lungs for transplantation. Ann Thorac Surg 1991; 52:1113-21.
14. Lechner J, Stoner G, Yoakum G et al. In vitro carcinogenesis studies with human tracheobronchial tissues and cells. In: Schiff LJ, ed. In vitro Models of Respiratory Epithelium. Boca Raton: CRC Press, 1986:143-59.
15. Egan T. Non-heart-beating donors in thoracic transplantation. J Heart Lung Transplant 2004; 23(1):3-10.
16. Hardy J, Webb W, Dalton M et al. Lung homotransplantation in man. JAMA 1963; 186:1065-74.
17. Magovern G, Yates A. Human homotransplantation of left lung: report of a case. Ann NY Acad Sci 1964; 120:710-28.
18. Love R, Stringham J, Chomiak P et al. First successful lung transplantation using a nonheart-beating donor (abstract). J Heart Lung Transplant 1995; 14(suppl):S88.
19. Steen S, Sjoberg T, Pierre L et al. Transplantation of lungs from a nonheart-beating donor. Lancet 2001; 357:825-9.
20. Jacquet L, Ziady G, Stein K et al. Cardiac rhythm disturbances early after orthotopic heart transplantation. J Am Coll Cardiol 1990; 16:832.
21. Gifford RPM, Weaver AS, Burg JE et al. Thyroid hormone levels in heart and kinder cadaver donors. J Heart Trans 1986; 5:249.

22. Novitzky D, Wicomb WN, Cooper DKC et al. Improved cardiac function following hormonal therapy in brain dead pigs: Relevance to organ donation. Cryobiology 1987; 24:1.
23. Novitzky D, Cooper DKC, Zuhdi N. The physiological management of cardiac transplant donors and recipients using triiodothyronine. Trans Proc 1988; 20:803.
24. Stringham H, Southard J, Hegge J et al. Limitations of heart preservation by cold storage. Transplantation 1992; 53:287-94.
25. Martin D, Scott D, Downes G et al. Primary cause of unsuccessful liver and heart preservation: cold sensitivity of the ATPase system. Ann Surg 1972; 175:111-7.
26. MacKnight A, Leaf A. Regulation of cellular volume. Physiol Rev 1977; 57:510-57.
27. Jamieson R. The role of cellular swelling in the pathogenesis or organ ischemia. West J Med 1974; 120:205-18.
28. Kelly R. Current strategies in lung preservation. J Lab Clin Med 2000; 6:427-440.
29. Kurihara S, Sakai T. Effect of rapid cooling on mechanical and electrical responses in ventricular muscle of guinea-pig. J Physiol 1985; 361:361-78.
30. Okada Y, Zuo XJ, Marchevsky A et al. Transient cold preservation alone stimulates Tumor Necrosis Factor-α gene expression in a model of rat syngeneic lung transplantation. Transplant Proc 2002; 35:1111-3.
31. Haniuda M, Dresler CM, Mizuta T et al. Free radical-mediated vascular injury in lungs preserved at moderate hypothermia. Ann Thorac Surg 1995; 60(5):1376-81.
32. Eppinger M, Deeb G, Bolling S et al. Mediators of ischemia-reperfusion injury of the rat lung. Am J Pathol 1997; 150:1773-84.
33. Mal H, Dehoux M, Sleiman C et al. Early release of proinflammatory cytokines after lung transplantation. Chest 1998; 113:645-51.
34. Rao V, Feindel CM, Cohen G et al. Is profound hypothermia required for storage of cardiac allografts? J Thorac Cardiovasc Surg 2001; 122(3):501-7.
35. Fitton TF, Wei C, Lin R et al. Impact of 24 h continuous hypothermic perfusion on heart preservation by assessment of oxidative stress. Clin Transplant 2004; 18(Suppl 12):22-7.
36. McCord J. Oxygen-derived free radicals in post ischemic tissue injury. N Engl J Med 1985; 94:428-32.
37. Gutteridge J, Richmond R, Halliwell B. Inhibition of the iron-catalyzed formation of hydroxyl radicals from superoxide and of lipid peroxidation by desferrioxamine. Biochem J 1979; 184:469.
38. Frank L, Massaro D. Oxygen toxicity. Am J Med 1980; 69:117-26.
39. Naidu BV, Farivar AS, Woolley SM et al. Enhanced peroxynitrite decomposition protects against experimental obliterative bronchiolitis. Exp Mol Pathol 2003; 75(1):12-7.
40. Hiratsuka M, Yano M, Mora BN et al. Heat shock pretreatment protects pulmonary isografts from subsequent ischemia-reperfusion injury. J Heart Lung Transplant 1998; 17(12):1238-46.
41. Hiratsuka M, Mora B, Yano M et al. Gene transfer of heat shock protein 70 protects lung grafts from ischemia-reperfusion injury. Ann Thorac Surg 1999; 67(5):1421-7.
42. Okada Y, Marchevsky AM, Zuo XJ et al. Accumulation of platelets in rat syngeneic lung transplants: A potential factor responsible for preservation-reperfusion injury. Transplantation 1997; 64(6):801-6.
43. Naka Y, Marsh HC, Seesney SM et al. Complement activation as a cause for primary graft failure in an isogenic rat model of hypothermic lung preservation and transplantation. Transplantation 1997; 64(9):1248-55.
44. Granger P. Role of xanthine oxidase and granulocytes in ischemia-reperfusion injury. Am J Physiol 1988; 255:H1269-75.
45. Klavsner J, Paterson I, Goldman G et al. Postischemic renal injury is mediated by neutrophils and leukotrienes. Am J Physiol 1989; 256:F794-802.
46. Schmid RA, Zollinger A, Singer T et al. Effect of soluble complement receptor type 1 on reperfusion edema and neutrophil migration after lung allotransplantation in swine. J Thorac Cardiovasc Surg 1998; 116(1):90-7.
47. Keshavjee S, Davis R, Zamora M et al. A randomized, placebo-controlled trial of complement inhibition in ischemia-reperfusion injury after lung transplantation in human beings. J Thorac Cardiovasc Surg 2005; 129(2):423-8.
48. Jahania M, Sanchez J, Narayan P et al. Heart Preservation for Transplantation: Principles and Strategies. Ann Thorac Surg 1999; 68:1983-87.
49. Okada K, Yamashita C, Okada M et al. Successful 24-hour rabbit heart preservation by hypothermic continuous coronary microperfusion with oxygenated University of Wisconsin solution. Ann Thorac Surg 1995; 60:1723-28.
50. Allison R, Kyle J, Adkins W et al. Effect of ischemia-reperfusion or hypoxia-reoxygenation on lung vascular permeability and resistance. J Appl Physiol 1990; 69:597-603.
51. Baumgartner W, Reitz B, Achuff S eds. Heart and Heart-Lung Transplantation. Philadelphia, WB Saunders 1990:319.

52. Hachida M, Morton D. Lung function after prolonged lung preservation. J Thorac Cardiovasc Surg 1989; 97:911-9.
53. Belzer F, Southard J. Principles of solid-organ preservation by cold-storage. Transplantation 1988; 45:673-6.
54. Keon KJ, Hendry PJ, Taichman GC et al. Cardiac transplantation: The ideal myocardial temperature for graft transport. Ann Thorac Surg 1988; 46:337.
55. The Toronto Lung Transplant Group. Unilateral lung transplantation for pulmonary fibrosis. N Engl J Med 1986; 314:1140-5.
56. Fujimara S, Handa M, Kindo T et al. Successful 48-hour simple hypothermic preservation of canine lung transplants. Transplant Proc 1987; 19:1334-6.
57. Steen S, Sjoberg T, Ingemansson R et al. Efficacy of topical cooling in lung preservation: Is a reappraisal due? Ann Thorac Surg 1994; 58:1658-63.
58. Hisatomi K, Moriyama Y, Yotsumoto G. Beneficial effect of additional cardioplegia flush during hypothermic static cardiac preservation. Transplantation 2000; 69(9):1950-3.
59. Hardesty R, Griffith B, Trento A et al. Improved myocardial and pulmonary preservation by metabolic substrate enhancement in the autoperfused working heart-lung preparation. J Heart Transplant 1985; 4:602.
60. Ladowski J, Kapelanski D, Teodori M et al. Use of autoperfusion for distant procurement of heart-lung allografts. J Heart Transplant 1985; 4:330-3.
61. Golding L, Stewart R, Molimoto T et al. Successful acute heart-lung transplantation after six hours preservation. J Heart Transplant 1985; 4:252-3.
62. Kontos G, Borkon A, Baumgartner W et al. Neurohormonal modulation of the pulmonary vasoconstrictor response in the autoperfused working heart-lung preparation during cardiopulmonary preservation. Transplantation 1988; 45:275-9.
63. Robicsek F. Cardiopulmonary preservation. J Heart Transplant 1988; 7:313-4.
64. Kontos G, Borkin A, Adachi H et al. Leukocyte depletion ameliorates the pulmonary vasoconstrictor response in the autoperfused working heart-lung preparation. Surg Forum 1986; 37:255-7.
65. Zang Z, Proffiff G, Salley R et al. Autoperfused heart-lung preparation: One reason for unsuccessful lung preservation. Acta Bio-Medica de "L'Ateneo Parmense" 1994; 65(mm. ¾):115-31.
66. Miyamoto Y, Lajos T, Bhayana J et al. Physiologic constraints in autoperfused heart-lung preservation. J Heart Transplant 1987; 6:261-6.
67. Robicsek F, Masters T, Duncan G et al. An autoperfused heart-lung preparation: metabolism and function. J Heart Transplant 1985; 4:334-8.
68. Hardesty R, Griffith B. Procurement for combined heart-lung transplantation. Bilateral thoracotomy with sternal transection, cardiopulmonary bypass and profound hypothermia. J Thorac Cardiovasc Surg 1985; 89:795-9.
69. Wahlers T, Haverich A, Fieguth H et al. Flush perfusion using Euro-Collins solutions vs cooling by means of extracorporeal circulation in heart-lung preservation. J Heart Transplant 1986; 5:89-98.
70. Baumgartner W, Williams G, Fraser C. Cardiopulmonary bypass with profound hypothermia: an optimal preservation method for multiple organ preservation. Transplantation 1987; 47:124-7.
71. Yacoub M, Khaghani A, Banner N et al. Distant organ procurement for heart and lung transplantation. Transplant Proc 1989; 21(1):2548-50.
72. Baumgartner W et al "Domino-donor" operation. JAMA 1989; 261(21):3123-5.
73. Peters R. Effect of cardiopulmonary bypass on lung function. In: Utley J, ed. Pathophysiology and Techniques of Cardiopulmonary Bypass. Baltimore: Williams and Wilkins, 1983:164-74.
74. Hammerschmidt D, Stroncek D, Bowers T et al. Complement activation an neutropenia occurring during cardiopulmonary bypass. J Thorac Cardiovasc Surg 1981; 81:370-7.
75. Gillinov A, DeValeria P, Winkelstein J et al. Complement inhibition with soluble complement receptor type 1 in cardiopulmonary bypass. Ann Thorac Surg 1993; 55:619-24.
76. Veith F, Deysine M, Nehlsen S et al. Preservation of pulmonary function, hemodynamics and morphology in isolated perfused canine lungs. J Thorac Cardiovasc Surg 1966; 52:437-41.
77. Poston R, Gu J, Prostein D et al. Optimizing donor heart outcome after prolonged storage with endothelial function analysis and continuous perfusion. Ann Thorac Surg 2004; 78(4):1362-70.
78. Baldwin J, Frist W, Starkey T et al. Distant graft procurement for combines heart and lung transplantation using a pulmonary artery flush and simple topical hypothermia for graft preservation. Ann Thorac Surg 1987; 42:670-3.
79. Starkey T, Sakakibara N, Hagberg R et al. Successful six-hour cardiopulmonary preservation with simple hypothermic crystalloid flush. J Heart Transplant 1986; 5:291-7.
80. Hopkinson D, Bhabra M, Hooper T. Pulmonary graft preservation: A worldwide survey of current clinical practice. J Heart Lung Transplant 1998; 15:169-74.

81. Haverich A, Scott W, Jamieson S. Twenty years of lung preservation: a review. J Heart Transplant 1985; 4:234-40.

82. Wittwer T, Fehrenbach A, Meyer D et al. Retrograde flush perfusion with low-potassium solutions for improvement of experimental pulmonary preservation. J Heart Lung Transplant 2000; 19(10):976-83.

83. Bitu-Moreno J, Francischetti I, Siemer R et al. Influence of different routes of flush perfusion on the distribution of lung preservation solutions in parenchyma and airways. Eur J Cardiothorac Surg 1999; 15(4):481-9.

84. LoCicero J, Massad M, Matano J et al. Contribution of the bronchial circulation to lung preservation. J Thorac Cardiovasc Surg 1991; 101:807-15.

85. Varela A, Montero F, Tendillo E et al. Retrograde lung preservation provides optimal pulmonary function after transplantation. J Heart Lung Transplant 1994; 13:547.

86. Chien C, Gallagher R, Ardery P et al. Retrograde vs antegrade flush in canine left allograft lung preservation for 6 hours. J Heart Lung Transplant 1995; 14:590.

87. Wittwer T, Franke U, Fehrenback A et al. Impact of retrograde graft perfusion in Perfadex-based experimental lung transplantation. J Surg Res 2004; 117(2):239-48.

88. Mayer E, Puskas J, Cardoso P et al. Reliable eighteen-hour lung preservation at 4 degrees and 10 degrees C by pulmonary artery flush after high-dose prostaglandin E1 administration. J Thorac Cardiovasc Surg 1992; 103(6):1136-42.

89. Wang L, Nakamoto K, Hsieh C et al. Influence of temperature of flushing solution on lung preservation. Ann Thorac Surg 1993; 55:771-5.

90. Wang L, Yoshikawa K, Miyoshi S et al. The effect of ischemic time and temperature on lung preservation in a simple ex vivo rabbit model used for functional assessment. J Thorac Cardiovasc Surg 1989; 98:333-342.

91. Shiraishi T, Igusi H, Shirakusa T. Effects of pH and temperature on lung preservation: A study with an isolated rat lung reperfusion model. Ann Thorac Surg 1994; 57:639-43.

92. Wallwork J, Jones K, Cavarocchi N et al. Distant procurement of organs for clinical heart-lung transplantation using a single flush technique. Transplantation 1987; 4:654-8.

93. DeLima N, Binns O, Buchanan S et al. Euro-Collins solution exacerbates lung injury in the setting of high-flow reperfusion. J Thorac Cardiovasc Surg 1996; 112:111-6.

94. Bhabra M, Hopkinson D, Shaw T et al. Controlled reperfusion protects lung grafts during a transient early increase in permeability. Ann Thorac Surg 1998; 65:187-92.

95. Halldorsson A, Kronon M, Allen B et al. Controlled reperfusion after lung ischemia: implications for improved function after lung transplantation. J Thorac Cardiovasc Surg 1998; 115:415-25.

96. 151Pirlo A, Benumo J, Trousdale F. Atelectatic lobe blood flow: open vs. closed chest, positive pressure vs. spontaneous ventilation. J Appl Physiol 1981; 50:1022-6.

97. Nicholas T, Barr H. Control of release of surfactant phospholipids in the isolated perfused rat lung. J Appl Physiol 1981; 51:90-8.

98. Puskas J, Hirai T, Christie N et al. Reliable thirty-hour lung preservation by donor lung hyperinflation. J Thorac Cardiovasc Surg 1992; 10:1075-83.

99. Christie N, Waddell T. Lung preservation. Chest Surg Clin North Am 1993; 329-47.

100. Demmy TL, Biddle JS, Bennett LE et al. Organ preservation solutions in heart transplantation: patterns of usage and related survival. Transplantation 1997; 63:262.

101. Michel P, Hadour G, Rodriguez C et al. Evaluation of a new preservation solution for cardiac graft during hypothermia. J Heart Lung Transplant 2000; 19(11):1089-97.

102. Stringham JC, Love RB, Welter D et al. Impact of University of Wisconsin solution on clinical heart transplantation. A comparison with Stanford solution for extended preservation. Circulation 1998; 98(Suppl 19):II157.

103. Collins G, Bravo-Shugarman M, Terasaki P. Kidney preservation for transportation: initial perfusion and 30 hour ice storage. Lancet 1969; 2:1219-22.

104. Trump B, Ginn F. Studies of cellular injury in isolated flounder tubules. II. Cellular swelling in high potassium media. Lab Invest 1968; 18:341-51.

105. Gordon E, Maeir D. Effect of ionic environment on metabolism and structure of rat kinder slices. Am J Physiol 1964; 207:71-6.

106. Tyers G, Todd G, Niebauer I et al. The mechanism of myocardial damage following potassium citrate (Melrose) cardioplegia. Surgery 1975; 78:45-53.

107. Green C, Pegg D, Mechanism of action of intracellular renal preservation solutions. World J Surg 1979; 3:115-20.

108. Keshavjee S, Yamazaki F, Cardoso P et al. A method for safe twelve hour pulmonary preservation. J Thorac Cardiovasc Surg 1989; 98:529-34.

109. Calhoon J, Groveer F, Gibbons W et al. Single lung transplantation: alternative indication and technique. J Thorac Cardiovasc Surg 1991; 101:816-25.

110. Fujimura S, Kono T, Handa M et al. Development of low potassium solution (EP4 solution) for ling-term preservation of a lung transplant: evaluation in primate and murine lung transplant model. Artif Organs 1996; 20(10):1137-44.
111. Keshavjee S, Yamazaki F, Yokomise H et al. The role of Dextran 40 and potassium in extended hypothermic lung preservation for transplantation. J Thorac Cardiovasc Surg 1992; 103:314-25.
112. Kirk A, Colquhoun I, Dark J. Lung preservation: a review of current practices and future directions. Ann Thorac Surg 1993; 56:990-1000.
113. Corcoran P, Wang Y, St. Louis J et al. Effects of lung preservation solution on differential lung function and pulmonary hemodynamics in an acute canine model. Surg Forum 1990; 41:719-22.
114. Hardesty R, Aeba R, Armitage J et al. A clinical trial of University of Wisconsin solution for pulmonary preservation. J Thorac Cardiovasc Surg 1993; 106:660-6.
115. Aeba R, Keenan R, Hardesty R et al. University of Wisconsin solution for pulmonary preservation in a rat transplant model. Ann Thorac Surg 1992; 53:240-6.
116. Corcoran P, St. Louis J, Wang Y et al. The effect or organ preservation solution on pulmonary function in canine single lung allotransplantation. Surg Forum 1990; 63:705-7.
117. Hirt S, Wahlers T, Jurman M et al. Improvement of currently used methods for lung preservation with prostacyclin and University of Wisconsin solution. J Heart Lung Transplant 1992; 11:656-64.
118. Kawahara K, Itoyangi N, Takashahi T et al. Transplantation of canine lung allografts preserved with UW solution for 24 hours. Transplantation 1993; 55:15-8.
119. Wilson C, Stansby G, Haswell M et al. Evaluation of eight preservation solutions for endothelial in situ preservation. Transplantation 2004; 78(7):1008-13.
120. Kober I, Obermayr R, Brull T et al. Comparison of the solutions of Bretschneider, St. Thomas' Hospital and the National Institutes of Health for cardioplegic protection during moderate hypothermic arrest. Eur Surg Res 1998; 30(4):243-51.
121. Cavallari A, Cillo U, Nardo B et al. A multicenter pilot prospective study comparing Celsior and University of Wisconsin preserving solutions for use in liver transplantation. Liver Transpl 2003; 9(8):814-21.
122. Wittwer T, Wahlers T, Cornelius J et al. Celsior solution for improvement of currently used clinical standards of lung preservation in an ex vivo rat model. Eur J Cardiothorac Surg 1999; 15(5):667-71.
123. Thabut G, Vinatier I, Brugiere O et al. Influence of preservation solution on early graft failure in clinical lung transplantation. Am J Respir Crit Care Med 2001; 164(7):1204-8.
124. Xiong L, Legagneux J, Wassef M et al. Protective effects of Celsior in lung transplantation. J Heart Lung Transplant 1999; 18(4):320-7.
125. Randes H, Albes J, Haas B et al. Influence of high molecular dextrans on lung function in an ex vivo porcine lung model. J Surg Res 2001; 101(2):225-31.
126. Peltz M, He T, Adams G et al. Pyruvate-modified Perfadex improves lung function after long-term hypothermic storage. J Heart Lung Transplant 2005; 24(7):896-903.
127. Gamex P, Cordoba M, Millan I et al. Improvements in lung preservation: 3 years' experience with a low-potassium Dextran solution. Arch Bronconeumol 2005; 41(1):16-9.
128. Rabanal J, Ibanez A, Mons R et al. Influence of preservation solution on early lung graft function (Euro-Collins vs Perfadex). Transplant Proc 2003; 35(5):1938-9.
129. Wittwer T, Wahlers T, Fehrenbach A et al. Improvement of pulmonary preservation with Celsior and Perfadex: impact of storage time on early post-ischemic lung function. J heart Lung Transplant 1999; 18(12):1198-201.
130. Sommer S, Warnecke G, Hohlfeld J et al. Pulmonary preservation with LPD and Celsior solution in porcine lung transplantation after 24 h of cold ischemia. Eur J Cardiothorac Surg 2004; 26(1):151-7.
131. Conte J, Baumgartner W. Overview and future practice patterns in cardiac and pulmonary preservation. J Card Surg 2000; 15(2):91-107.
132. Williams A, Riise G, Anderson B et al. Compromised antioxidant status and persistent oxidative stress in lung transplant recipients. Free Radic Res 1999; 30(5):383-93.
133. Halliwell B, Gutteridge J eds. Free Radicals in Biology and Medicine. 2nd ed Oxford: Clarendon Press, 1989.
134. Magovern G, Bolling S, Casale et al. The mechanism of mannitol in reducing ischemic injury: hypermolarity or hydroxyl scavenger? Circulation 1984; 40(suppl 1):91-5.
135. Alvarez-Ayuso L, Calero P, Granado F et al. Antioxidant effect of gamma-tocopherol supplied by propofol preparations (Diprivan) during ischemia-reperfusion in experimental lung transplantation. Transpl Int 2004; 17(2):71-7.
136. Long S, Laubach V, Tribble C et al. Pyrrolidine dithiocarbamate reduces lung reperfusion injury. J Surg Res 2003; 112(1):12-8.
137. Kozower B, Christofidou-Solomidou M, Sweitzer T et al. Immunotargeting of catalase to the pulmonary endothelium alleviates oxidative stress and reduces acute lung transplantation injury. Nat Biotechnol 2003; 21(4):392-8.

138. Inci I, Inci D, Dutly A et al. Melatonin attenuates posttransplant lung ischemia-reperfusion injury. Ann Thorac Surg 2002; 73(1):220-5.
139. Ross S, Kron I, Gangemi J et al. Attenuation of lung reperfusion injury after transplantation using an inhibitor of nuclear factor-kappaB. Am J Physiol Lung Cell Mol Physiol 2000; 279(3):L528-36.
140. Baker C, Longoria J, Gade P et al. Addition of a water-soluble alpha-tocopherol analogue to University of Wisconsin solution improves endothelial viability and decreases lung reperfusion injury. J Surg Res 1999; 86(1):145-9.
141. Wilund L, Nilsson F, Schertsen H et al. Treatment with an antioxidant inhibits vascular changes caused by circulating lymphocytes during acute lung rejection in dogs. Transplantation 1997; 64(6):807-11.
142. Shiraishi T, Kuroiwa A, Shirakusa T et al. Free radical-mediated tissue injury in acute lung allograft rejection and the effect of superoxide dismutase. Ann Thorac Surg 1997; 64(3):821-5.
143. Oayumi A, Godin D, Jamieson W et al. Correlation of red cell antioxidant status and heart-lung function in swine pretreated with allopurinol (a model of heart-lung transplantation). Transplantation 1993; 56(1):37-43.
144. Katz A, Coran A, Oldham K et al. Decreased oxidized glutathione with aerosolized cyclosporine delivery. J Surg Res 1993; 54(6):597-602.
145. Detterbeck F, Keagy B, Paull D et al. Oxygen free radical scavengers decrease reperfusion injury in lung transplantation. Ann Thorac Surg 1990; 50(2):204-9; discussion 209-10.
146. Kelly R, Murar J, Hong Z et al. Low potassium Dextran lung preservation solution reduces reactive oxygen species production. Ann Thorac Surg 2003; 75(6):1705-10.
147. Serrick C, Jamjoum A, Reis A et al. Amelioration of pulmonary allograft injury by administering a second rinse solution. J Thorac Cardiovasc Surg 1996; 112(4):1010-6.
148. Hausen B, Mueller P, Ramsamooj R et al. Donor treatment with lazaroid U74389G reduces ischemia-reperfusion injury in a rat lung transplant model. Ann Thorac Surg 1997; 64:814-20.
149. Tanoue Y, Morita S, Ochiai Y et al. Successful twenty-four-hour canine lung preservation with lazaroid U74500A. J Heart Lung Transplant 1996; 15(1 Pt 1):43-50.
150. Kuwaki K, Komatsu K, Sohma H et al. Improvement of ischemia-reperfusion injury by lazaroid U74389G in rat lung transplantation model. Scand Cardiovasc J 2000; 34(2):209-12.
151. Kuwaki K, Komatsu K, Sohma H et al. The effect of various doses of lazaroid U74389G on lung ischemia reperfusion injury. J Thorac Cardiovasc Surg 1999; 47(2):67-72.
152. Kuwaki K, Komatsu K, Sohma H et al. Lazaroid U74389G ameliorates ischemia-reperfusion injury in the rat lung transplant model. J Thorac Cardiovasc Surg 1999; 5(1):11-7.
153. Hillinger S, Schmid R, Stammberger U et al. Lazaroid donor pretreatment does not improve lung allograft reperfusion injury in swine. Transplant Proc 1998; 30(7):3382-4.
154. Hillinger S, Schmid R, Stammberger U et al. Donor and recipient treatment with the Lazaroid U-74006F do not improve posttransplant lung function in swine. Eur J Cardiothorac Surg 1999; 15(4):475-80.
155. Date H, Matsumura A, Manchester J et al. Evaluation of lung metabolism during successful twenty-four-hour canine ling preservation. J Thorac Cardiovasc Surg 1993; 105:480-91.
156. Modry D, Jirsch D, Boehme G et al. Hypothermic perfusion preservation of the isolated dog lung. Ann Thorac Surg 1973; 16:583-97.
157. Bryan C, Cohen D, Dew A et al. Glutathione decreases the pulmonary reimplantation response in canine lung autotransplants. Am Rev Respir Dis 1989; 139:A45.
158. Moncada S, Flower R, Van J. Prostaglandins, prostacyclin, thromboxane A_2 and leukotrienes. In: Gilman A, Goodman S, Rall T et al, eds. Pharmacological Basis for Therapeutic Interventions. 7th ed. New York: Macmillan; 1985:660-73.
159. Gorman R, Buntin S, Miller O. Modulation of human platelet adenylate cyclase by prostacyclin (PGX). Prostaglandins 1977; 13:377-88.
160. Bolanowski P, Bauer J, Machiedo G et al. Prostaglandin influence on pulmonary intravascular leukocytic aggregation during cardiopulmonary bypass. J Thorac Cardiovasc Surg 1977; 73:221-4.
161. Jones G, Hurley J. The effect of prostacyclin on the adhesion of leukocytes to injured vascular endothelium. J Pathol 1984; 142:51-9.
162. Starling M, Neutze J. Prostaglandins and extracorporeal circulation. Prostaglandin Leuko Essential Fatty Acids 1990; 40:1-8.
163. Fantone J, Marasco W, Elgas L et al. Stimulus specificity of prostaglandin inhibition of rabbit polymorphonuclear leukocyte lysosomal enzyme release and superoxide anion production. Am J Pathol 1984; 15:9-16.
164. Higgins R, Hammond G, Baldwin J et al. Improved ultrastructural lung preservation with prostaglandin E_1 as donor pretreatment in a primate model of heart-lung transplantation. J Thorac Cardiovasc Surg 1993; 105:965-71.

165. Okada Y, Marchevsky A, Kass R et al. A stable prostacyclin analogue, beraprost sodium, attenuates platelet accumulation and preservation-reperfusion injury of isografts in a rat model of lung transplantation. Transplantation 1998; 66(9):1132-6.

166. Jurmann M, Dammenhayn L, Schafers H et al. Prostacyclin as an additive to single crystalloid flush: Improved pulmonary preservation in heart-lung transplantation. Transplant Proc 1987; 19:4103-4157.

167. Mulvin D, Hones K, Howard R et al. The effect of prostacyclin as a constituent of a preservation solution in protecting lungs from ischemic injury because of its vasodilator properties. Transplantation 1990; 49:828-30.

168. Bonser R, Fragomeni L, Jamieson S et al. The use of PGE1 in 12 hour lung preservation. J Heart Transplant 1990; 9:70.

169. Bonser R, Fragomeni L, Jamieson S et al. Deleterious affects of prostaglandin E_1 in 12 hour lung preservation. Br Heart J 1989; 61:463.

170. Veno T, Yokomise H, Oka T et al. The effect of PGE1 and temperature on lung function following preservation. Transplantation 1991; 52:626-30.

171. Hooper T, Fetherston G, Flecknell P et al. The use of a prostacyclin analog, iloprost, as an adjunct to pulmonary preservation with Euro-Collins solution. Transplantation 1990; 49:495-9.

172. Novick R, Reid K, Denning L et al. Prolonged preservation of canine lung allografts: the role of prostaglandins. Ann Thorac Surg 1991; 51:853-9.

173. Harjula A, Baldwin J, Shumway N. Donor deep hypothermia or donor pretreatment with prostaglandin E1 and single pulmonary artery flush for heart-lung graft preservation: an experimental primate study. Ann Thorac Surg 1988; 46(5):553-5.

174. Pepke-Zabba J, Higenbottam T, Dinh-Xuan A et al. Inhaled nitric oxide as a cause of selective pulmonary vasodilation in pulmonary hypertension. Lancet 1991; 338:1173-4.

175. Kayano K, Toda K, Naka Y et al. Superior protection in orthotopic rat lung transplantation with cyclic adenosine monophosphate and nitroglycerin containing preservation solution. J Thorac Cardiovasc Surg 1999; 118:135-44.

176. Bhabra M, Hopkinson D, Shaw T et al. Low-dose nitric oxide inhalation during initial reperfusion enhances rat lung graft function. Ann Thorac Surg 1997; 63:339-44.

177. Naka Y, Roy D, Smerling A et al. Inhaled nitric oxide fails to confer the pulmonary protection provided by distal stimulation of the nitric oxide pathway at the level of cyclic guanosine monophosphate. J Thorac Cardiovasc Surg 1995; 110:1434-41.

178. Eppinger M, Ward P, Jones M et al. Disparate effects of nitric oxide on lung ischemia-reperfusion injury. Ann Thorac Surg 1995; 60:1169-76.

179. Okabayashi K, Triantafillou A, Yamashita M et al. Inhaled nitric oxide improves lung allograft function after prolonged storage. J Thorac Cardiovasc Surg 1996; 112(2):293-9.

180. Aitchison J, Orr H, Flecknell P et al. Nitric oxide during perfusion improves posttransplant function of nonheart-beating donor lung. Transplantation 2003; 75(12):1960-4.

181. Pinsky D, Naka Y, Chowdhury N et al. The nitric oxide/cyclic GMP pathway in organ transplantation: critical role in successful lung preservation. Proc Natl Acad Sci USA 1994; 91(25):12086-90.

182. Hillinger S, Sandera P, Carboni G et al. Survival and graft function in a large animal lung transplant model after 30 h preservation and substitution of the nitric oxide pathway. Eur J Cardiothorac Surg 2001; 20(3):508-13.

183. Schmid R, Hillinger S, Walter R et al. The nitric oxide synthase cofactor tetrahydrobiopterin reduces allograft ischemia-reperfusion injury after lung transplantation. J Thorac Cardiovasc Surg 1999; 118(4):726-32.

184. Yamashita M, Schmid R, Ando K et al. Nitroprusside ameliorates lung allograft reperfusion injury. Ann Thorac Surg 1996; 62(3):791-6.

185. Adkins W, Barnard J, May S et al. Compounds that increase cAMP prevent ischemia-reperfusion pulmonary capillary injury. J Appl Physiol 1992; 32:492-7.

186. Siflinger-Birnboim A, Bode D, Malik A. Adenosine 3'5'-cyclic monophosphate attenuates neutrophil-mediated increase in endothelial permeability. Am J Physiol 1993; 264:H370-5.

187. Ogawa S, Koga S, Kuwabara K et al. Hypoxia-induced increased permeability of endothelial monolayers occurs through lowering of cAMP levels. Am J Physiol 1992; 262:C546-54.

188. Nakamura T, Hirata T, Fukuse T et al. Dibutyrl cyclic adenosine monophosphate attenuates lung injury caused by cold preservation and ischemia-reperfusion. J Thorac Cardiovasc Surg 1997; 114:635-42.

189. Bleisweis M, Jones D, Hoffman S et al. Reduced ischemia-reperfusion injury with rolipam in rat cadaver lung donors: effect of cyclic adenosine monophosphate. Ann Thorac Surg 1999; 67:194-200.

190. Spedding M, Paoletti R. Classification of calcium channels and the sites of action of drugs modifying channel function. Pharmacol Rev 1992; 44:363-76.

191. Cardella J, Keshavjee S, Bai X et al. Increased expression of nitric oxide synthase in human lung transplants after nitric oxide inhalation. Transplantation 2004; 77(6):886-90.

192. Kontos G, Adachi H, Borkon A et al. A no-flush core-cooling technique for successful cardiopulmonary preservation in heart-lung transplantation. J Thorac Cardiovasc Surg 1987; 94:836-42.
193. Hachida M, Morton D. The protection of ischemic lung with verapamil and hydralizine. J Thorac Cardiovasc Surg 1988; 95:178-83.
194. Hachida M, Morton D. A new solution (UCLA formula) for lung preservation. J Thorac Cardiovasc Surg 1989; 97:513-20.
195. Sasaki S, Yasuda K, McCully J et al. Does PGE1 attenuate potassium-induced vasoconstriction in initial pulmonary artery flush on lung preservation? J Heart Lung Transplant 1999; 18:139-42.
196. Sasaki S, Yasuda K, McCully J et al. Calcium channel blocker enhances lung preservation. J Heart Lung Transplant 1999; 18:127-32.
197. Balkhi H, Peterson M, Connolly R et al. Comparison of Euro-Collins and University of Wisconsin solution in single flush preservation of the ischemic reperfused lung: an in vivo rabbit model. Transplantation 1995; 59:1090-5.
198. Vercelloti G, Huh P, Yin H et al. Enhancement of PMN-mediated endothelial damage by PAF: PAF primes PMN response to activation stimuli. Clin Res 1986; 34:917-22.
199. Chesney C, Pifer D, Byers L et al. Effect of PAF on human platelets. Blood 1982; 59:582-5.
200. Foegh M, Chambers E, Khirabadi B et al. PAF in organ transplant rejection. Adv Prostaglandin Thomboxane Leukotriene Res 1989; 19:377-81.
201. Haverich A, Aziz S, Scott W et al. Improved lung preservation using Euro-Collins solution for flush-perfusion. J Thorac Cardiovasc Surg 1986; 34:368-76.
202. Bienviste J. Characteristics and semi-purification of platelet-activating factor form human and rabbit leukocytes. Fed Proc 1974; 33:797.
203. Casals-Stenzel J, Heuer H. Pharmacology of PAF-antagonists. Prog Biochem Pharmacol 1988; 22:58-65.
204. Wahlers T, Hirt S, Haverich A et al. Future horizons of lung preservation by application of a platelet-activating factor antagonist compared with current clinical standards. J Thorac Cardiovasc Surg 1992; 103:200-5.
205. Corcoran P, Wang Y, Katz N et al. Platelet activating factor antagonist enhances lung preservation in a canine model of single lung allotransplantation. J Thorac Cardiovasc Surg 1992; 104:66-72.
206. Conte J, Foegh M. Lung preservation: A new indication for a PAF antagonist, BN 52021. In: O'Flaherty J, Rampwell P, eds. PAF Antagonists: New Developments for Clinical Application. The Woodlands, Texas: Portfolio Publishing, 1990:129-37.
207. Conte J, Katz N, Wallace R et al. Long-term lung preservation with PAF antagonist BN 52021. Transplantation 1991; 51:1152-6.
208. Corcoran P, Tse S, Katz N et al. Reduction of conjugated dienes in lung transplantation: effect of BN 52021. Ann Thorac Surg 1993; 56:1279-84.
209. Qayumi K, English J, Duncan S et al. Extended lung preservation with platelet-activating factor-antagonist TCV-309 in combination with prostaglandin. J Heart Lung Transplant 1997; 16:946-54.
210. Qayumi K, English J, Feeley E et al. A new platelet-activating factor antagonist (CV-6209) in preservation of heart and lung for transplantation. Cardiovasc Drug Ther 1997; 11:777-85.
211. Tang W, Weil M, Gazzmuri R et al. Reversible impairment of myocardial contractility due to hypercarbic acidosis in the isolated perfused rat heart. Crit Care Med 1991; 19:218-24.
212. Ross S, Tribble C, Linden J et al. Selective adenosine-A2A activation reduces lung reperfusion injury following transplantation. J Heart Lung Transplant 1999; 18(10):994-1002.
213. Fiser S, tribble C, Kaza A et al. Adenosine A2A receptor activation decreases reperfusion injury associated with high-flow reperfusion. J Thorac Cardiovasc Surg 2002; 124(5):973-8.
214. Reece T, Ellman P, Maxey T et al. Adenosine A2A receptor activation reduces inflammation and preserves pulmonary function in an in vivo model of lung transplantation. J Thorac Cardiovasc Surg 2005; 129(5):1137-43.
215. Rivo J, Zeira E, Galun E et al. Activation of A3 adenosine receptor provides lung protection against ischemia-reperfusion injury associated with reduction in apoptosis. Am J Transplant 2004; 4(12):1941-8.
216. Reece T, Laubach V, Tribble C et al. Adenosine A2A receptor agonist improves cardiac dysfunction from pulmonary ischemia-reperfusion injury. Ann Thorac Surg 2005; 79(4):1189-95.
217. Caldwell-Kenkel J, Currin R, Thurman R et al. Damage to endothelial cells after reperfusion of rat livers stored for 24 hours: protection by mild acidic pH and lack of protection by antioxidants. FASEB J 1989; 3:A625.
218. Bond J, Harper I, Chacon E et al. The pH paradox in the pathophysiology of reperfusion injury to rat neonatal cardiac myocytes. Ann NY Acad Sci 1994; 723:25-37.
219. Naccache P, Showell H, Becker E et al. Transport of sodium potassium and calcium across rabbit polymononuclear leukocyte membranes: effect of chemotactic factor. J Cell Biol 1977; 73:428-44.

220. Ryan J, Hicks M, Cropper J et al. Sodium-hydrogen exchanger inhibition, pharmacologic ischemic preconditioning, or both for extended cardiac allograft preservation. Transplantation 2003; 76(5):766-71.
221. Meldrum D, Cleveland J, Mitchell M et al. Constructive priming of myocardium against ischemia-reperfusion injury. Shock 1996; 6(4):238-42.
222. Gasparri R, Jannis N, Flameng W et al. Ischemic preconditioning enhances donor lung preservation in the rabbit. Eur J Cardiothorac Surg 1999; 16(6):639-46.
223. Li G, Chen S, Lou W et al. Protective effects of ischemic preconditioning on donor lung in canine lung transplantation. Chest 1998; 113(5):1356-9.
224. Du Z, Hicks M, Winlaw D et al. Ischemic preconditioning enhances donor lung preservation in the rat. J Heart Lung Transplant 1996; 15(12):1258-67.
225. Tang D, Pavot D, Mouria M et al. Warm ischemia lung protection with pinacidil: an ATP regulated potassium channel opener. Ann Thorac Surg 2003; 76(2):385-9.
226. Yamashita M, Schmid R, Fujino S et al. Nicorandil, a potent adenosine triphosphate-sensitive potassium-channel opener, ameliorates lung allograft reperfusion injury. J Thorac Cardiovasc Surg 1996; 112(5):1307-14.
227. Ryan J, Hicks M, Cropper J et al. Sodium-hydrogen exchanger inhibition, pharmacologic ischemic preconditioning, or both for extended cardiac allograft preservation. Transplantation 2003; 76(5):766-71.
228. Stevens G, Sanchez M, Chappell G. Enhancements of lung preservation by prevention of lung collapse. J Surg Res 1973; 14:400-5.
229. Stevens G, Rangel D, Yakeishi R et al. The relationship of ventilation to preservation of ischemic canine lung graft. Curr Topics Surg Res 1969; 1:51-66.
230. Kao S, Wan D, Yeh D et al. Static inflation attenuates ischemia/reperfusion injury in an isolated rat lung in situ. Chest 2004; 126(2):552-8.
231. Faridy E. Effect of distension on release of surfactant in excised dogs' lungs. Respir Physiol 1976; 27:99-114.
232. Oyarzun M, Stevens P, Clements J et al. Effect of lung collapse on alveolar surfactant in rabbits subjected to unilateral pneumothorax. Exp Lung Res 1989; 15:909-24.
233. Weder W, Harper B, Shimokawa S et al. Influence of intra-alveolar oxygen concentration on lung preservation in a rabbit model. J Thorac Cardiovasc Surg 1991; 101:1037-43.
234. Koyama I, Toung T, Rogers M et al. O_2 radicals mediate reperfusion lung injury in ischemia O_2-ventilated canine lobe. J Appl Physiol 1987; 63(1):111-5.
235. Heiniuda M, Dresler C, Mizuta T et al. Free radical-mediated vascular injury in lungs preserved at moderate hypothermia. Ann Thorac Surg 1995; 60:1376-81.
236. Shiraishi Y, Lee J, Laks H et al. Use of leukocyte depletion to decrease injury after lung preservation and rewarming ischemia: an experimental model. J Heart Lung Transplant 1998; 17(3):250-8.
237. Breda M, Hall T, Stuart R et al. Twenty-four hour lung preservation by hypothermia and leukocyte depletion. J Heart Transplant 1985; 4(3):325-9.
238. Schueler S, DeValeria P, Hatanaka M et al. Successful twenty-four hour lung preservation with donor core cooling and leukocyte depletion in an orthotopic double lung transplantation model. J Thorac Cardiovasc Surg 1992; 104(1):73-82.
239. Suzuki M, Inaven W, Kvietys M et al. Superoxide mediates reperfusion-induced leukocyte endothelial cell interactions. Am J Physiol 1989; 257:H1740-5.
240. Bando K, Pillai R, Cameron D et al. Leukocyte depletion ameliorates free radical mediated lung injury after cardiopulmonary bypass. J Thorac Cardiovasc Surg 1990; 99(5):873-7.
241. Pillai R, Bando K, Schueler S et al. Leukocyte depletion results in excellent heart-lung function after 12 hours of storage. Ann Thorac Surg 1990; 50:211-4.
242. Byrne J, Smith W, Murphy M et al. Complete prevention of myocardial stunning, contracture, low-reflow and edema after heart transplantation by blocking neutrophil adhesion molecules during reperfusion. J Thorac Cardiovasc Surg 1992; 104:1589-96.
243. Zehr K, Herskowitz A, Lee P et al. Neutrophil adhesion inhibition prolongs survival of cardiac allografts with hyperacute rejection. J Heart Lung Transplant 1993; 12:837-45.
244. Kapelanski D, Iguchi A, Niles S et al. Lung reperfusion injury is reduced by inhibiting a CD 18-dependant mechanism. J Heart Lung Transplant 1993; 12:294-307.
245. Clark S, Rao J, Flecknell P et al. Pentoxifylline is as effective as leukocyte depletion for modulating pulmonary reperfusion injury. J Thorac Cardiovasc Surg 2003; 126(6):2052-7.
246. Bando K, Schueler S, Cameron D et al. Twelve-hour cardiopulmonary preservation using donor core cooling, leukocyte depletion and liposomal superoxide dismutase. J Heart Lung Transplant 1991; 10:304-9.
247. Ardehali A, Laks H, Russell H et al. Modified reperfusion and ischemia-reperfusion injury in human lung transplantation. J Thorac Cardiovasc Surg 2003; 126:1929-34.
248. Jones M, Hsieh C, Yoshikawa K et al. A new model for assessment of lung preservation. J Thorac Cardiovasc Surg 1988; 96:608-14.

249. Ehrie M, Prop J, Crapo J et al. A successful model of lung transplantation in the rat. Am Rev Respir Dis 1981; 123(suppl):58.
250. Marck K, Wildevuur C. Lung Transplantation in the rat: I. Technique and survival. Ann Thorac Surg 1982; 31:74-80.
251. Cohen R, Schenkel R, Barr M et al. Post-explantation preservation of living related donor lobes for bilateral pulmonary transplantation. J Heart Lung Transplant 1995; 14:S42.
252. Fiser S, Kron I, Kaza A et al. Controlled perfusion decreases reperfusion injury after high-flow reperfusion. J Heart Lung Transplant 2002; 21(6):687-91.
253. Clark SC, Sudarshan C, Khanna R et al. Controlled reperfusion and pentoxifylline modulate reperfusion injury after single lung transplantation. J Thorac Cardiovasc Surg 1998; 115:1335-1341.
254. Lick SD, Brown PS, Kurusz M et al. Technique of controlled reperfusion of the transplanted lung in humans. Ann Thorac Surg 2000; 69:910-912.
255. Gunther A, Balser M, Schmidt R et al. Surfactant abnormalities after single lung transplantation in dogs: impact of bronchoscopic surfactant administration. J Thorac Cardiovasc Surg 2004; 127(2):344-54.
256. Koletsis E, Chatzimichalis A, Fotopoulos V et al. Donor lung pretreatment with surfactant in experimental transplantation preserves graft hemodynamics and alveolar morphology. Exp Biol Med (Maywood) 2003; 228(5):540-5.
257. Hohlfeld J, Struber M, Ahlf K et al. Exogenous surfactant improves survival and surfactant function in ischemia-reperfusion injury in minipigs. Eur Respir J 1999; 13(5):1037-43.
258. Buchanan S, deLima N, Mauney M et al. Intratracheal surfactant administration preserves airway compliance during reperfusion in an isolated perfused rabbit lung model. J Heart Lung Transplant 1995; 14:S43.
259. Novick R, Veldhuizen R, Possmayer F et al. Exogenous surfactant therapy in thirty-eight hour lung graft preservation for transplantation. J Thorac Cardiovasc Surg 1994; 108:259-68.

CHAPTER 13

Future of Organ Procurement and Preservation

Luis H. Toledo-Pereyra*

As we advance with the use of better organs for transplantation and as we identify better means of recovering and preserving organs, we need to characterize the molecular and metabolic landscape of maximally protected organs prior to, during and after transplantation.

A fundamental question for the uninitiated would be: What are the most important subjects to address in improving organ recovery and preservation? The answer, as you would imagine, is not simple, due to the complexity associated with organ function and survival. Multiple factors participate in these processes and, therefore, it is difficult to isolate the origin of the mechanisms associated with injury development or protective pathways.

As I review the second edition of this book,[1] originally published in 1997, I clearly realize that one of my predictions for the future totally missed the mark:

"The procurement and preservation of xenografts will be more frequently utilized. The problem here will be double, one is associated with successful preservation of the organ and the second, with successful treatment of rejection after transplantation. These two apparently different issues will be addressed and moderate success will be achieved."

Obviously not, as xenograft organ preservation did not appear on the horizon of preservation technology. With today's understanding, we need to extend the time in which we expect to accomplish this goal. It is possible that advances in xenograft preservation will be delayed until xenograft-positive results are more consistently obtained and this might take several decades. This more conservative prediction aligns with the current progress in xenotransplantation.[2]

On another front, namely organ procurement and preservation, several topics of particular interest must be addressed, including the "molecular basis of preservation injury, biomolecular protection of damaged organs and maximal utilization of suboptimal organ donors,"[1] as critical components in advancing the field.

Serious progress has been made, as recently analyzed in three excellent reviews by Mc Laren and Friend,[3] Maathius, Leuvenink and Ploeg[4] and Jamieson and Friend.[5] They comprehensively cover the most crucial issues dealing with advances in procurement and preservation.

The molecular biology and biomolecular protection of recovered and preserved organs have been approached before, especially as pertains to donor and/or organ preconditioning. Better organ recovery has been achieved in this area, from pharmacological management with hormonal replacement, energy supplementation, antioxidant therapy and extracellular calcium channel blockade to genetic manipulation using adenovirus transfection with various protective genes (e.g., Bcl-2, HO-1, etc).[3-5]

*Corresponding Author: Luis H. Toledo-Pereyra—Departments of Research and Surgery Michigan State University, Kalamazoo Center for Medical Studies, Kalamazoo, Michigan 49008, USA. Email: toledo@kcms.msu.edu

Organ Preservation for Transplantation, Third Edition, edited by Luis H. Toledo-Pereyra.
©2010 Landes Bioscience.

Anaya-Prado from our group[6] critically reviewed the role of ischemia and reperfusion insofar as the use of new organ-protective agents is concerned. Interesting observations revealed the importance of characterizing lesions associated with organ recovery and preservation. They emphasized the role of nitric oxide (NO) in the ischemic state as well as the upregulation of inflammatory mediators such as cytokines, chemokines and adhesion molecules. More recently, Lopez-Neblina from our group[7] focused on the utilization of a relatively new approach, the use of intracellular calcium channel blockade with dantrolene, in protecting severely ischemic rat livers. This particularly important research will permit us to identify more specifically the responsible pathways related to ischemia and reperfusion and therefore to organ preservation injury. Identifying these metabolic roads to injury will facilitate the discovery of precise therapeutic approaches.

New and well-crafted anti-ischemic therapy—to be given to the donor, or to the organ directly through a preservation solution and/or to the recipient—will therefore be a specific way of potentially improving the function of marginal recovered organs, frequently from non-heart beating donors (NHBDs). In this newly emerging field, ischemia represents a significant focus of dedicated study. It is reasonable to believe that the more organs obtained from NHBD, the more protection from ischemia will be required. Thus, is necessary to emphasize more laboratory work on ischemia, as already mentioned. I anticipate that extensive anti-cytokine therapy and/or anti-chemokine therapy or anti-adhesion molecule therapy, will be introduced in the years to come. Some anti-ischemic agents have already been applied experimentally, but now we need to move to the preservation and cold ischemia arena.

Compounds characterized as anti-apoptotic substances, which can upregulate Bcl-2 and downregulate p53 and caspase 3, represent new means of managing ischemic and preservation lesions. Since both of these lesions produce an overwhelming inflammatory response, it is logical to assume that anti-inflammatory products will overcome these pathological injuries, especially because many common mediators are intimately associated with types of lesions.

Another topic of considerable interest is the provision of normothermic machine perfusion (NMP) or normothermic preservation (NP) in general. Even though NMP is technically more complicated than hypothermic machine perfusion (HMP) and certainly cold preservation, there are some potential advantages as discussed by Jamieson and Friend.[5] In brief, they believe that avoiding ischemia, preventing cold injury and permitting viability studies are significant enough to overcome the complexity in the utilization of NMP.[5]

Stubenitsky in Koostra's group[8] introduced in 2001 an exsanguinous metabolic support solution for the resuscitation of NHBDs. After 30 minutes of warm ischemia and 24 hours of cold storage, canine kidneys were subjected to normothermic resuscitation. The results were exceptional, with recipient survival of 90% observed in the metabolically supported group compared to 73% otherwise.[8] Metcalfe from Nicholson's group[9] utilized a blood-based perfusion system for NMP, which later demonstrated improved results with the use of leukocyte-depleted blood.[10] A few years ago, the Valero and Garcia-Valdecasas group applied normothermic recirculation and NP to ischemichally injured porcine livers. Better survival was noted but necrosis of the biliary tree was also evident.[11] Other attempts at NMP by Schon and his group[12] confirmed the opportunity of using this method in pig livers and demonstrated substantial improvement after transplantation. New avenues of research with NMP show the promise of better results in the future.

Many worthwhile areas of organ preservation have lagged, such as the use of perfluorocarbons (PFC) for pancreas and islet cell preservation, the role of persufflation in organ protection, the effect of microdialysis on liver viability, the possibility of modifying immunogenicity by maintaining ischemic organs for long periods of NMP (the organ culture concept) and the improbable likelihood of organ freezing. There are substantial differences among these opportunities. First, PFC is already being successfully applied and only fine tuning is needed. Second, persufflation and microdialysis have been applied clinically in small studies.[5] Third, NMP is showing an initial positive role and more improvements are on the horizon. Finally, organ freezing—the most difficult because of crystal formation and thawing—is the Achilles tendon of this enterprise. Organ freezing has proven extremely difficult to understand, as reflected by persistent challenges to successful transplantation after any freezing event.

What about other organs, such as the heart, lungs, small bowel? In truth, preservation research has not been as vigorous with these organs. Preservation and transplant researchers have not focused on these organs as much, even though the heart has the shortest preservation time, which mandates many clinical accommodations when performing heart transplantation. Despite the differences in tissue function and biochemistry of these organs compared to kidneys, livers and pancreases, for example, improved preservation should apply in principle and with certain limitations to many of the forgotten organs. I hope new interest arises in these most intriguing regions of anatomical and physiological challenge.

Finally, let me offer some specific predictions for organ procurement, preservation and transplantation in the next decade:

1. Clinical application of donor and/or organ preconditioning.
2. Acceptable drug manipulation of ischemic injury in the clinical setting.
3. Identification of molecular injury after preservation.
4. Modification of molecular injury by inhibiting the inflammatory response.
5. Advances of HMP for organ preservation.
6. Improvements of NMP for organ preservation.
7. Evidence of modification of immunogenicity through extended preservation or development of solutions with immunosuppressive effect.
8. Advance persufflation studies and further exploration of the beneficial role of O_2 in HMP.
9. Donor pretreatment with CO and NO for better preservation and transplantation.

My predictions are generalizations, but at least they offer some guideposts for progress in the next 10 years or so. Specific predictions for specific organs are omitted due to time constraints and perhaps to limited usefulness. Let me complete this list of predictions by paraphrasing what I said in the same chapter of the previous edition.[1] "I hope that five to ten years from now, when we are ready to organize the fourth edition of this book, we will have accomplished all the goals and realizations predicted here today." History will judge our advancement and the rational approach to organ procurement and preservation presented in this book.

References

1. Toledo-Pereyra LH. Future of organ procurement and preservation. In: Toledo-Pereyra LH, ed. Organ Procurement and Preservation for Transplantation. Second Edition. Austin/New York: Landes Bioscience/ Chapman & Hall, 1997.
2. Toledo-Neblina LH, Lopez-Neblina F. Xenotransplantation: A view to the past and an unrealized promise to the future. Exp Clin Transpl 2003; 1:1.
3. McLaren AJ, Friend PJ. Trends in organ preservation. Transpl Int 2003; 16:701.
4. Maathius MJ, Leuvenink HGD, Ploeg RJ. Perspectives in organ preservation. Transplantation 2007; 83:1289.
5. Jamieson RW, Friend PJ. Organ reperfusion and preservation. Frontiers in Bioscience 2008; 13:221.
6. Anaya-Prado R, Toledo-Pereyra LH, Lentsch A et al. Organ ischemia reperfusion injury. J Surg Res 2002; 105:248.
7. Lopez-Neblina F, Toledo-Pereyra LH, Toledo AH et al. Ryanodine receptor antagonism protects the ischemic liver and modulates TNF-alpha and IL-10. J Surg Res 2007; 140:121.
8. Stubenitsky BM, Booster MH, Brasile L et al. Exsanguinous metabolic support perfusion—A new strategy to improve graft function after kidney transplantation. Transplantation 2000; 70:1254.
9. Metcalfe MS, Waller JR, Hosgood SA et al. A paired study comparing the efficacy of renal preservation by normothermic autologous blood perfusion and hypothermic pulsatile perfusion. Transpl Proc 2002; 34:1473.
10. Harper S, Hosgood S, Kay M et al. Leucocyte depletion improves renal function during reperfusion using an experimental isolated haemoperfused organ preservation system. Brit J Surg 2006; 93:623.
11. Valero R, Garcia-Valdecasas JC, Rull R et al. Hepatic blood flow and oxygen extraction ratio during normothermic recirculation and total body cooling as viability predictors in non-heart beating donor pigs. Transplantation 1998; 66:170.
12. Schon MR, Kollmar O, Wolf S et al. Liver transplantation after organ preservation with normothermic extracorporeal perfusion. Ann Surg 2001; 233:114.

INDEX

Printed and bound by CPI Group (UK) Ltd, Croydon, CR0 4YY

21/10/2024

01777085-0005